THE
SODIUM SULFUR BATTERY

THE

SODIUM SULFUR BATTERY

J. L. SUDWORTH and A. R. TILLEY

of Beta Research & Development Ltd

LONDON NEW YORK

CHAPMAN and HALL

First published in 1985 by
Chapman and Hall Ltd
11 New Fetter Lane, London EC4P 4EE
Published in the USA by
Chapman and Hall
29 West 35th Street, New York NY 10001

Printed in Great Britain
at the University Press, Cambridge

ISBN 0 412 16490 6

British Library Cataloguing in Publication Data

The Sodium Sulfur battery.
1. Sodium Sulphur batteries
I. Sudworth, J. L. II. Tilley, A. R.
621.31'2424 TK2941

ISBN 0-412-16490-6

Library of Congress Cataloging in Publication Data

Sudworth, J. L.
The Sodium sulfur battery
Bibliography: p.
Includes index.
1. Sodium sulphur batteries. I. Tilley, A. R., 1944- . II. Title.

TK2941.S84 1985 621.31'2424 85-10933
ISBN 0-412-16490-6

Contents

Contributors

R. J. Bones — Materials Development Division
Harwell Laboratory
Oxfordshire
OX11 0RA
England

D. S. Demott — Beta Research & Development Ltd
50 Goodsmoor Road
Sinfin
Derby
DE2 9GN
England

J. H. Duncan — Beta Research & Development Ltd
50 Goodsmoor Road
Sinfin
Derby
DE2 9GN
England

R. S. Gordon — Ceramatec Inc.
163 W. 17th Street South
Salt Lake City
Utah 84115
USA

G. R. Miller — Ceramatec Inc.
163 W. 17th Street South
Salt Lake City
Utah 84115
USA

P. T. Moseley — Materials Development Division
Harwell Laboratory
Oxfordshire
OX11 0RA
England

R. W. Powers
General Electric Company
Corporate Research & Development
PO Box 8
Schenectady
New York
USA

F. M. Stackpool
Chloride Silent Power Ltd
Davy Road
Astmoor
Runcorn
Cheshire
WA7 1PZ
England

Foreword

In correspondence with us beginning in early 1967, J. L. Sudworth expressed his interest in our experiments with a sodium sulfur cell. Before long he had established an active programme at British Rail, the first of several centres in Britain dedicated to the development of a large scale battery system. Well supported by fundamental research, their combined efforts soon earned them a pre-eminent position in the field. There now exists a large sum of experience scattered through numerous reports and the literature of several disciplines. Some excellent review papers have issued, but there has been a salient need for a work of wider scope addressed to specialists in the field of energy storage and to prospective users of electrochemical storage systems.

It is scarcely necessary for us to affirm that the editors of this book have succeeded admirably in fulfilling this need by distilling the essence from nearly two decades of work in Britain and from laboratories around the globe. Their success must of course be shared with the contributing authors, all of whom have (to paraphrase Brande) earned distinction as great architects in the science through service as labourers in the quarry. The editors have asked us for an account of the earliest work at Ford. Briefly, one of us (J. T. Kummer) believed that a secondary cell could be made with a sodium sulfur couple if a stable ionically conducting membrane could be found. With the urging and enthusiastic administrative support of T. W. Dewitt, we began a search for conductors among glasses in the soda aluminosilicate and aluminoborate systems. Using the guideposts erected by earlier workers, especially Izard, Rawson and Blodget, we prepared many glasses and a few did show promise. At this time, some primitive electrochemical cells were built to test the concept. Because the conductivities of the glasses increased with increasing content of sodium meta-aluminate, an attempt was made to measure the conductivity of this crystalline compound. A sample pressed and sintered, unencapsulated, at high temperature was unusually conductive. Analysis by X-ray diffraction and weight loss revealed that this sample had lost soda and was composed of a phase with a layer structure known as beta alumina. Experiments to increase the alkali concentration and conductivity of beta alumina by doping with mono and divalent metals gave unexpected results. Additions of oxides known to be spinel formers with alumina could stabilize, above sintering temperatures, a phase now referred to as beta″ alumina, which had even greater conductivity than beta alumina. With the notable exception of their large alkali ion mobilities, some of the properties of beta alumina and even of binary beta″ alumina were, to our surprise, already known. The confusion in some of the literature at the time regarding these phases was resolved here by Bettman who went on to lead X-ray crystallographic studies to determine accurate structures for these layered alkali polyaluminates.

Early tests with sintered discs of beta alumina in sodium sulfur cells were plagued by progressive increases in the ceramic electrolyte resistance. The effect was eventually proved to be a manifestation of a subtle corrosion of the solid electrolyte by liquid sodium. The reaction was strongly suspected of involving an impurity rich phase at grain boundaries, but a totally convincing explanation was never found. The problem was bypassed, as it were, and a crucial impetus was given to the programme, when Dzieciuch prepared ceramics of both beta alumina and beta″ alumina from pure materials which were stable in sodium. The programme gained further momentum from the results of an extensive study led by Tennenhouse that included the preparation and testing of lithia stabilized beta″ alumina ceramics. This work generated the formulations and practical processing techniques that permitted fabrication of virtually single phase dense tubular membranes. These were used for the first demonstration cells and became standard in our laboratory.

So ended the interlude when progress was relatively easy and inexpensive. The accomplishments of our colleagues at Ford more closely identified with the later challenging years are represented elsewhere in this book.

J. T. Kummer
N. Weber

Preface

The publication in 1967 of a paper by J. T. Kummer and N. Weber of the Ford Motor Company, in which they described a new secondary battery – the sodium sulfur battery – set in train development programmes at centres throughout the world. At almost the same time, Levine and Brown published details of a cell with the same electrodes, but using conducting glass electrolyte instead of the beta alumina used by Kummer and Weber. Today the literature on sodium sulfur batteries is extensive and growing rapidly.

Comprising as it does a solid electrolyte, a molten salt electrolyte, a liquid metal electrode and a redox electrode, and operating between 300 and 400°C, the cell is of interest to workers in many disciplines and consequently papers covering the various aspects of the cell are scattered throughout the literature. No comprehensive publication covering all aspects of the cell has been attempted, and it is the aim of this book to provide a critical account of the progress which has been made in the development of this novel electrochemical system. Where the subject matter lies within our direct experience and has been of particular interest to us, we have written those chapters or sections of chapters ourselves. Where the subject matter was not within our direct experience we have invited others to contribute.

Thus the chapter on electrolyte properties is contributed by Dr P. T. Moseley. In the chapter on production of the solid electrolyte, Dr R. W. Powers has contributed a section on electrophoretic deposition, R. J. Bones a section on the sol gel process of powder preparation, and Professor R. S. Gordon sections on continuous sintering. The remaining sections were contributed by Mr J. H. Duncan. In the chapter on the sodium electrode, Mr D. S. Demott has contributed the section on the sodium/beta alumina interface. In the chapter on cell design R. J. Bones and F. M. Stackpool have contributed the sections on diffusion bond seals, brazed seals and glass seals. In the chapter on the current collector, Professor G. R. Miller has contributed the sub-section on doped rutile as a current collector.

By adopting this somewhat unorthodox approach, we hope to provide the reader with a first hand account of the many facets of the sodium sulfur cell.

The sodium sulfur cell is still developing into its final form, or forms; many different approaches to cell design are being pursued and many different theories exist to explain the same phenomena. We have tried as far as possible, and we have asked our contributors to try, to provide a critical appraisal of all aspects of the cell. Our conclusions cannot be the last word on the subject, and only time will tell to what extent we have identified the best design concepts or the correct theories. We believe, however, that this approach will be of more value to the reader than an uncritical description of all theories and design concepts.

The glass electrolyte cell developed by C. Levine and his colleagues at Dow Chemicals has presented some difficulties. In many respects it is completely different from the beta alumina electrolyte cell, but it also has many problems in common with that cell. We did consider asking Charles Levine to contribute a chapter devoted solely to this cell, but he has recently contributed such a chapter to another book (*The Sulfur Electrode*, ed. R. P. Tischer, Academic Press, 1983) and we could not justify a repetition here. We have decided, therefore, to deal with this cell in the appropriate chapters and in doing so to highlight the differences and similarities between the two types of cells.

Acknowledgements

We gratefully acknowledge our debt to all our colleagues in the sodium sulfur battery group at British Rail, Derby, on whose work much of this book is based.

We would also like to thank B. Cleaver for providing us with an advance copy of his chapter on alkali metal polysulfides in *The Sulfur Electrode*, edited by R. P. Tischer, and P. Furness for drawing to our attention the technique of hazard plotting.

We would especially like to thank K. G. A. Pankhurst, the former Deputy Director of the British Rail R & D Division, for his enthusiastic backing and unstinting support from the inception of the programme until his retirement in 1978; E. D. Henley, his successor; and A. H. Wickens, Director of Research of the British Rail R & D Division. We would also like to thank Dorothy Kashefi-Mofrad, Ruth Sykens and Linda Demott who typed the chapters contributed by us.

Finally, we would like to acknowledge the numerous discussions over a number of years with members of the sodium sulfur battery community.

P. T. Moseley gratefully acknowledges helpful discussions with A. E. Hughes, R. M. Dell, R. J. Bones and G. Tappin. R. S. Gordon acknowledges the support of the US Department of Energy under contracts EY-76-C-02-2566 and DE-AM-02-79CH10012.

Introduction

In this, the last quarter of the twentieth century, the realization that fossil fuel resources are finite, and that oil and gas reserves in particular could be largely depleted within the lifetime of many people, has prompted a search for alternative energy sources.

Whichever energy source is utilized – nuclear, direct solar radiation, or renewable sources derived from the energy of the sun such as wind, wave and ocean thermal – it is likely to be converted to electrical energy. Whilst enormous progress has been made in the generation of electricity, the development of methods of storing electrical energy has lagged far behind.

The most convenient way of storing electrical energy is by converting it to chemical energy in a secondary battery. It is interesting to note that the first secondary battery – the lead acid battery – was discovered by Gaston Planté in 1859 only two years after the development by Werner von Siemens of a practical electric generator. The Planté cell, which uses plates of lead coated with lead and lead dioxide by an electrochemical forming process, is still extensively used for stationary storage batteries, particularly in power stations, but for most other applications it has been replaced by the pasted plate battery developed by Camille Fauré in 1881. In the Fauré battery a paste of lead oxide and sulfuric acid is applied to a lead grid resulting in a much higher specific energy. Pasted plate batteries are widely used today for starting, lighting and ignition duties on automobiles, trucks etc. and as the power source for electric vehicles. For the latter application, a modified positive electrode is often used. This consists of resin impregnated synthetic textile fibre tubes which are porous to sulfuric acid; inside each tube a central lead rod is located, and the resulting annular gap is filled with lead oxide. Batteries with this type of electrode can withstand over 1000 deep cycles without displaying signs of degradation. The other secondary battery in commercial use is the alkaline battery, normally nickel cadmium but occasionally nickel iron. Alkaline batteries were developed between 1895 and 1905 by Waldemar Jungner and Thomas Edison, and are used today where a rugged battery is required and for stand-by power supplies, where the ability of the battery to retain its charge for long periods is an advantage.

Lead acid and nickel cadmium secondary batteries are the only ones in general use today, but they suffer from a number of disadvantages. Both batteries have a relatively low specific energy – 25 to 30 Wh kg^{-1} is usual – and both batteries use raw materials which will increasingly be in short supply. It is unlikely that these batteries will be displaced from most of their present applications by new secondary batteries, and therefore they will undoubtedly continue to be used for starting, lighting and ignition in automobiles, trucks and locomotives. However, for traction applications, where weight is important, and for large scale energy storage, where cost and availability of materials are significant, there is a need for the development of new batteries.

Several ambient temperature batteries are being actively developed for these applications. For traction applications, nickel zinc, nickel iron and iron air batteries are receiving most attention. For stationary batteries, zinc chloride and zinc bromine batteries are the only ambient temperature batteries being considered. (In the case of the zinc chloride battery, refrigeration is required to keep the chlorine hydrate from decomposing too rapidly, so strictly speaking this is not an ambient temperature battery.)

The shortcomings of ambient temperature aqueous batteries led some workers to consider batteries with molten salt electrolytes. Cairns and Shimotake developed a group of batteries with lithium negative electrodes and sulfur, selenium and tellurium positive electrodes. Because of the low electrical resistance of the molten salt electrolyte and the highly reversible electrode processes at the operating temperature of 350 °C, very high specific powers were obtained with these cells as well as high specific energies. Development was concentrated on the lithium sulfur cell: however, the high vapour pressure of sulfur resulted in an extremely short cell life, and this together with other problems caused the elemental couple to be dropped in favour of a cell in which both lithium and sulfur were combined with other elements. The lithium was alloyed with aluminium, an idea first used by Rightmire in 1969 in an effort to make a durable lithium chlorine cell. The lithium chlorine cell as originally developed by Heitbrink operated at about 650 °C and used a fused lithium chloride electrolyte. The highly corrosive nature of lithium and chlorine, combined with the high operating temperature, resulted in very short lifetimes, and it was this problem that Rightmire attempted to overcome by alloying the lithium, absorbing the chlorine in active charcoal and lowering the operating temperature by using a mixed salt electrolyte. The resulting cell, although more durable than the original lithium chlorine cell, still suffered from a number of drawbacks including a considerably reduced performance, and the development was stopped. The lithium aluminium alloy, however, was used in conjunction with a positive electrode of iron sulfide (either FeS or FeS_2) by Vissers at the Argonne National Laboratory. This battery was developed for traction applications, together with variants such as lithium silicon/iron sulfide. The operating temperature of these batteries is approximately 450 °C and specific energies of 150 Wh kg^{-1} for the individual cells are said to be obtainable. The separator used in the cells – boron nitride felt – is very expensive at present, and alternatives such as magnesium oxide powder have been investigated. The major uncertainty about this battery, assuming that the technical problems can be solved, is the price and availability of lithium metal in the long term. Clearly there is a requirement for a secondary battery with a high specific energy which can be fabricated from low cost materials which are widely available. None of the batteries described, with the possible exception of the lithium alloy/iron sulfide battery, satisfies these requirements.

The sodium sulfur battery, which is the subject of this book, could satisfy the requirements described, and it is the editors' hope that readers unfamiliar with this battery will find in the following pages enough information to decide for themselves whether it will be a suitable power source for the applications of interest to them.

In 1966, when details of the sodium sulfur cell were first released by the Ford Motor Company, work had already been in progress for three years on cells with beta alumina solid electrolytes, and in the Foreword to this book Weber and Kummer describe how their discovery arose. Independently, Levine and Brown at the Dow Chemical Company also constructed a sodium sulfur cell but with a glass electrolyte instead of beta alumina. It is interesting to hear the story of the conception of the glass electrolyte cell as related by Charles Levine.

> The story of the beginnings of the Dow sodium sulfur effort is interesting. I had been working on making hydrogen air fuel cells using fine hollow fibres of ion exchange material as the electrolyte. In a configuration similar to a multitube boiler, the 100 μm ~ 200 μm diameter hollow fibres were bundled together, the insides and the outsides of the fibres metal coated (for current pickup) and plasticized. Hydrogen was passed through the fibre lumen and air passed over the outsides. This made a 'General Electric' type ambient temperature ion exchange fuel cell but with a higher power density due to the area of the ion exchange fibres.
>
> When we finally determined the amount of platinum needed at the electrodes, we quickly decided that there would be no non-space or non-military application for the device. We held a discussion to see if any of the technology could be useful. Bob Heitz, our then Director of Research, mentioned a high school science class in which they repeated the experiment of Burt (*J, Opt. Soc.*, 1925, **11**, 87–91). Heitz asked, 'Can we fill thin wall capillaries of glass with sodium and have a sodium electrode?' On Christmas morning, 1964, we started the first sodium in glass electrode using Cu^{+}–Cu^{2+} for the cathode.
>
> In searching for a viable cathode, William Brown asked about using sulfur. 'Nonsense', I replied, 'sulfur is a good insulator.' He said, 'With sulfide in it?' Near the end of January 1965 we had shown the sodium in glass/sulfur sulfide cell to be reversible.
>
> In 1966 Ford Motor Company announced their sodium sulfur battery. In a meeting with Ford, it was decided that the glass and beta alumina systems were different enough so that joining forces would not be advantageous. 'May the best man win!'
>
> During the first 5–6 years we surveyed silicate glass structures but never found a silicate based glass that would be stable to both sodium and

sulfur. Finally, we decided on the borate glass system. The rest is slow, plodding progress.

Since that early work the potential of the sodium sulfur cell has been realized and specific energies in excess of 200 Wh kg^{-1} have been obtained for single cells. Several prototype batteries have been built, making the sodium sulfur system the most advanced of the high specific energy batteries.

1 The cell

J. L. SUDWORTH

1.1 INTRODUCTION

The sodium sulfur cell is unusual in a number of ways. Not only does it have a solid electrolyte and a liquid metal negative electrode, but the extent of the cell reaction varies according to the temperature of operation. At first sight it would appear that the cell reaction would be

$$2Na + S \rightarrow Na_2S$$

but examination of the phase diagram (Fig. 1.1) shows that the cell would have to operate above 900 °C if the sodium sulfide was to remain molten. In fact, the normal operating temperature is in the range 300 to 400 °C, and to avoid precipitation of solids the discharge is normally terminated when the polysulfide melt reaches the composition Na_2S_3. In order to discuss the cell EMF it is necessary to choose the standard states of the reactants and products. Here we will follow Cleaver and Davies' treatment (1973), in which they chose the standard states of sodium, sulfur and Na_2S to be pure liquids at the operating temperature, and in which they represented the polysulfide melts as mixtures of $Na_2S(l)$ with $S(l)$. They presented the following expressions for the EMF of the sodium sulfur cell:

$$Na(l)|\text{beta alumina}|Na_2S_x(l)|C \qquad (\text{cell 1; EMF } E_1)$$

or

$$Na(l)|\text{beta alumina}|Na_2S_y(l)+S(l)|C \qquad (\text{cell 2; EMF } E_2)$$

These apply equally well to the glass electrolyte cell.

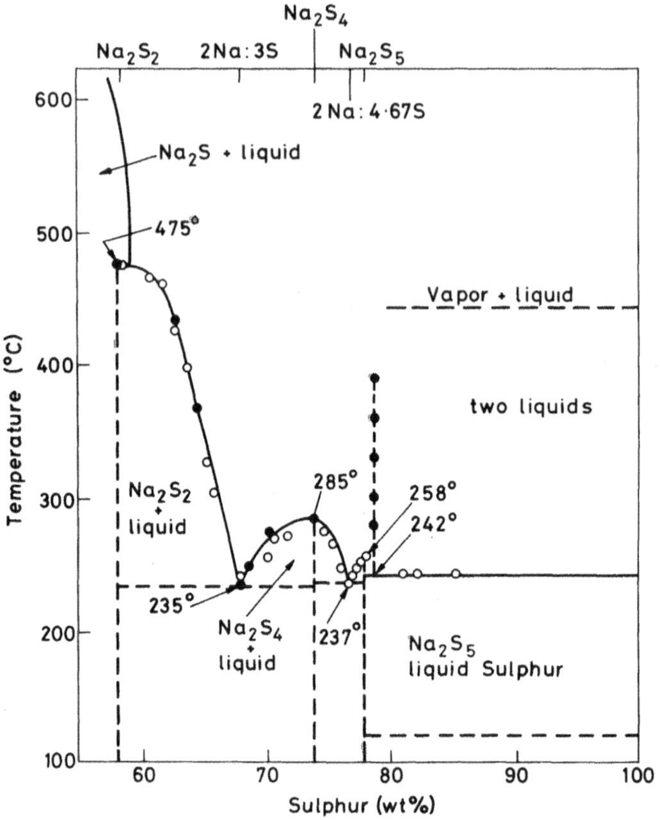

Fig. 1.1 The Na_2S/S phase diagram according to Gupta and Tischer (1972) and Oei (1973).

From the phase diagram it can be seen that for melts containing between 78 and 100 wt% sulfur two immiscible liquids are formed: a sulfur rich phase which is in fact almost pure sulfur, and an ionically conducting melt of approximate composition $Na_2S_{5.2}$. The EMF of cell 2 is invariant as long as the two phases are present, whilst at 350 °C the EMF of cell 1 declines steadily as the cell is discharged and the melt composition changes from $Na_2S_{5.2}$ to $Na_2S_{2.7}$. Thereafter solid Na_2S_2 is present and the EMF of the cell

$$Na(l)|\text{beta alumina}|Na_2S_x(l) + Na_2S_2(s)|C \qquad (\text{cell 3; EMF } E_3)$$

is again invariant because the composition of $Na_2S_x(l)$ is fixed (see Fig. 1.1) and thus the activity of Na_2S in this liquid is also fixed.

The expressions for the EMFs of these cells contain terms involving Na_2S activities which have not been determined directly, and the values shown in

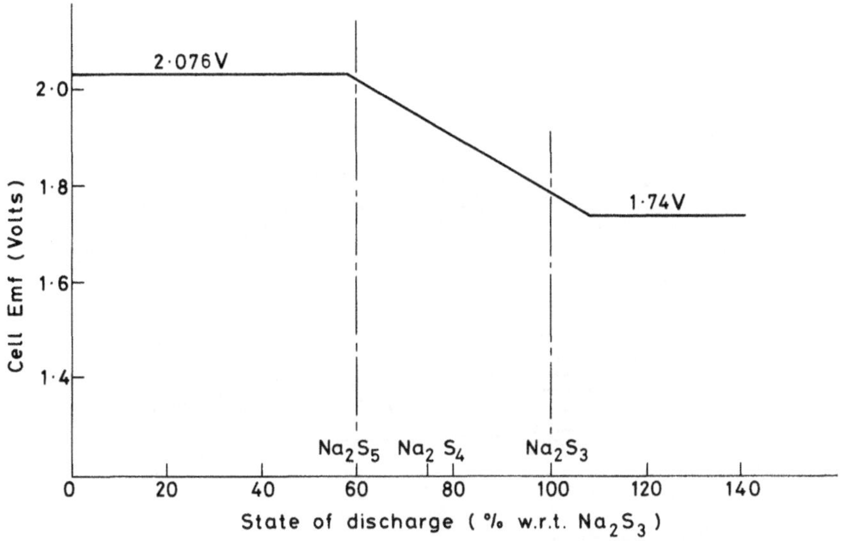

Fig. 1.2 EMF of the sodium sulfur cell at 350 °C.

Fig. 1.2 have been determined experimentally. At 350 °C the EMF of the cell in the two phase region is 2.076 V, and this declines to 1.78 V at a polysulfide composition corresponding to Na_2S_3. Further discharge to the composition corresponding to $Na_2S_{2.7}$ results in the EMF declining to 1.74 V, at which value it remains as long as the liquid phase is present.

The specific energy of the cell reaction

$$2Na + 3S \rightarrow Na_2S_3$$

calculated from Fig. 1.2 is 760 Wh kg^{-1}, and this is the value normally quoted for the sodium sulfur cell. It can be seen from the analysis that this is an arbitrary value; it is quite possible to discharge the cell further, in which case the specific energy should be recalculated from Fig.1.2.

So far only reversible operation of the cell has been considered. In practice the cell will be discharged and recharged irreversibly and the cell voltage will be a function of the current and the state of charge. In fact the cell voltage mirrors the cell EMF over a wide range of currents during discharge and charge, and this is very well illustrated by the charge/discharge characteristics of the cell described by Kummer and Weber (1967) in their first publication (Fig. 1.3). Of note in Fig. 1.3 are the high current densities on charge as well as discharge, the relative insensitivity of the discharge capacity to current density, and the low cell resistance.

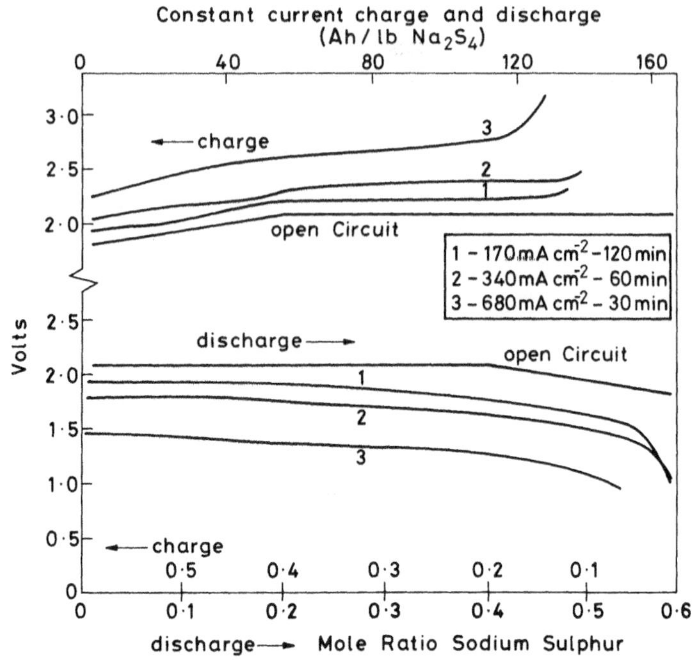

Fig. 1.3 Charge/discharge characteristics of the sodium sulfur cell described by Kummer and Weber (1967).

1.2 DEFINITION OF TERMS

Because it is so different from other secondary cells, the sodium sulfur cell has to some extent generated its own terminology. Furthermore, different groups of workers have used different terminology. To avoid confusion, the terms used in this book will be defined here.

The terms used to describe the parameters of cell performance are illustrated in Fig. 1.4 (Piercy and Sudworth, 1977). The *charge acceptance* refers to the percentage of the theoretical capacity which can be regenerated during charge starting from the sulfur electrode composition Na_2S_3. Thus, if it were possible to convert all the sodium sulfide to sulfur, the charge acceptance would be 100%. The *depth of discharge*, sometimes abbreviated to DoD, is the fraction of the elemental sulfur which has been converted to Na_2S_3. Thus, the *theoretical capacity of the cell* is the capacity calculated from the weight of sulfur in the cell assuming complete conversion to Na_2S_3. The weight of sulfur may not be known if, for example, it is added as a mixture with other materials. In this case the theoretical cell capacity can be determined by discharging the cell until it is in the single phase region, allowing the EMF to reach a steady value, then passing a known quantity of current and allowing the EMF to reach a steady

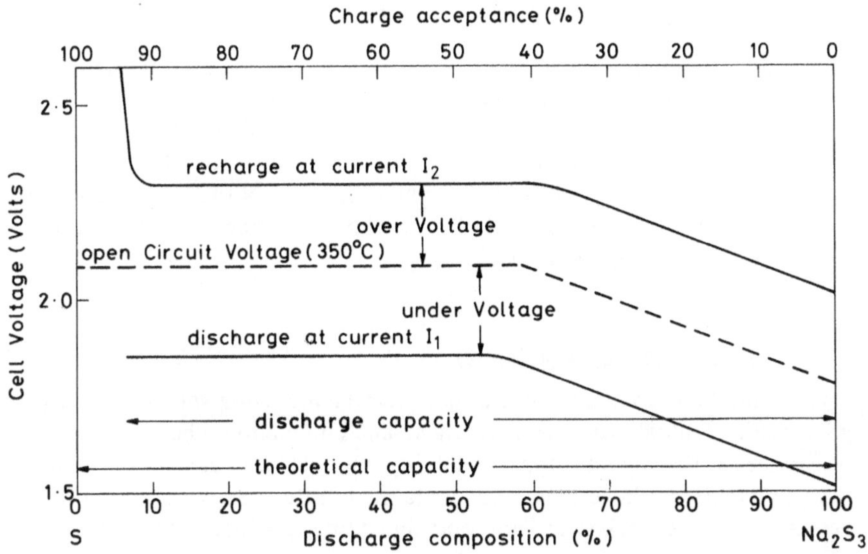

Fig. 1.4 Parameters for assessing cell performance. R discharge = under voltage/I_1. (Piercy and Sudworth, 1977.) Reproduced with permission of the Electrochemical Society Inc.

value once more. By using the known variation of cell EMF with composition in this region, the capacity can easily be calculated.

One term which has been used for different cell components in sodium sulfur cells with beta alumina electrolyte is the *current collector*. Some workers use this to refer to the carbon felt matrix in the sulfur electrode whilst others use it to refer to the massive electronically conducting component by means of which current is drawn from the sulfur electrode. In this book the latter meaning will be used. Another term sometimes used instead of current collector is *pole*. This is often preferred for the current collector in plate cells and in tubular cells in which the sulfur electrode is contained within the electrolyte tube.

The *resistance* of cells with beta alumina electrolyte is often referred to the *area of the solid electrolyte*, as is the cell *current*. This is convenient when discussing cells which have different electrolyte areas, and the unit $\Omega\ cm^2$ obtained by multiplying the measured cell resistance by the electrolyte area has been widely adopted for the cell resistance. The electrolyte area used to calculate this parameter should be the area in contact with the sulfur electrode. Similarly, the current density is referred to the same area. In the glass electrolyte cell the current density is referred to the area of the foil electrode.

For the *resistivity* of the solid electrolyte the unit Ω cm will be used. The area resistance of the electrolyte is, of course, this value multiplied by the thickness of the electrolyte.

1.3 THE MAIN FEATURES OF THE BATTERY

Subsequent chapters will be devoted to the various aspects of the cell and battery, and for those readers who are familiar with the field the remaining sections of this chapter will probably be superfluous. Those readers, however, who are interested in this battery but who do not have first hand experience of it might profit from a brief outline of the main features of the sodium sulfur cell before tackling the necessarily detailed descriptions of the various components of the battery in Chapters 2 to 13.

1.3.1 The solid electrolyte

The essential requirements for the sodium ion conducting solid electrolyte are good ionic conductivity with zero electronic conduction, chemical corrosion resistance to the reactants, impermeability and adequate mechanical strength. Only two materials have been found suitable: borate glass, which has been used exclusively by the workers at Dow, and beta alumina, which has been used in all other cell designs.

The resistivity of the glass (about 10^3 Ω cm) is higher than that of beta alumina by nearly three orders of magnitude, and for that reason it is used in the form of thin walled capillary tubes. Little information has been published by Levine and his co-workers on the manufacturing process for the glass, but it is known that chloride is added to improve conductivity and that the glass is drawn into hollow fibres. The fibres have a diameter of only 90 μm and many thousands are required for a cell with a useful capacity. Even with a wall thickness of only 10–15 μm, most of the resistance of the cell arises from the glass electrolyte and the current density at the surface of the fibres is kept to a value of 2 mA cm^{-2}. Degradation of the glass in service can be caused by the presence of impurities in the sodium metal. Oxides and hydroxides of sodium are particularly deleterious to the glass, and it has been found necessary to remove oxygen from the sodium by means of getters such as zirconium titanium alloys. The glass is fairly tolerant to the presence of other alkali metals and the potassium level of commercial grade sodium causes no problems. Calcium, however, can cause degradation of the glass and the amount present in the sodium must be kept to a minimum.

The presence of minute particles of platinum has been found to give rise to cracking of the glass fibres and it is necessary to ensure that the level of oxygen in the glass melt is sufficient to prevent the dissolved platinum oxide precipitating as platinum.

The solid electrolyte which is most widely used in sodium sulfur cells is a crystalline solid known as beta alumina. In fact, there are two distinct materials, beta alumina and beta″ alumina. The chemical composition of the two materials is very similar but the crystal structure is quite distinct. Both contain slabs of close packed oxide which are four oxygen layers deep, but beta alumina has a twofold screw axis whilst beta″ alumina has a threefold screw axis. For this reason the two structures are often referred to as two block and three block material. From the point of view of utility as solid electrolytes the features which differentiate these materials are their electrical resistivity and their thermal stability. The three block material, beta″ alumina, has a considerably lower resistivity than the two block material at the temperature at which the cell operates (typically 5 Ω cm at 350 °C for the former compared with 15 Ω cm for the latter) but it is unstable at temperatures in excess of 1550 °C, and to fabricate dense polycrystalline ceramic from this material it is necessary to stabilize it with dopants such as magnesia and lithia. The amount of these dopants is usually below the 2% level and they may be used alone or in combination with each other. The level of dopants does seem to affect the stability of the beta″ alumina as an electrolyte and only a few compositions have been found to be useful. For many years some workers believed that the two block material was inherently more stable as an electrolyte, but it is now generally recognized that this is not the case and the lower resistivity of the three block material has made it the preferred electrolyte.

Although the material is stable to sodium and sulfur at 350 °C, passage of current can lead to electrolyte degradation. Electrolyte removed from failed cells may be seen to be penetrated by black filaments, often referred to as dendrites. These appear to be sodium filled cracks in the ceramic. The formation of these dendrites has been explained in terms of generation of sodium in pre-existing microcracks during cell charging and growth of these microcracks due to the Poiseuille pressure resulting from the flow of sodium along the crack. To prevent the formation of these dendrites the population of flaws is reduced to a minimum, the strength of the ceramic is maximized and the grain size is minimized. It does appear, however, that other mechanisms of electrolyte failure exist, since ceramic which is fine grained and strong may still develop dendrites and relatively coarse grained ceramic may have a very long life in a cell. Whilst some workers emphasize the importance of fine grain size, others take a different view. This division of opinion is reflected in the views of two of our contributors: P. T. Moseley, who believes that fine grain size is important for long life, and R. S. Gordon, who considers the point not proven. The chemical composition of the electrolyte is an important factor, but once a stable composition has been identified it is almost certainly advantageous to aim for a strong ceramic (greater than 200 MN m^{-2}).

Fabrication of a ceramic with these properties is not straightforward, as the material has a propensity towards exaggerated grain growth during sintering. The first requirement seems to be a powder preparation method which will result in the soda and the dopants being uniformly distributed. This can be achieved by dry milling the components of the powder for long periods followed by calcination to convert the powder at least partially to beta alumina. For better homogeneity the minor constituents such as lithia and soda may be prereacted separately with the alpha alumina, and these materials may then be milled together to form a homogeneous powder. Alternatively, advantage may be taken of the solubility of sodium and lithium hydroxides and aqueous solutions of these can be milled with alpha alumina to produce a slurry of alpha alumina in a solution of the alkali metal hydroxides. This slurry solution may then be spray dried to produce a powder, each particle of which will have the same chemical composition.

Once a homogeneous powder has been produced it must be formed into a tube or disc and two forming methods have been widely adopted. A convenient way of forming ceramic powders is isostatic pressing and beta alumina artefacts can easily be formed by this method. It has the advantage of being a well established and rapid method of producing green shapes, which can be easily automated, given a powder with the correct rheological properties. A more unconventional method which has also been used is electrophoretic deposition. This requires much closer control of the powder properties and of the environment in which forming takes place, but deposition of the powder can be reasonably fast and the final product has a very good microstructure.

To sinter these powder compacts into dense polycrystalline ceramics requires temperatures around 1600 °C and the time/temperature profile during firing must be closely controlled if exaggerated grain growth is to be avoided. Both batch firing and pass through firing methods are in use but in both cases it is necessary to encapsulate the artefact to prevent soda loss. An exception to this is zone sintering where the first few artefacts to be fired generate a soda rich atmosphere which allows subsequent artefacts to sinter without excessive soda loss occurring. The refractory used for encapsulation must resist attack by the sodium oxide vapour and only two materials have been found to be satisfactory – platinum and magnesia. The amount of platinum may be minimized by using it to line an alpha alumina tube, but it is obviously much more expensive than magnesia. It has the advantage, however, that it can be used for an indefinite number of firings whereas magnesia tubes must be expected to have a finite life if only because of breakages.

Pass through firing can be shown to be economically more desirable than batch firing but at present most beta alumina is manufactured by batch firing. It may well be that the final manufacturing route will be a compromise between the two methods and tunnel kilns have been suggested in this respect. Whether these would be compatible with the close control of firing conditions necessary for sintering beta alumina remains to be seen.

1.3.2 Cell configuration

Cell configuration is largely determined by the geometry of the solid electrolyte. In the glass electrolyte cell, the need to use the glass in the form of thin walled capillaries results in a configuration in which many thousands of glass capillaries are sandwiched between sheets of aluminium foil in such a way that the effective volume of sulfur around each capillary corresponds Faradaically to the volume of sodium in the capillary. The capillaries are fused at their open ends to a sheet of glass referred to as the tube sheet and this forms the base of the sodium reservoir. The tube assembly is encased in a stainless steel container which contains the sulfur/polysulfide melt.

Beta alumina is generally available in the form of tubes or discs and three cell configurations are possible. In plate cells a beta alumina disc is sealed to an alpha alumina ring and the positive and negative compartments are closed with an inert current collector disc. To make a battery it is convenient to pack these plate cells in a bipolar array, thus eliminating the problem of intercell connections. Two problems have prevented the plate cell being widely adopted. The first is the change in volume in the two electrode compartments during discharge. This can lead to pressure differences across the beta alumina disc causing fracture of the ceramic. Although solutions, such as flexing bipoles, have been suggested this still remains a severe problem. The problem of sealing the cell has also defied solution: the ratio of seal area to electrolyte area is necessarily high and the seals must be immersed in the reactants. Consequently the plate cell has found application only as a laboratory tool.

With tubular electrolyte geometry there are two possible cell configurations – central sodium (sometimes referred to as sodium core cells), in which the sodium is contained within the electrolyte tube, and central sulfur (sulfur core cells). The cell described by Weber and Kummer was of the central sodium type as were most early cells. As the problem of corrosion of the current collector came to be recognized as a serious impediment to cell longevity, and as ceramic technology advanced to the point where tubes of several centimetres diameter could be produced, some groups decided that turning the cell inside out would have a number of advantages. Thus the two types of cell have been developed in parallel and as yet neither has shown sufficient advantages over the other to make it the preferred configuration. The advantage of the central sodium cell is that the cell case also acts as the current collector and this gives it a better specific energy than the central sulfur cell, especially on a volumetric basis. The advantage of the central sulfur cell is that aluminium can be used as a passive substrate for the current collector and such current collectors can show virtually no corrosion after over 1000 charge/discharge cycles. In contrast many of the central sodium cells use expensive stainless steel as the cell case and this has to be coated to prevent corrosion. Any defects in that coating can lead to corrosion of the stainless steel, adversely affecting cell performance. Mild steel protected by chromizing is used in some central sodium cells but again any defects in the coating will lead to corrosion of the mild steel and possibly spalling of the chromized layer.

With tubular cells it is possible to orientate them vertically or horizontally. Vertical orientation with the seals uppermost has the advantage of keeping the seals out of contact with the polysulfide melt and thus reducing the possibility of chemical or electrochemical attack, but horizontal orientation may be necessary if the cell is long and space is limited.

1.3.3 The sulfur electrode

In the sulfur electrode elemental sulfur must be reduced to sodium polysulfides on discharge and reformed by oxidation of these compounds on charge. We have seen from the phase diagram that for much of the discharge two liquid phases, elemental sulfur and sodium pentasulfide, are present; the former is a viscous liquid with a very high resistance whilst the latter, like the lower polysulfides, is a mobile, ionically conducting molten salt. In the glass electrolyte cell reduction of the sulfur takes place at the surface of the metal foil electrode and the distance of this from the electrolyte tubes is small enough that all the sulfur in the vicinity can be reduced without polarization of the electrode. In cells with beta alumina electrolyte the distance between the current collector and the beta alumina is much greater and the electrode volume must be filled with an electronically conducting matrix with a high surface area. Initially, of course, the only electrolyte present is the beta alumina, but as sodium pentasulfide is formed this can be transported throughout the electrode and also act as an electrolyte allowing sulfur to be reduced over the whole surface of the matrix material. The distribution of reaction is determined partly by the relative resistances of the melt and the matrix material and can be varied by changing the properties of the latter. Carbon felt or mat is usually used as the matrix in the sulfur electrode and this is available in a range of resistivities. Whichever matrix material is chosen there is usually little difficulty in discharging the cell fully: the problem arises on recharge when insulating sulfur is generated in the two phase region. The electrochemistry of the anodic and cathodic processes in polysulfide melts has been extensively studied and at the current densities normally used in cells there should be little polarization arising from the electrode reaction during charge and discharge. Despite this it is commonly observed that, in cells in which the sulfur electrode consists of a uniform carbon felt matrix, charge can be terminated at relatively low current densities by severe electrode polarization as soon as the sulfur electrode enters the two phase region. There now seems little doubt that this is due to blocking of the solid electrolyte surface by elemental sulfur. Many electrode designs have been suggested to overcome this problem, and the most successful of these allow the cell to be recharged well into the two phase region at current densities of up to half an ampere per square centimetre.

The sodium polysulfides are extremely corrosive melts and selection of a material for the current collector in the sulfur electrode requires some care. In the glass electrolyte cell, aluminium is used as the electrode, but as this forms a film of insulating aluminium sulfide it must be coated with another material. Both carbon and molybdenum are satisfactory and these can be present as very thin films at the low current densities used in this cell. In cells with beta alumina electrolyte aluminium can also be used but the coating must be more substantial. Coated aluminium current collectors are particularly suitable for central sulfur cells where the current collector is a small diameter rod which can easily be coated. For central sodium cells, where the cell case is also the current collector, aluminium is less acceptable because of its relatively low melting point. This could compromise cell integrity when cell failure occurs, and also means that the metal is close to its yield point at the cell operating temperature, an undesirable feature for a structural component such as the cell case. Coating the inside of a tube is also more difficult than coating the outside surface of a rod, and for these reasons aluminium has not been widely used in the central sodium cell. For these cells, steel cases with some kind of protective coating are normally used.

1.3.4 The sodium electrode

There are two requirements for satisfactory operation of the sodium electrode: the sodium metal must remain in contact with the whole of the solid electrolyte surface during charge and discharge, and there must be little or no interfacial polarization at the surface of the solid electrolyte. There are a variety of methods for ensuring the first condition is satisfied. However, interfacial polarization is a complex phenomenon which can manifest itself in a number of ways, inducing resistance rise with cycling and asymmetry of cell resistance. In glass electrolyte cells it is necessary to purify the sodium to prevent attack of the glass by oxides and hydroxides and to prevent calcium ions from entering the glass. In beta alumina cells the extent to which interfacial polarization occurs is also dependent upon the purity of the alpha alumina used in the fabrication process. Electrolyte made from high purity alpha alumina is much less likely to exhibit interfacial polarization effects than that made from commercial grade material. The main impurities in alpha alumina are calcium and silica and there is good evidence that calcium, which is mobile in the beta″ alumina structure, can migrate to the beta″ alumina/sodium interface and affect the electrode processes occurring there. Other factors involved include the dopant level in the beta alumina and the oxygen content of the sodium metal. It is possible, however, to control this phenomenon to the extent that cells containing beta alumina manufactured from low purity alpha alumina can retain a stable resistance for the life of the cell.

1.3.5 Cell design and construction

The dimensions of the cell will be determined by the application for which it is intended. However, within the limits set by this there will be optimum dimensions for the electrolyte and the sulfur electrode and, because the current efficiency of the cell is 100% and the resistance is ohmic, it is possible to calculate the optimum dimensions rather easily. For the glass electrolyte cell the dimensions of the glass fibres are to all intents and purposes fixed and these determine the spacing of the fibres in the cell. The only variables are the number of fibres in the cell and the length of the fibres. For the beta alumina electrolyte cells the dimensions of the electrolyte can vary over a wide range and optimization of cell dimensions for particular applications is possible.

It can be shown that, for a specific discharge rate (i.e. energy : power ratio) and an electrolyte of given dimensions, the specific energy passes through a maximum as the thickness of the sulfur electrode increases. This arises simply because to discharge a cell in a given time will require higher current densities as the amount of sulfur (and thus the capacity) is increased, and there will be a point at which the increase in weight or volume of the cell and the reduction in voltage efficiency balance out the increase in energy. The point at which this maximum is reached will depend upon the cell resistance and the dimensions of the cell components. Similarly there will be an optimum electrolyte diameter and an optimum length for a given set of conditions. Whilst these calculations are a useful guide, practical considerations may dictate that other dimensions are chosen. Thus the scrap rate in the production of electrolyte tubes may reach prohibitive levels at tube lengths considerably less than those corresponding to the theoretical maximum.

Once the cell dimensions, the sulfur electrode structure, the current collector material and the method for retaining full coverage of the electrolyte by sodium have been chosen, it only remains to seal the cell. Direct sealing of the metal cell components to the solid electrolyte is best avoided as this can lead to generation of sodium metal in the seal during charge. In all practical cell designs the electrolyte tube is sealed to an insulating collar or header: in the glass electrolyte cells the fibres are sealed to a glass header or tube sheet, and in beta alumina cells the electrolyte is sealed to an alpha alumina collar with a glass. It is important to achieve a seal in which tensile stresses are low both at the operating temperature and at room temperature, and careful matching of expansion coefficients over the relevant temperature range is essential. Once the insulating collar is sealed to the electrolyte tube there are a number of ways in which the metal cell case can be joined to the insulating collar, but those adopted for use in cells often involve the use of aluminium metal as an intermediate material between the metal and the insulating collar. Its passivity to sulfur and its low yield point make it in many ways an ideal sealing material.

Sodium and sulfur are very reactive elements and it is essential to build in design features which will ensure that the cell is safe even under conditions of extreme abuse. This means ensuring that chemical reaction of sodium and sulfur is controlled sufficiently for the heat of reaction to be dissipated without the cell reaching a dangerously high temperature, and protecting the metal components from corrosion by the sodium polysulfides at temperatures which may be encountered during cell failure. With low capacity cells safety can be ensured by simple design features, but as the cell capacity increases these safety features become more complex and can lead to problems with cell performance unless great care is taken to prevent this.

1.3.6 Cell performance

The main factor determining cell performance is the internal resistance and the main contributor to this is the ceramic electrolyte. A low cell resistance is desirable on two counts – to achieve a high power : weight ratio and to minimize the amount of heat produced on discharge. The cell resistance can be reduced by increasing the temperature of operation but there is little advantage in operating at temperatures much above 350 °C. As well as achieving a low initial cell resistance it is essential that it should remain low during the life of the cell. Provided that the interfacial polarization mentioned earlier is eliminated and corrosion of the current collector does not occur, there is no reason why the cell resistance should not remain stable.

The sodium sulfur cell is unique in that the only cell reaction is formation and decomposition of the various polysulfides and this determines the way in which the cell is charged and discharged. A fully charged cell will develop an extremely high resistance and if current is forced through such a cell degradation of the solid electrolyte is unavoidable. Thus it is essential to have good charge control for sodium sulfur cells. On discharge, on the other hand, the end of discharge is an arbitrary point and the cell will continue to discharge beyond this point. Depending upon the cell design, this can result in fracture of the ceramic electrolyte or rupture of the cell seals or the cell case as the volume of material in the sulfur electrode exceeds the design limits of the cell. Alternatively, current concentration effects resulting from a shortage of sodium in the sodium electrode can cause fracture of the solid electrolyte. Thus, severe overdischarge must be prevented.

A very important feature of cell performance is the life of the cell. A cell can fail suddenly or it can degrade slowly. Sudden failure is most often associated with failure of the solid electrolyte and there is a tendency for this kind of cell failure to fall into the pattern of infant mortality failure, i.e. a decreasing failure rate. As ceramic technology develops and more reproducible and reliable material is produced these early failures should be eliminated. Slow degradation of cell performance is associated with capacity loss and resistance rise and it is possible to eliminate these by correct choice of the current collector material and elimination of interfacial polarization at the sodium/solid electrolyte interface. This still leaves one failure mechanism: seal leakage. This almost invariably falls into the pattern of wearout failure, i.e. increasing failure rate, and in some cells may be the major determinant of cell life. Sodium sulfur cells have achieved lives of several thousand cycles and there seems no fundamental reason why even longer lives should not be achieved. The problem is more one of reliability and elimination of early failures than one of longevity.

Subsequent chapters will cover in depth all the points outlined in this chapter, and in addition will consider the problems which arise when cells are connected together to form batteries. At the end of the book is an appendix containing data which we have found to be useful and which we hope will prove equally useful to others working on sodium sulfur batteries.

1.4 REFERENCES

Cleaver, B. and Davies, A. J. (1973) *Electrochimica Acta*,**18**, 733–39.
Kummer, J. T. and Weber N. (1967) *Proc. SAE Congr.*, paper 670179, 1–6.
Piercy R. and Sudworth J. L. (1977) *Proc. Symp. Electrode Materials Processes Energy Conv. Storage*, p. 950.
Gupta, N. K. and Tischer, R. P. (1972) *J. Electrochem. Soc.*, **119**, 1033.
Oei, D.-G. (1973) *Inorg. Chem.*, **12**, 435.

2 The solid electrolyte – properties and characteristics

P. T. MOSELEY

2.1 INTRODUCTION

The concept of the sodium sulfur cell was almost certainly a direct consequence of the discovery of the remarkably high conductivity of sodium ions in the ceramic oxide, beta alumina (Weber and Kummer, 1967). The requirements for the material to act as separator and electrolyte in the cell are extremely demanding, including total retention of integrity in aggressive environments at high temperatures. Also, in view of the market demands for a device to offer a reliable capability for at least 1000 deep charge/discharge cycles, neither the mechanical properties nor the electrical specification for a high ionic and zero electronic conductivity can be allowed significant shortfall or degradation over long periods.

Few materials are known to be capable of meeting these demanding needs and the majority of designs of sodium sulfur battery still incorporate an electrolyte separator which is a polycrystalline artefact of some form of beta alumina.

Beta alumina was first reported as long ago as 1916 (Rankin and Merwin) but initially it was not appreciated that sodium is an essential constituent of the material. That beta alumina is actually a sodium aluminium oxide was established during the first crystal structure analysis (Bragg *et al.*, 1931) when beta alumina was assigned the formula $Na_2O \cdot 11Al_2O_3$. It has since become clear that beta alumina is never stoichiometric as prepared and the formula $(Na_2O)_{1+x}11Al_2O_3$ is more appropriate, but the original name remains in general use.

One other material receiving some attention as a possible sodium conducting membrane in sodium sulfur batteries is a sodium ion conducting glass developed by the Dow Company (Levine *et al.*, 1972). The present survey, dealing with electrolyte properties, properly reflects the dominant position of beta alumina as the material most widely used and most deeply scrutinized, but comparison is made with the conducting glass wherever possible.

There are a number of oxides now known to belong to the beta alumina group by virtue of sharing common features of crystal structure. The group may be subdivided into those members containing a twofold screw axis and those containing a threefold screw axis. The archetypes of the two subgroups are designated beta alumina and beta″ alumina respectively, and there are fundamental differences between the properties of the two forms. Both forms are used in electrolytes.

2.2 CRYSTAL STRUCTURE

2.2.1 The parent structures

The beta alumina group of oxides is characterized by structures composed of alternating slabs of close packed oxide and layers with a low atom density containing mobile cations (typically sodium). The close packed oxide slabs accommodate small metal cations (typically Al^{3+}) in both octahedral and tetrahedral interstices but not in the same proportion as in spinel ($MgAl_2O_4$).

As mentioned earlier, there are two main subgroups which differ in the stacking sequence of layers up the unique axis. The first subgroup (typified by beta) is stacked according to a twofold screw axis, contains a mirror plane through the layers of mobile cations and results in hexagonal crystal structures. The idealized structure of the unit cell of beta alumina is shown in perspective in Fig. 2.1. Fig. 2.2 shows a $11\bar{2}0$ section through the structure highlighting the stacking sequence of layers up to the *c* axis. The two figures show how the slabs of close packed oxide extend normally to the hexagonal *c* axis and adjacent slabs are held apart by rigid Al–O–Al spacer units. Ionic diffusion occurs exclusively within the open planes perpendicular to the *c* axis.

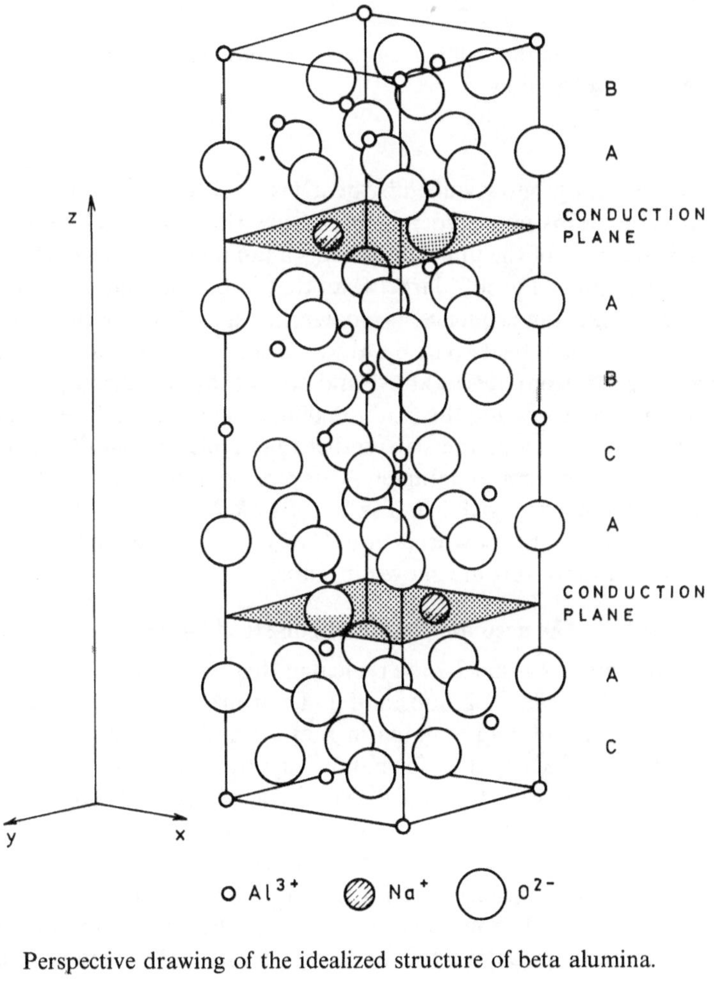

Fig. 2.1 Perspective drawing of the idealized structure of beta alumina.

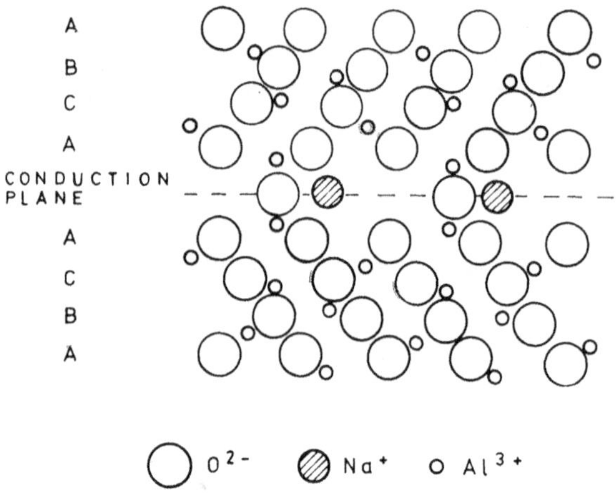

Fig. 2.2 $11\bar{2}0$ section through the idealized structure of beta alumina showing stacking sequence up to the c axis.

The second subgroup (typified by beta″) is stacked according to a threefold screw axis, contains no mirror plane and has rhombohedral structures. The idealized structure of the unit cell of beta″ alumina is shown in perspective in Fig. 2.3. The unit cell is 50% larger than that of beta alumina by virtue of the difference in stacking sequence, as shown in Fig. 2.4. Adjacent close packed oxide slabs are again held apart by Al–O–Al spacer units, but in this structure alternate sodium atom sites lie above and below the plane through the centre of the oxide spacer atoms and the Na^+ ion diffusion path actually encompasses a finite volume (the conduction slab) rather than a plane as in the beta alumina structure. Sets of structural parameters for representative compositions of beta and beta″ alumina are given in Tables 2.1 and 2.2.

The early parts of the powder X-ray ($CuK\alpha_1$) diffraction patterns calculated from these two data sets are shown in Tables 2.3 and 2.4.

2.2.2 The accommodation of non-stoichiometry

The data in Tables 2.1 and 2.2 are typical in that both beta and beta″ alumina characteristically exhibit a substantial departure from stoichiometry. The idealized structure of beta alumina in Figs 2.1 and 2.2 only accounts for the material represented by the formula $(Na_2O)_{1+x}11Al_2O_3$ where $x=0$. In practice x ranges between 0.25 and 0.55 in undoped material (Tofield, 1982) and reaches 0.66 when magnesium is present in the crystal (Baffier *et al.*, 1981). The accommodation of the excess $x(Na_2O)$ in interstitial sites of the parent lattice has important consequences both for the structure and for the conductivity of beta alumina. Fig. 2.5 shows a section through the conduction plane of beta alumina with the various sites that have been considered for sodium occupancy

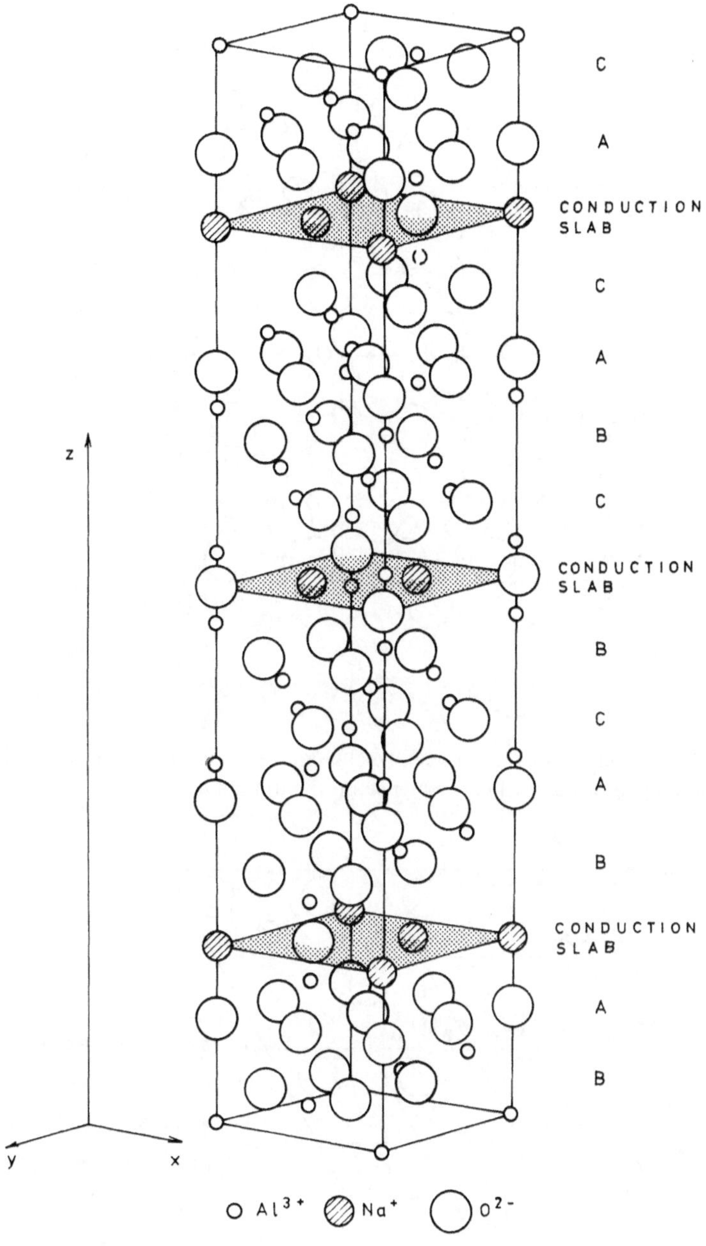

Fig. 2.3 Perspective drawing of the idealized structure of beta″ alumina.

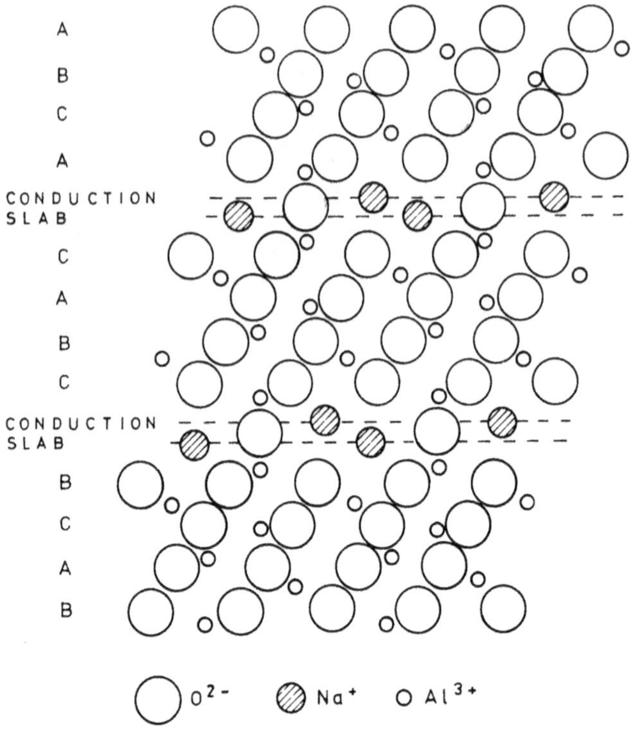

Fig. 2.4 $11\bar{2}0$ (referred to hexagonal axes) section through the idealized structure of beta″ alumina showing stacking sequence up the unique axis.

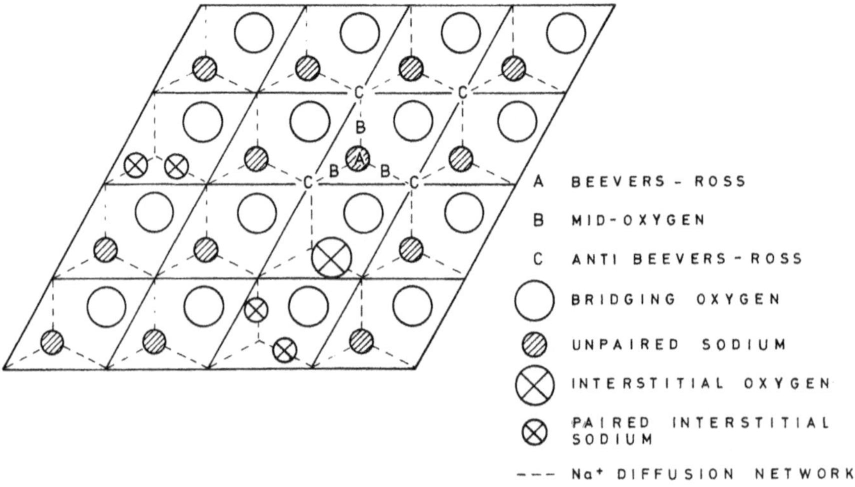

Fig. 2.5 Section through the conduction plane of beta alumina.

Table 2.1 Representative crystallographic information for the beta alumina structure

Hexagonal P 6_3/mmc

Lattice parameters for a specimen with 6.6 wt% Na_2O, i.e. $(Na_2O)_{1.29}11Al_2O_3$ (Peters *et al.*, 1971): $a = 5.594$ Å, $c = 22.53$ Å

Calculated density = 3.25 g cm^{-3}

Atomic positional parameters

Atom	*Number of equivalent positions, Wyckoff notation*	*Type of site*	*x*	*y*	*z*
O(1)	12(k)	tetrahedral	0.157	$\bar{x}$	0.050
O(2)	12(k)	tetrahedral	0.503	$\bar{x}$	0
O(3)	4(f)	tetrahedral	1/3	2/3	0.056
O(4)	4(e)	tetrahedral	0	0	0.143
O(5)	2(c)*	linear	1/3	2/3	1/4
Al(1)	12(k)†	octahedral	0.832	$\bar{x}$	0.106
Al(2)	4(f)	tetrahedral	1/3	2/3	0.025
Al(3)	4(f)	tetrahedral	1/3	2/3	0.176
Al(4)	2(a)	octahedral	0	0	0
Na(1)	2(c)*	BR‡	2/3	1/3	1/4
Na(2)	6(h)	m0§	0.873	$\bar{x}$	1/4

* These sites refine best when split into 6h sites with 1/3 occupancy on each.
† This site is only partly occupied. The balance of the Al atoms switch to a Frenkel defect site near 0.832 $\bar{x}$ 0.176 and stabilize an interstitial oxygen near to an m0 site.
‡ The Beevers-Ross (1937) site. Here occupied ~1.5 per unit cell.
§ The mid-oxygen site. Here occupied ~1.0 per unit cell.

Mean aluminium–oxygen distances

Atom	*Coordination number*	*Mean Al–O (Å)*
Al(1)	6	1.92
Al(2)	4	1.80
Al(3)*	4	1.75
Al(4)	6	1.90

* Constituting part of the Al–O–Al spacer column.

Table 2.2 Representative crystallographic information for the beta″ alumina structure

Rhombohedral $R\bar{3}m$

Lattice parameters referred to hexagonal axes for a specimen with 8.85 wt% Na_2O, 0.75 wt% Li_2O, i.e. $Na_{1.72}Al_{10.66}Li_{0.30}O_{17}$ (Jorgensen *et al.*, 1981): $a = 5.610$ Å, $c = 33.463$ Å

Calculated density = 3.28 g cm^{-3}

Atomic positional parameters

Atom	*Number of equivalent positions, Wycoff notation*	*Type of site*	*x*	*y*	*z*
O(1)	18(h)	tetrahedral	0.155	$\bar{x}$	0.034
O(2)	18(h)	tetrahedral	0.164	$\bar{x}$	0.235
O(3)	6(c)	tetrahedral	0	0	0.296
O(4)	6(c)	tetrahedral	0	0	0.098
O(5)	3(b)	linear	0	0	1/2
Al(1)	18(h)	octahedral	0.167	$\bar{x}$	0.929
Al(2)	6(c)*	tetrahedral	0	0	0.350
Li	6(c)*	tetrahedral	0	0	0.350
Al(3)	6(c)	tetrahedral	0	0	0.450
Al(4)	3(a)	octahedral	0	0	0
Na(1)	6(c)	BR/aBR†	0	0	0.828
Na(2)	18(h)	m0‡	0.104	$\bar{x}$	0.834

* Site occupied by Al/Li in the ratio 1.71/0.29.
† In the beta″ structure the Beevers-Ross (BR) and anti-Beevers-Ross (aBR) sites are crystallographically equivalent. Here occupied ~1.32 per unit cell.
‡ The second set of sodium positions occurs in pairs (only one of which can be occupied) centred on the mid-oxygen site. Here occupied ~3.69 per unit cell.

Mean aluminium–oxygen distances

Atom	*Coordination number*	*Mean Al–O (Å)*
Al(1)	6	1.91
Al(2)	4	1.85
Al(3)*	4	1.74
Al(4)	6	1.89

* Constituting part of the Al–O–Al spacer column.

Table 2.3 The early part of the X-ray ($CuK\alpha_1$) powder diffraction pattern of beta alumina (calculated using the data in Table 2.1)

2 theta (degrees)	*D (Å)*	*h*	*k*	*l*	*Relative intensity*
7.84	11.265	0	0	2	100.0
15.72	5.633	0	0	4	25.8
18.30	4.845	1	0	0	0.1
18.72	4.736	1	0	1	11.2
19.93	4.450	1	0	2	10.4
21.81	4.071	1	0	3	1.9
23.68	3.755	0	0	6	0.1
24.21	3.673	1	0	4	1.0
27.00	3.299	1	0	5	13.9
30.09	2.968	1	0	6	0.6
31.75	2.816	0	0	8	1.5
31.97	2.797	1	1	0	8.5
32.97	2.715	1	1	2	0.0
33.40	2.681	1	0	7	33.5
35.82	2.505	1	1	4	14.7
36.89	2.435	1	0	8	1.8
37.08	2.422	2	0	0	1.3
37.31	2.408	2	0	1	10.4
37.96	2.368	2	0	2	4.0
39.04	2.305	2	0	3	6.8
39.99	2.253	0	0	10	1.5
40.17	2.243	1	1	6	7.0
40.51	2.225	2	0	4	2.7
40.53	2.224	1	0	9	1.2
42.33	2.134	2	0	5	3.7
44.30	2.043	1	0	10	1.3
44.47	2.036	2	0	6	12.4
45.68	1.985	1	1	8	0.0
46.91	1.935	2	0	7	3.2
48.20	1.887	1	0	11	0.5
48.44	1.878	0	0	12	0.0
49.60	1.836	2	0	8	2.2
49.76	1.831	2	1	0	0.2
49.93	1.825	2	1	1	1.5
50.45	1.807	2	1	2	0.6
51.32	1.779	2	1	3	0.1
52.08	1.755	1	1	10	0.4
52.21	1.751	1	0	12	0.1
52.51	1.741	2	1	4	0.5
52.53	1.741	2	0	9	0.4
54.01	1.696	2	1	5	1.8
55.67	1.650	2	0	10	1.4

Table 2.3 (cont.)

2 theta (degrees)	*D (Å)*	*h*	*k*	*l*	*Relative intensity*
55.81	1.646	2	1	6	0.1
56.34	1.632	1	0	13	0.1
56.98	1.615	3	0	0	0.9
57.14	1.611	3	0	1	0.0
57.20	1.609	0	0	14	0.8
57.62	1.599	3	0	2	0.0
57.89	1.592	2	1	7	8.9
58.41	1.579	3	0	3	0.0
59.01	1.564	2	0	11	3.6
59.23	1.559	1	1	12	1.0
59.50	1.552	3	0	4	2.0
60.24	1.535	2	1	8	1.0
60.58	1.527	1	0	14	0.3

Table 2.4 The early part of the X-ray ($CuK\alpha_1$) powder diffraction pattern of beta″ alumina (calculated using the data in Table 2.2)

2 theta (degrees)	*D (Å)*	*h*	*k*	*l*	*Relative intensity*
7.92	11.154	0	0	3	100.0
15.88	5.577	0	0	6	35.2
18.44	4.808	1	0	1	0.0
19.01	4.666	0	1	2	7.3
21.13	4.201	1	0	4	13.6
22.60	3.932	0	1	5	1.7
23.91	3.718	0	0	9	0.0
26.13	3.408	1	0	7	0.1
28.13	3.170	0	1	8	0.0
31.88	2.805	1	1	0	10.5
32.07	2.789	0	0	12	2.0
32.46	2.756	1	0	10	13.9
32.90	2.720	1	1	3	0.0
34.77	2.578	0	1	11	17.4
35.80	2.506	1	1	6	20.1
37.08	2.423	0	2	1	11.3
37.38	2.404	2	0	2	3.1
38.56	2.333	0	2	4	11.1
39.43	2.283	2	0	5	4.2
39.59	2.275	1	0	13	0.0
40.24	2.239	1	1	9	11.6

Table 2.4 (cont.)

2 theta (degrees)	D (Å)	h	k	l	Relative intensity
40.40	2.231	0	0	15	2.3
41.67	2.166	0	2	7	7.2
42.10	2.145	0	1	14	2.5
43.02	2.101	2	0	8	0.4
45.85	1.978	1	1	12	0.2
46.14	1.966	0	2	10	32.5
47.28	1.921	1	0	16	0.7
47.88	1.898	2	0	11	0.1
48.96	1.859	0	0	18	0.0
49.68	1.834	2	1	1	0.5
49.92	1.825	1	2	2	0.7
49.95	1.824	0	1	17	0.0
50.87	1.794	2	1	4	0.8
51.57	1.771	1	2	5	0.2
51.70	1.767	0	2	13	0.6
52.36	1.746	1	1	15	0.4
53.41	1.714	2	1	7	0.1
53.76	1.704	2	0	14	0.4
54.53	1.681	1	2	8	0.0
55.45	1.656	1	0	19	0.4
56.80	1.619	3	0	0	1.0
57.17	1.610	2	1	10	3.7
57.45	1.603	3	0	3	0.0
57.45	1.603	0	3	3	0.1
57.82	1.593	0	0	21	1.1
58.16	1.585	0	2	16	4.7
58.28	1.582	0	1	20	0.9
58.68	1.572	1	2	11	7.5
59.38	1.555	0	3	6	1.9
59.38	1.555	3	0	6	0.8
59.62	1.550	1	1	18	2.0
60.49	1.529	2	0	17	3.4
62.03	1.495	2	1	13	0.1
62.50	1.485	3	0	9	1.1

marked in one unit cell. The Beevers-Ross (1937) site (ideally 2d: 2/3, 1/3, 1/4) accommodates the greater part of the sodium for beta alumina near the centre of the composition range (i.e. $x \sim 0.20$–0.3). The remainder of the sodium is found (Peters *et al.*, 1971) slightly displaced from the mid-oxygen position (6h, 0.833–0.833 1/4). As a result of the electrostatic interactions the sodium site is displaced towards the anti-Beevers-Ross site (with the result $x \sim 0.873$) and the

Na_{mO}-Na_{mO} distance becomes 3.47 Å instead of the 2.8 Å implied by the ideal site. The ideal anti-Beevers-Ross sites are not normally found to be occupied to any great extent.

Originally there appeared to be two possible ways of accommodating the charge of excess sodium ions in the plane: either by aluminium vacancies in the spinel-like layers, or by oxygen interstitials in the conducting plane. The detection of a significant underpopulation of the Al(1) site (Table 2.1) by Peters *et al.* (1971) appeared to settle the matter, until a neutron study (Roth *et al.*, 1976) revealed the presence of an interstitial aluminium atom displaced from the site of Al(1) to a position (Fig. 2.6) where it is able to bind an interstitial oxygen ion in one of the three equivalent mid-oxygen sites in the conduction plane. The result is a complex linear defect in which each unoccupied Al(1) site allows one extra interslab 'spacer' to form. The presence of each interstitial oxygen allows two extra sodium ions into the plane, and potential energy calculations (Wang *et al.*, 1975) indicate that these may remain paired and form the basis of an 'interstitialcy' diffusion mechanism.

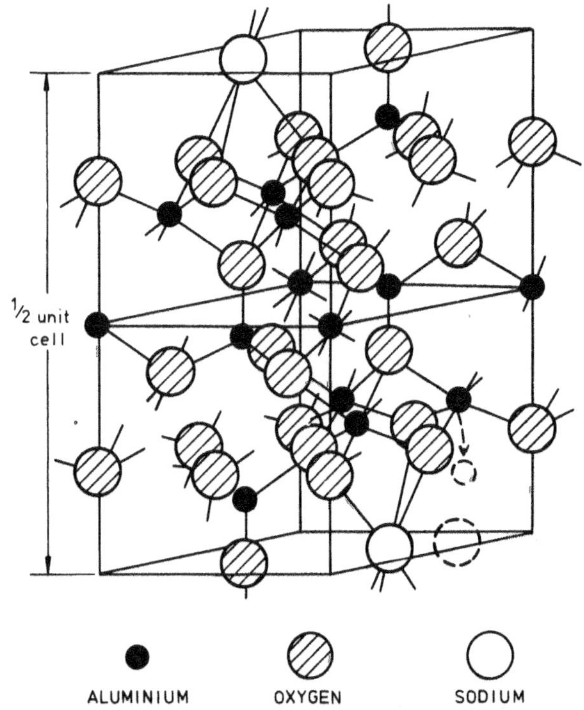

Fig. 2.6 The interstitial aluminium atom displaced from an Al(1) site in one of the three crystallographically equivalent positions in the structure of beta alumina. The interstitial aluminium and a paired atom across the mirror plane stabilize an interstitial oxygen ion in the plane.

The interstitial aluminium ion has been observed in a number of crystallographic refinements of undoped beta aluminas (Tofield, 1982) and in each case it is matched by vacancies at the Al(1) site. Thus the $V_{Al}Al_i$–O_i^{2-}–Al_iV_{Al} defect appears to be the principal means of accommodating excess sodium in undoped beta alumina.

In the case of potassium beta alumina containing magnesium no evidence for this type of defect is found (Collin *et al.*, 1980) and the charge compensation mechanism is exclusively provided by the substitution of Mg^{2+} for Al^{3+} in the spinel-like blocks.

The accommodation of increased sodium levels is accompanied by a contraction in the *c* axis of the crystal structure. The lattice parameters are substantially larger when the charge compensation involves the construction of additional interstitial spacer units than when charge compensation is achieved by Mg^{2+} doping in the block (see Table 2.5).

Table 2.5 Lattice parameter contractions with increase in sodium content of $(Na_2O)_{1+x}11Al_2O_3$

Charge compensation	*x*	*c axis (Å)*	*Reference*
Interstitials	0.29	22.54	Bourke *et al.*
	0.93	22.38	(1980)
Magnesium doping	0.30	22.45	Collin *et al.*
	0.60	22.29	(1980)

The two charge compensation mechanisms for accommodating sodium beyond the stoichiometric formula $Na_2O \cdot 11Al_2O_3$ allow a straightforward calculation of the upper limit of the sodium content of the material in the two cases by consideration of the number of vacancies required in order to hold electrostatic interaction within tolerable limits (Newsam, 1982). In the case of magnesium doped beta alumina each m0 cation is assumed to require two associated vacancies and each mirror plane containing some m0 occupation has a vacancy at the Beevers-Ross site. The maximum value of *x* in the formula $(Na_2O)_{1+x}11Al_2O_3$ is 0.667, and it is clear that this composition requires the formation of a two dimensional superlattice $\sqrt{(3)}a \times \sqrt{(3)}a$ where *a* is the parameter of the parent cell. This composition is close to the maximum value found by Boilot *et al.* (1981) who observed diffuse X-ray scattering evidence for a $\sqrt{(3)}a \times \sqrt{(3)}a$ superlattice.

In the absence of dopant cations, the analysis is modified by the need to accommodate interstitial oxygen ions in the conduction plane. In this case the maximum value of *x* is 0.571 and there is no requirement for a superlattice to form.

The maximum compositions for beta alumina with and without dopant cation such as Mg^{2+} are summarized in Table 2.6.

A corollary of this type of analysis indicates (Tofield, 1982) that the proportion of sodium ions on Beevers-Ross sites will decrease as the soda content of the crystal increases. Thus a crystal with $x=0.25$ may have 81% occupation of Beevers-Ross sites, while for a crystal with $x=0.50$ this occupancy will fall to 62.

Table 2.6 Compositions of beta alumina with maximum soda content

Material	*Formula*	*Soda content (wt%)*
Undoped	$(Na_2O)_{1.57}11Al_2O_3$	8.0
Doped with magnesium	$(Na_2O)_{1.67}(MgO)_{1.33}10.33Al_2O_3$	8.6

The maximum soda compositions given in Table 2.6 might be significantly exceeded in two sets of circumstances. The first would be by sodium ions taking up positions of Al(1) atoms which had moved to the interstitial aluminium site shown in Fig. 2.6. In this case a limiting composition of $Na_{2.0}Al_{10.667}O_{17}$ would be reached. There has been a preparation of a sodium rich beta alumina (Bourke *et al.*, 1980) for which chemical analysis indicates a composition close to this formula, but structural details are not yet available. Nuclear magnetic resonance (NMR) data for other preparations of highly soda enriched beta alumina (Dubin *et al.*, 1979) suggest a 'second type of sodium site' but the definitive demonstration of this mechanism has yet to be performed.

The second means of exceeding the limits in Table 2.6 would be by siting sodium on both Beevers-Ross and anti-Beevers-Ross sites, which would also lead to a value of $x_{max}=1.0$. This is not thought likely to occur however in view of the rather marked steric instability of the anti-Beevers-Ross site, which is likely to be the reason for sodium not occupying this site in structures with compositions nearer the middle of the range (Table 2.1).

In the case of doped materials a change in structure to beta″ alumina offers the means of accommodating higher levels of sodium. It has been found that with Mg doping of the spinel block $Al_{11-y}Mg_y$ the beta alumina structure is obtained for $0 \leqslant y \leqslant 0.6$ and the beta″ structure is found for $0.64 \leqslant y \leqslant 0.74$ (Baffier *et al.*, 1981).

There is no realistic stoichiometric formula to which the composition of beta″ alumina may be referred. The idealized structures drawn in Figs 2.3 and 2.4 correspond to a composition $Na_6Al_{32}[Vac]_{Al^{3+}}O_{51}$ but, as in the case of beta alumina, material with the idealized formula cannot be stably accommodated in the structure. The beta″ structure is only formed when stabilized by the presence of ions with a valency of less than three which substitute for aluminium in the spinel-like blocks. Because these defects have a charge of +1 or +2 rather than zero as in the $[Vac]_{Al^{3+}}$ of the idealized formula, the full complement of two sodium ions per unit cell conducting slab (Figs 2.3 and 2.4) is never achieved. Hence the beta″ structure is most conveniently described in terms of sodium vacancies (Fig. 2.7) and involves a somewhat different approach to diffusion mechanisms than does the interstitial system in beta alumina.

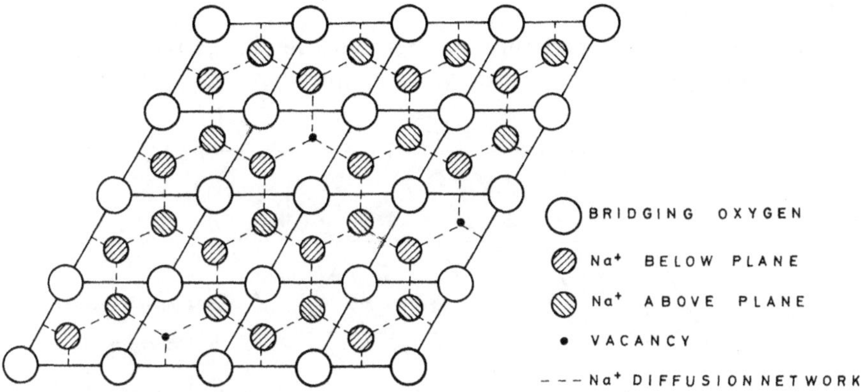

Fig. 2.7 Section through the conduction slab of beta″ alumina.

Dopant ions that stabilize the beta″ structure include Li^+, Co^{2+}, Mg^{2+} and Zn^{2+}. These dopants are all found to substitute for the tetrahedrally coordinated aluminium on site Al(2) (Jorgensen *et al.*, 1981; Dernier and Remeika, 1976; Brown *et al.*, 1981) where the Al^{3+} ions are sustaining rather longer bond lengths than in the Al–O–Al spacer column (see Table 2.2). It is likely (Roth *et al.*, 1976) that the larger dopant ions occupy the Al(2) site with less strain. Ideal compositions would be: for Mg^{2+} stabilized beta″, $Na_{1.67}Al_{10.33}Mg_{0.67}O_{17}$; and for Li^+ stabilized beta″, $Na_{1.67}Al_{10.67}Li_{0.33}O_{17}$. A maximum value for the soda content would be achieved when the two sites marked in Fig. 2.7 were totally occupied, as in the formula $Na_2MgAl_{10}O_{17}$ discussed by Bettman and Peters (1969), but the absence of vacancies would severely hamper ionic diffusion and high levels of magnesia doping lead to a product which is prone to attack by atmospheric moisture.

As in the case of beta alumina the length of the *c* axis of beta″ alumina appears to be inversely related to the sodium content (May and Henderson, 1979).

2.2.3 Ordering

The structural information described so far is the product of X-ray and neutron diffraction experiments and thus only pertains to the average atomic arrangements. Both beta and beta″ alumina possess structures which are highly defective in the sodium containing zones and this provides ample opportunity for both short and long range ordering, which is conveniently studied by X-ray diffuse scattering (Boilot *et al.*, 1980).

Two ordering schemes – one for the case of $x=0$ with all sodium near Beevers-Ross sites only, and the other for $x=0.67$ based on a $\sqrt{(3)}a \times \sqrt{(3)}a$ superlattice which orders five sodium ions and one vacancy – can be invoked to describe the diffuse X-ray scattering from magnesium doped beta alumina (Boilot *et al.*, 1981).

The $\sqrt{(3)}a \times \sqrt{(3)}a$ superlattice, which is shown in Fig. 2.8, is also adopted by the magnesia doped beta″ structure with a composition near to 1.66 Na ions per

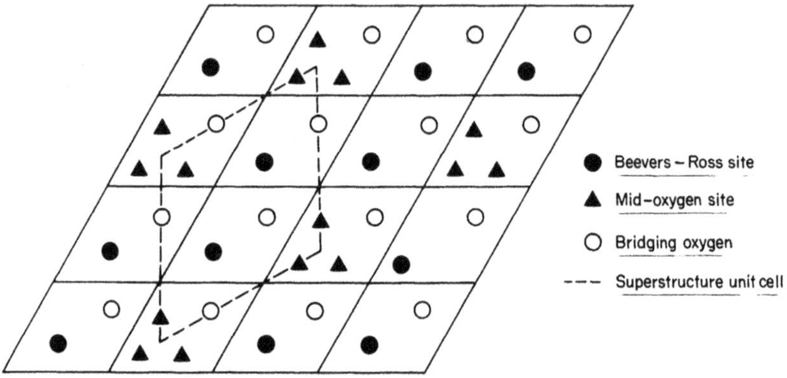

Fig. 2.8 The superlattice ordering of sodium atoms adopted by magnesia doped beta and beta″ alumina with a composition of 1.66 Na atoms per unit cell plane. The unit cell of the superlattice has sides $\sqrt{(3)}a$ long where a is the hexagonal a lattice parameter of the parent structure.

unit plane cell (Boilot *et al.*, 1980). This composition represents the maximum sodium content for the beta structure and the minimum for the beta″ structure. At this composition both structures exhibit a maximum in mid-oxygen site occupancy as a result of the superlattice ordering. As the soda content increases above 1.66 per unit plane cell, only the beta″ structure is stable and the higher levels of sodium can be accommodated by the involvement of a greater proportion of BR and aBR sites. The soda rich limit would place 2.0 sodium ions in each unit plane cell with total occupancy of both BR and aBR sites, but as mentioned previously this composition is not stable.

The extent to which the sodium containing planes are ordered defines the ion–ion correlation length which can be monitored by making measurements of the half-widths of the X-ray diffuse scattering features (Bates *et al.*, 1981). For materials of the same composition near the soda rich limit of the beta structure the correlation length is longer for beta than for beta″. This difference, together with the tight constriction between adjacent spinel block oxygen ions at the aBR site, gives rise to higher resistivities in the beta phase (Boilot *et al.*, 1981).

Ordering of the mobile cation in the conducting zones is clearly a factor which will influence the facility of ionic diffusion and conductivity, and the order/disorder transition of the beta″ superlattice which takes place in the region of 200 °C (Boilot *et al.*, 1980) is probably the origin of the non-Arrhenian conductivity behaviour of the material (Wang *et al.*, 1981).

2.2.4 Other sodium beta alumina structures

The two most common beta alumina structures, beta and beta″, both contain slabs of close packed oxide which are four oxygen layers deep. However, several other structures have been detected in which the number of oxygen layers is greater than four. These structures all possess either a twofold screw axis or a threefold screw axis and thus fall naturally into the hexagonal (beta) or rhombohedral (beta″) subdivisions of the beta alumina group of oxides. For instance, the compound $Na_2O \cdot 4MgO \cdot 15Al_2O_3$, referred to as beta‴ (read as beta triple prime) alumina (Bettman and Terner, 1971) has a spinel-like slab made up of six close packed oxygen layers separated by mirror planes 15.9 Å apart. This structure is hexagonal with $a = 5.62$ Å and $c = 31.8$ $(= 2 \times 15.9)$ Å and is clearly an analogue of beta alumina. The rhombohedral compound analogous to beta″ alumina and comprising three spinel-like blocks of six oxygen layers each is also known (Weber and Venero, 1970). This compound, designated beta⁗ (read as beta four prime) has a lattice parameter of 47.7 (3×15.9) Å.

The proliferation of superscripts to describe these materials is clearly absurd, and following the recognition of two further rhombohedral structures related to beta alumina (Bovin and O'Keefe, 1980) it is proposed that a modified Ramsdell notation be adopted in which is given (a) the number of blocks in the repeat distance up the unique axis, (b) the symmetry of the structure (H or R) and (c) the number of cation layers between conducting planes. Thus the beta structure is written as 2H(5) and beta″ as 3R(5). The longer period variants are as given in Table 2.7.

The six oxygen layer structures 2H(7) and 3R(7) are not found in the binary

Table 2.7 Long period variants of the beta alumina structure

*c axis** *(Å)*	*Original notation*	*Bovin-O'Keefe notation*
31.8	β'''	2H(7)
47.7	β''''	3R(7)
169.5	—	15R(5)
339	—	30R(5)

* The hexagonal *a* axes of all these structures are approximately 5.6 Å.

Na_2O–Al_2O_3 system but only in the MgO containing ternary. The less stable compound $Na_2O \cdot 2MgO \cdot 13Al_2O_3$ has a spinel-type structure with five oxygen layers between the conducting planes (Takabatake and Fukatsu, 1973). The significance of these structures in the context of solid electrolyte construction is that they offer a lower density of conducting planes per unit volume than either beta or beta″ and thus are undesirable.

2.2.5 Theoretical considerations

Although a large volume of information has been obtained from diffraction studies of beta alumina there remain a number of theoretical questions about the structure which need to be addressed. Among these are the origins of the driving force for the non-stoichiometry of the crystal and the reasons why the preferred number of oxygen layers per unit close packed slab is four and not more or less.

Some insight into these problems is provided by consideration of the local electroneutrality of the structure (West, 1979). Thus, the application of Pauling's rules (1963) to the stoichiometric structure (Fig. 2.1) indicates that there is an aggregation of positive charge at the centre of the spinel slabs and negative charge at its outer limits (i.e. the walls of the conduction planes). The observed departure from stoichiometry in $(Na_2O)_{1+x}11Al_2O_3$ will improve the local electroneutrality position regardless of whether charge compensation is achieved by complex Al–O–Al interstitials or by Mg^{2+} doping for Al^{3+} near the centre of the block. It is scarcely surprising therefore that stoichiometric beta alumina $(Na_2O)_{1.0}11Al_2O_3$ can only be prepared by an indirect route through silver beta alumina and hydrogen beta alumina as intermediate steps (Tofield, 1982) and seems to be metastable. The local electroneutrality concept provides a sensible explanation of the apparent preference of beta alumina for a nonstoichiometric formulation with partial occupancy of sites in its crystal structure.

Computer simulation studies of the energetics of the structure of beta alumina (Walker and Catlow, 1980) successfully predict that the Al–O–Al interstitial defect with excess sodium should be more stable than the stoichiometric formulation. Such techniques prove of particular value when applied to predicting the details of ionic motion through the structure. Thus results of two separate studies (Walker and Catlow, 1982; Wang *et al.*, 1981) both indicate that sodium in the beta alumina structure moves by an interstitialcy mechanism, i.e. by an interstitial ion pushing a neighbouring ion out of a regular site into a further interstitial position.

2.3 CRYSTAL DEFECTS

2.3.1 Transmission electron microscopy

Local aspects of the crystal structure of beta alumina which are not visible to X-ray or neutron diffraction can be viewed directly in the transmission electron microscope (TEM). The reliable interpretation of micrographs can be supported by comparison with a computed image as shown in Fig. 2.9.

The several variants of beta alumina have such closely related structures that, as revealed by several TEM studies (Bevan *et al.*, 1974; Gratias *et al.*, 1976; De Jonghe, 1977a), they can intergrow syntactically. An example of the intergrowth of beta and beta″ alumina is shown in Fig. 2.10 (Bovin, 1979b). For magnesium doped material this type of local inhomogeneity appears (Sato and Hirotsu, 1976) to represent an additional means of varying the sodium inventory of beta aluminas, although it is unlikely that syntactic intergrowth could provide the sole means of accommodating the observed non-stoichiometry (Bevan *et al.*, 1975). Defects found in the crystal structure of Mg stabilized beta″ alumina (for

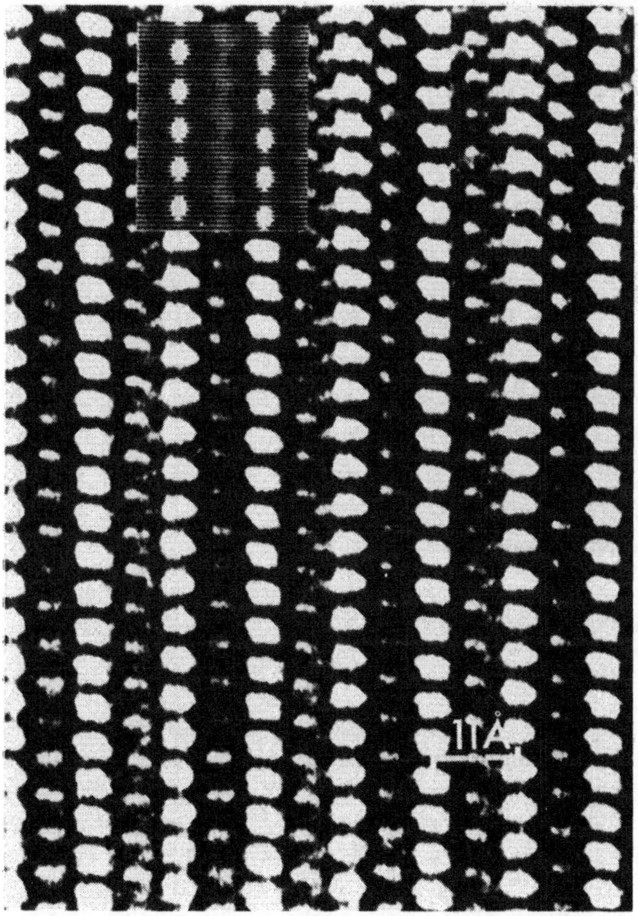

Fig. 2.9 Electron micrograph of the structure of beta alumina, beam along [11$\bar{2}$0] (after Humphreys *et al.*, 1980). Inset is the computed image.

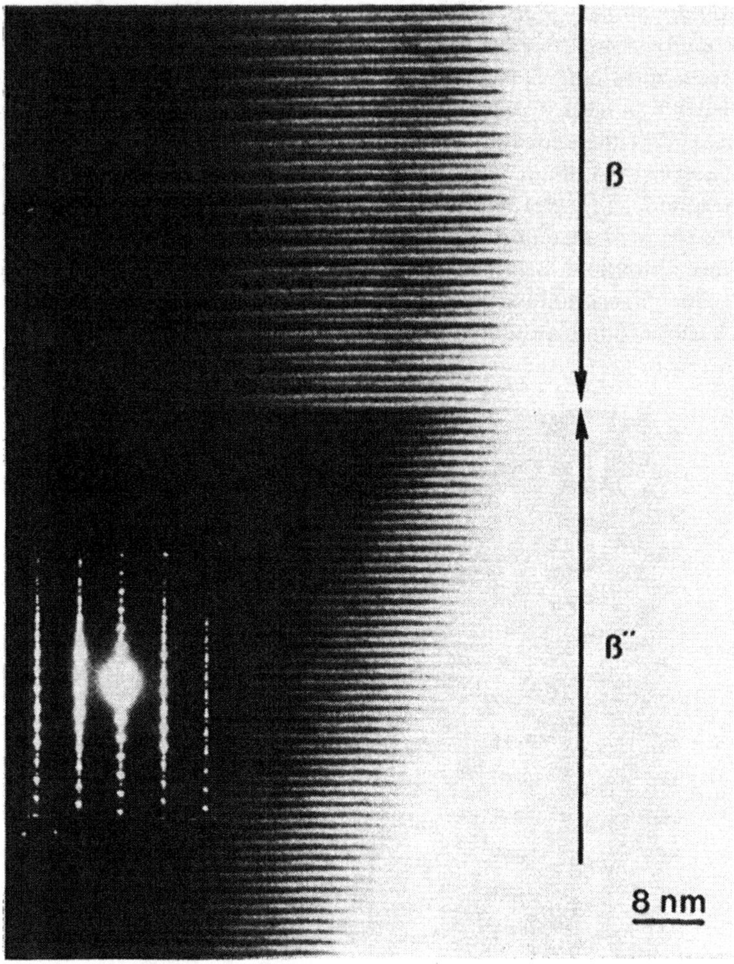

Fig. 2.10 Syntactic intergrowth of beta and beta″ alumina (after Bovin, 1979b). Inset diffraction pattern.

instance spinel blocks of seven or eleven oxygen layers) are effective at blocking the sodium ion conducting planes (Bovin, 1979a). A complicating factor in the study of such material is provided by the demonstration (De Jonghe, 1977b; Hull *et al.*, 1981; Matsui and Horiuchi, 1981) that irradiation by the electron beam can actually create such defects by causing Na_2O expulsion from the structure followed by migration of dislocation half-loops (Fig. 2.11) and coalescence of the adjacent spinel blocks.

Electron microscopy has also provided information about dislocations in undoped beta alumina. Low angle [*hk*0] tilt boundaries (De Jonghe, 1976) are

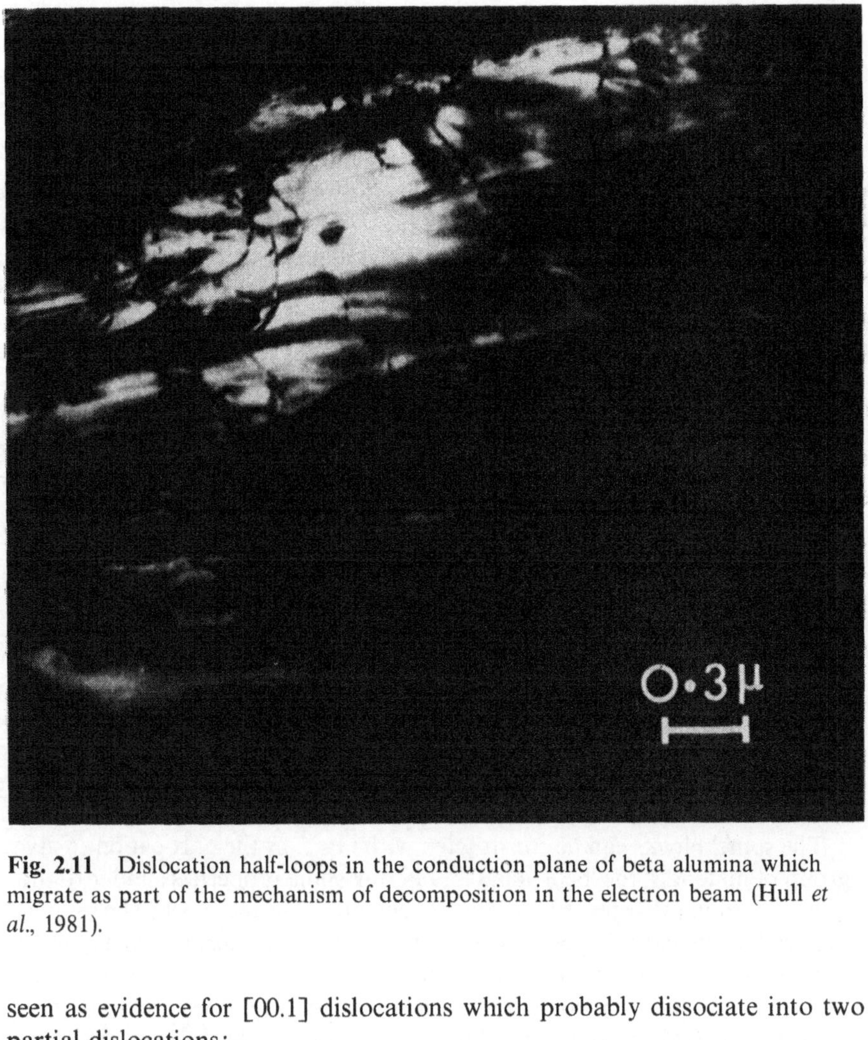

Fig. 2.11 Dislocation half-loops in the conduction plane of beta alumina which migrate as part of the mechanism of decomposition in the electron beam (Hull *et al.*, 1981).

seen as evidence for [00.1] dislocations which probably dissociate into two partial dislocations:

$$C_0[00.1] \rightarrow \tfrac{1}{2}C_0[00.1] + \tfrac{1}{2}C_0[00.1]$$

and estimates of the stacking fault energy have been made from measurements of the separation of partial dislocations (Stevens and Miles, 1976). Two methods of analysis lead to stacking fault energies of 1.65×10^{-3} J m^{-2} and 0.6×10^{-3} J m^{-2} respectively for beta alumina. These are lower than for other oxide ceramics but of the same order as for layered materials such as graphite.

The grain boundary structure of beta alumina can influence the resistivity of the material (De Jonghe, 1979). A key factor determining the effect of grain boundaries is whether or not screw dislocations with b equal to one spinel block can accommodate the orientational mismatch. This type of dislocation has been observed (De Jonghe, 1979).

2.3.2 Point defects

X-irradiation of undoped crystals of beta alumina produces four different paramagnetic defects which can be observed in the electron nuclear double resonance (ENDOR) spectrum (Barklie *et al.*, 1980). One of the defects, probably an F^+ centre (one electron trapped at an oxygen vacancy), is located at the O(5) site, and a second is thought to be an O_2^- ion at three equivalent sites in the mirror plane. The remaining two defects are probably F^+ centres at oxygen sites within the spinel-like block.

X-irradiation of beta″ crystals also produces point defects. Thermoluminescence glow curves of specimens of beta″ (Demott and Redfern, 1976) show two peaks, both giving emission with a wavelength of 505 nm. The peaks are thought to correspond to separate F-type centres with electron trap depths of 0.8 and 1.1 eV respectively.

2.4 COMPOSITION AND PHASE RELATIONSHIPS

The beta alumina structures are remarkable not only for their ionic conductivities but also for their versatility in isomorphous replacement. There is little of the structure of $(Na_2O)_{1+x}11Al_2O_3$ which cannot be substituted, at least in part, by an alternative ion. Components in the spinel block are adjusted by a suitable choice of starting materials when the crystal is grown, and the mobile cations can be replaced at a later stage as described in Section 2.6.1.

The spinel blocks can be constructed with Ga^{3+} as the only cation to give a group of materials, the 'beta gallates', offering some properties similar to those of beta alumina (Foster, 1979). The oxygen atoms in the blocks can be partly replaced by nitrogen if AlN is used as a nitrogen source during the preparation (Harata *et al.*, 1981), and the oxynitride beta alumina product shows signs of improved conductivity compared with undoped beta, presumably since the extra charge on the nitride ion stabilizes excess sodium levels in the conduction plane.

Also, as described in Section 2.2.2, Li^+ and a wide range of divalent cations can substitute for a fraction of the aluminium ions in the spinel block. The limitation on cationic substitution for Al^{3+} in the block is one of size. Ions with a radius <0.97 Å can make the substitution (Boilot and Théry, 1976) while larger ions may not. All the divalent cations of size <0.97 Å are transition metals of potentially variable valence, except for Mg^{2+}; it is for this reason that MgO and Li_2O are preferred additives to beta alumina in the preparation of a good ionic conduction with no electronic contribution.

The functions of additives in stabilizing one beta alumina phase or another depend upon their incorporation into the structure at the sintering stage. If lithium, for example, is inserted into the beta alumina structure at a later stage by ion exchange, then it is accommodated in the conduction planes rather than in the spinel block (Tofield and Farrington, 1979) and does not bring about the formation of beta″ alumina. Impurity cations larger than 0.97 Å radius, which substitute into the conduction plane, impede the migration of Na^+ ions and can cause a serious reduction in ionic conductivity. The alkaline earth cations, in particular, are deleterious to beta alumina in this respect (Buechele and De Jonghe, 1979).

The formulations chosen for practical electrolytes have evolved around 8 wt% Na_2O, 2 wt% MgO for Mg^{2+} doping (Tan and May, 1977a) and 8.9 wt% Na_2O, 0.7 wt% Li_2O for Li^+ doping (Bugden and Duncan, 1977). These compositions arise quite reasonably out of an appreciation of the charge compensation mechanism in the crystal structure. The resistivity passes through a minimum when the soda content corresponds to that in the undoped beta alumina *plus* that which corresponds to the doping level (Kennedy and Sammells, 1972). Thus 2 wt% MgO is equivalent to 0.59 mol Mg^{2+} per mole of beta alumina, and this should stabilize an additional 0.59 mol of Na^+ ion (equivalent to 1.5 wt% Na_2O); if the undoped beta alumina contained 6.1 wt% Na_2O, then the material doped with 2 wt% MgO will show a minimum resistivity at $6.1+1.5=7.6$ wt% Na_2O. The experimental results confirm these predictions; samples containing larger amounts of soda than the optimum value show higher resistivity because they require additional oxygen interstitials in the conduction plane.

Reliable thermodynamic and phase diagram data are only available for undoped beta alumina. Even for the binary system the stoichiometry range of the material and the intervention of the metastable beta″ phase pose problems for experimental determinations. The beta″ phase forms at around 1100 °C when preparations of beta alumina in the composition range $Na_2O \cdot 5.33Al_2O_3$–$Na_2O \cdot 7Al_2O_3$ are fired initially. The beta″ phase disproportionates to beta alumina and delta $NaAlO_2$ when the temperature is raised to 1550 °C (De Vries and Roth, 1969) and does not reappear on cooling. Beta″ alumina is in fact metastable in the absence of dopants. The presence of MgO or Li_2O dopants modifies the phase stability profoundly, rendering the beta″ structure stable up to 1700 °C (Imai and Harata, 1972). The instability of the stoichiometric formulation of beta alumina is acknowledged in the phase diagram of Fig. 2.12.

The sodium oxide activity over beta alumina is conveniently measured by means of electromotive force measurements. In the temperature range 550–925 °C the activity is given by (Elrefaie and Smeltzer, 1981)

$$\log a_{Na_2O} = \frac{-16\,400}{T} + 3.60$$

This expression can be used to calculate the standard Gibbs free energy of

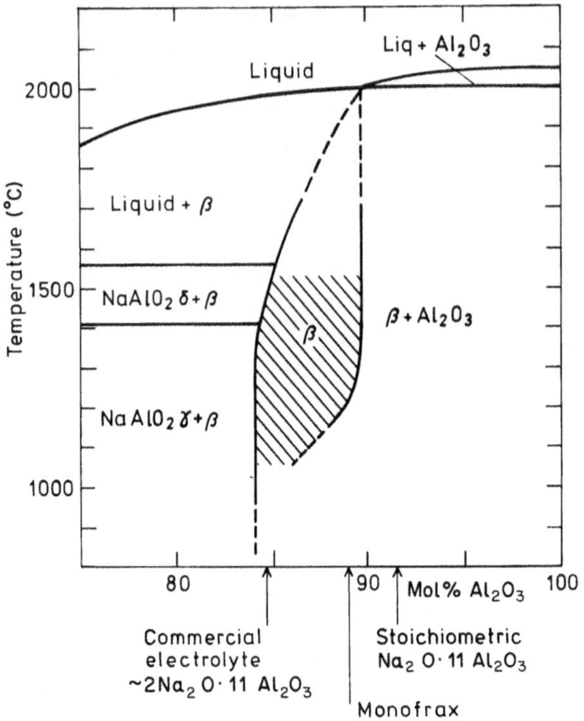

Fig. 2.12 The phase diagram of the Na_2O/Al_2O_3 system. After Le Cars *et al.* (1972) reproduced with permission, Masson, S. A., Paris). The hatched region corresponds to the coexistence of beta and beta″.

formation of beta alumina from Na_2O (solid) and alpha Al_2O_3 according to

$$Na_2O(s)+11Al_2O_3=Na_2O\cdot 11Al_2O_3 \tag{2.1}$$

or from the elements

$$2Na+22Al+17O_2(g)=Na_2O\cdot 11Al_2O_3 \tag{2.2}$$

Hence, from Equation 2.1,

$$\Delta G_f^\circ \text{ (beta } Al_2o_3)=-314\,000+68.920T$$

and from Equation 2.2,

$$\Delta G_f^\circ \text{ (beta } Al_2o_3)=-19.279\,200+3795.607T$$

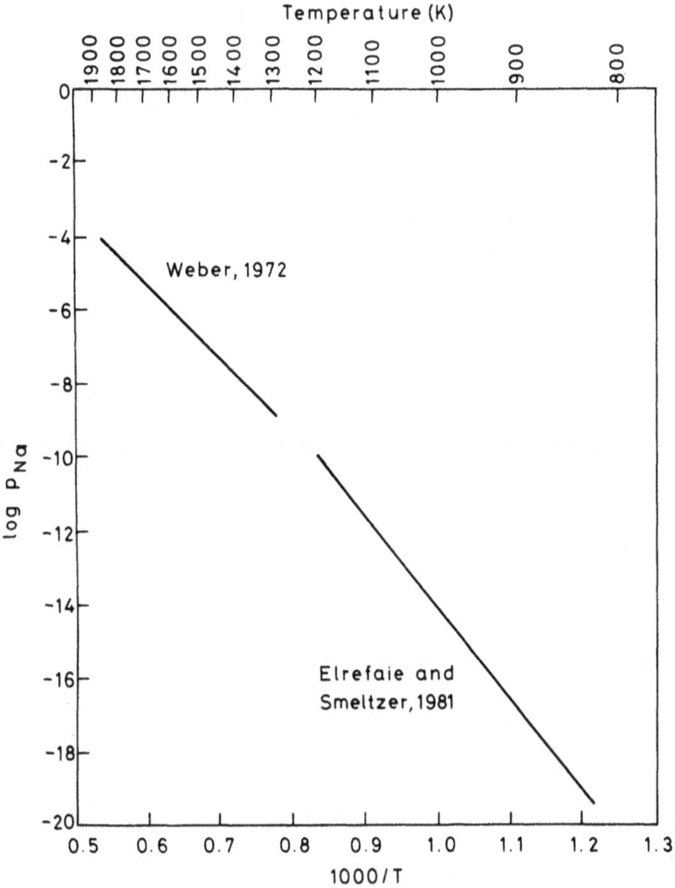

Fig. 2.13 Variation of sodium vapour pressure (in atmospheres) over beta alumina with temperature in air (Elrefaie and Smeltzer, 1981). Reproduced with permission of the publisher, The Electrochemical Society, Inc.

The same activity data can also be used to calculate the variation of sodium vapour pressure over beta alumina with temperature in air (Elrefaie and Smeltzer, 1981) and this is shown in Fig. 2.13.

Preparations of beta alumina containing more than the equilibrium level of sodium oxide are expected to be rather labile, with an enhanced tendency to lose sodium to the environment (Ewart *et al.*, 1983).

Both beta and beta″ alumina appear to be less thermally stable when subjected to high pressures (Roth and De Vries, 1977). In each case the decomposition reaction is accompanied by a volume decrease.

2.5 DIFFUSION AND CONDUCTIVITY

2.5.1 Beta alumina

The beta alumina type of crystal structure is ideally equipped for good ionic transport in the conduction planes, having a high proportion of defect sites for the mobile ions linked by networks of easy access (Figs 2.5 and 2.7). As a result the ionic conductivity of beta alumina at 350 °C can be almost as high as that of the electrolyte in aqueous batteries. At the same time the material remains virtually an electronic insulator.

The diffusion of ions through beta alumina is a thermally activated process and the jump frequency of the Na^+ ion is given by

$$\nu = \nu_0 \exp(-E/\mathrm{k}T) \tag{2.3}$$

where ν_0 is the frequency with which the particle approaches the barrier to diffusion (the 'attempt frequency'), E is the activation energy of the jump, k is the Boltzmann constant and T is the absolute temperature.

The diffusion coefficient is then given by

$$D = D_0 \exp(-E/\mathrm{k}T) \tag{2.4}$$

and this may be combined with the simple Nernst-Einstein equation

$$D = \frac{\sigma \mathrm{k}T}{nq^2} \tag{2.5}$$

to give the following expression for the temperature dependence of the ionic conductivity:

$$\sigma T = \sigma_0 \exp(-E/\mathrm{k}T) \tag{2.6}$$

where $\sigma_0 = nq^2D_0/\mathrm{k}$, n is the number of particles per unit volume and q their effective charge. Hence values of E and σ_0 are obtained from plots of log σT against $1/T$. It is clear that ionic conductivity is influenced both by the barrier to diffusion (E) offered by the local crystal structure and by the vibration of the mobile ion (ν_0) which influences the frequency with which it approaches the barrier. Values of E occur in the range 0–1 eV and values of ν_0 lie in the range 10^{11}–10^{14} Hz (Bates *et al.*, 1979; Bailey *et al.*, 1979; Hayes and Ailion, 1979).

Measurements of ionic conductivity of beta alumina are conveniently made using AC impedance techniques with standard blocking electrodes (e.g. Hooper, 1977) which overcome electrode polarization problems posed by DC methods.

(a) *Single crystal conductivity*

The conductivity σ of beta alumina single crystals depends on composition.

Table 2.8 Conductivity of beta and beta″ alumina single crystals, composition $(Na_2O)_{1+x}11Al_2O_3$

Phase	*Stoichio-metry* *x*	σ_0 $(\Omega^{-1}\ cm^{-1}\ K)$	*E* *(eV)*	*Reference*
β	0	—	0.62	Hayes *et al.* (1980)
β(i)	0.25	2.2×10^3	0.16	Whittingham and Huggins (1971)
β	0.4	4.8×10^3	0.20	Allen *et al.* (1978)
β(i)	1.0	—	0.30	Bourke *et al.* (1980)
β(Mg)	0.66	8×10^3	0.17	Baffier *et al.* (1981)
β''(Mg)	0.66	1.2×10^5 3×10^4	0.22 (25–250 °C) 0.17 (250–650 °C)	Baffier *et al.* (1981)

(i) Excess soda stabilized by interstitials.
(Mg) Excess soda stabilized by magnesium.

Some values of σ and E are given in Table 2.8. Undoped beta alumina shows a maximum conductivity and minimum activation energy roughly in the middle of its stoichiometric range. The decrease of conductivity with increasing soda content above around $(Na_2O)_{1+x}11Al_2O_3$, $x=0.2$–0.3 clearly reveals the effect of the additional interstitial oxygen defects on binding and blocking the motion of sodium ions. By contrast, beta alumina crystals doped with Mg^{2+} have a much higher conductivity than do undoped crystals. In both cases however the conductivity obeys the Arrhenius law predicted by Equation 2.6 and shown in Fig. 2.14.

The conductivity of beta″ alumina also depends on composition (Roth *et al.*, 1981; Engstrom *et al.*, 1981) but additionally beta″ alumina exhibits a change in the slope of the log σT against $1/T$ plot in the region of 200 °C as shown in Fig. 2.14. The change in slope coincides with a substantial change in the correlation length in the conduction zone (Boilot *et al.*, 1980) and thus a two dimensional order/disorder transition may be responsible for this behaviour. Two dimensional ordering phenomena are well established for both beta and beta″ structures and are likely to exert subtle influences on the conduction mechanisms of the two materials.

In general the diffusion coefficient obtained from ionic conductivity measurements with the simple Nernst-Einstein relation is not the same as that measured by tracer diffusion (Ormrod and Kirk, 1977). The ratio of the tracer diffusion coefficient to the diffusion coefficient derived from σ is defined as the Haven ratio H_R. The Haven ratio for beta alumina decreases with temperature as shown in Table 2.9. This is thought to be due to high correlation effects and the onset of defect association (ordering) at low temperatures. The Haven ratio for Li^+ containing beta″ varied very little over the temperature range 95–600 °C.

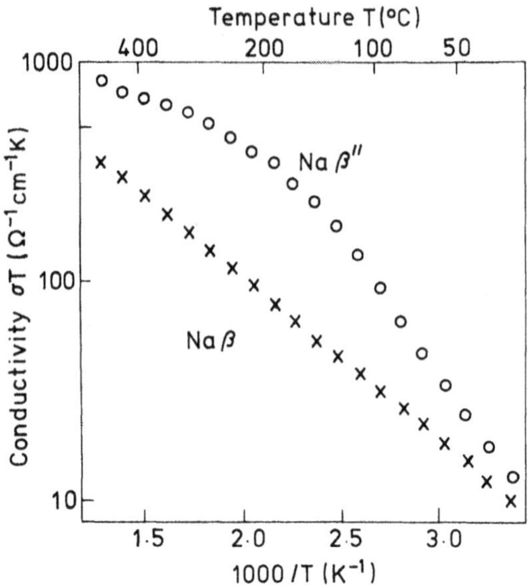

Fig. 2.14 Arrhenius plot (log σT against $1/T$) of the conductivities of beta and beta″ alumina. Reproduced with permission from Bates *et al.* (1982).

Table 2.9 Variation of Haven ratio H_R with temperature for beta and beta″ alumina

Phase	*Temperature (°C)*	H_R	*Reference*
β	600	0.45	Kim *et al.* (1979)
β	100	0.35	
β	−79	0.18	Chandrashekhar and Kim (1981)
β'' (two Li^+ containing compositions)			
	600	0.16, 0.15	Ormrod and Kirk (1977)
	95	0.18, 0.18	

This uniformity may be a manifestation of the difference in diffusion mechanism compared with beta.

A model applied to the vacancy mechanism which is thought to hold for beta″ alumina (Wang *et al.*, 1981) predicts (Ormrod and Kirk, 1977) a value of H_R of 0.17, in good agreement with the measured values. Application of the same model to the interstitialcy mechanism which is favoured for beta alumina (Wang *et al.*, 1981; Wolf, 1979) predicts a value for H_R of 0.3 which may represent confirmation of the mechanism in the region of 100 °C but with departures at high or low temperatures.

(b) *Polycrystalline conductivity*

AC measurements are particularly useful in allowing the bulk and grain boundary components of the resistivity to be separated. Fig. 2.15 illustrates the application of the technique (Hooper, 1977). The polycrystalline specimens (beta plus beta″) measured have an overall resistance which contains a contribution from the bulk (R_B) and the grain boundaries (R_{GB}). By analysing the impedance and corresponding admittance plots (Fig. 2.15 inset) it is possible to separate out the bulk phase conductivity G_2. The measured activation energy (0.16 eV) is close to that for single crystals but the resistivity at 300 °C is considerably higher (15.4 Ω cm). As can be seen in Fig. 2.15 the grain boundary contribution to the total electrolyte resistance is negligible above 250 °C. By contrast, for specimens of pure beta″ polycrystalline material the bulk resistance is still less than that of the grain boundaries, even at 350 °C (Virkar *et al.*, 1978).

The resistivity of polycrystalline beta alumina ceramic of near theoretical density is a function of (1) the relative proportion of beta and beta″ phases present, (2) the composition, and (3) the microstructure and crystallite size (May and Hooper, 1978). Typical values of the ionic conductivities of polycrystalline and single crystal beta and beta″ alumina at 300 °C are given in Table 2.10. Fig. 2.16 shows that there is an approximately linear relation between the resistivity and the proportion of the two phases present; the scatter of points arises from variations in density, microstructure etc. From these results it appears that the pure beta″ phase would be most desirable for use as a solid electrolyte. However, it is more sensitive to moisture than the beta phase, resulting in a deterioration of mechanical properties and conductivity. In practice therefore the ceramic employed as electrolyte in sodium sulfur batteries is usually a two phase beta plus beta″ composition with a conductivity in the range 0.1–0.2 Ω^{-1} cm^{-1} at 350 °C. This can be manufactured to have both excellent strength characteristics and resistance to attack by moisture.

Two further factors that influence the practical conductivity of beta alumina ceramics are the presence of impurities and the development of pronounced texture (preferred orientation of grains). Two of the more likely contaminants of ceramic electrolytes are calcium and silicon, and the effects of both on microstructure and resistivity are known. The presence of calcium at concentrations greater than 400 ppm promotes abnormal grain growth and allows the formation of intergranular calcium aluminate phases which are likely to impede the passage of mobile ions (Buechele and De Jonghe, 1979). The measured resistivity at 300 °C increases exponentially with calcium content and is particularly sensitive to sintering conditions. Beta alumina doped with a few weight per cent of SiO_2 also exhibits increased resistivity (Sieh and De Jonghe, 1978).

The huge anisotropy in conductivity of beta alumina crystals (by a factor of at least 10^{15} at 300 °C (Kummer, 1972)) might be expected to manifest itself in the

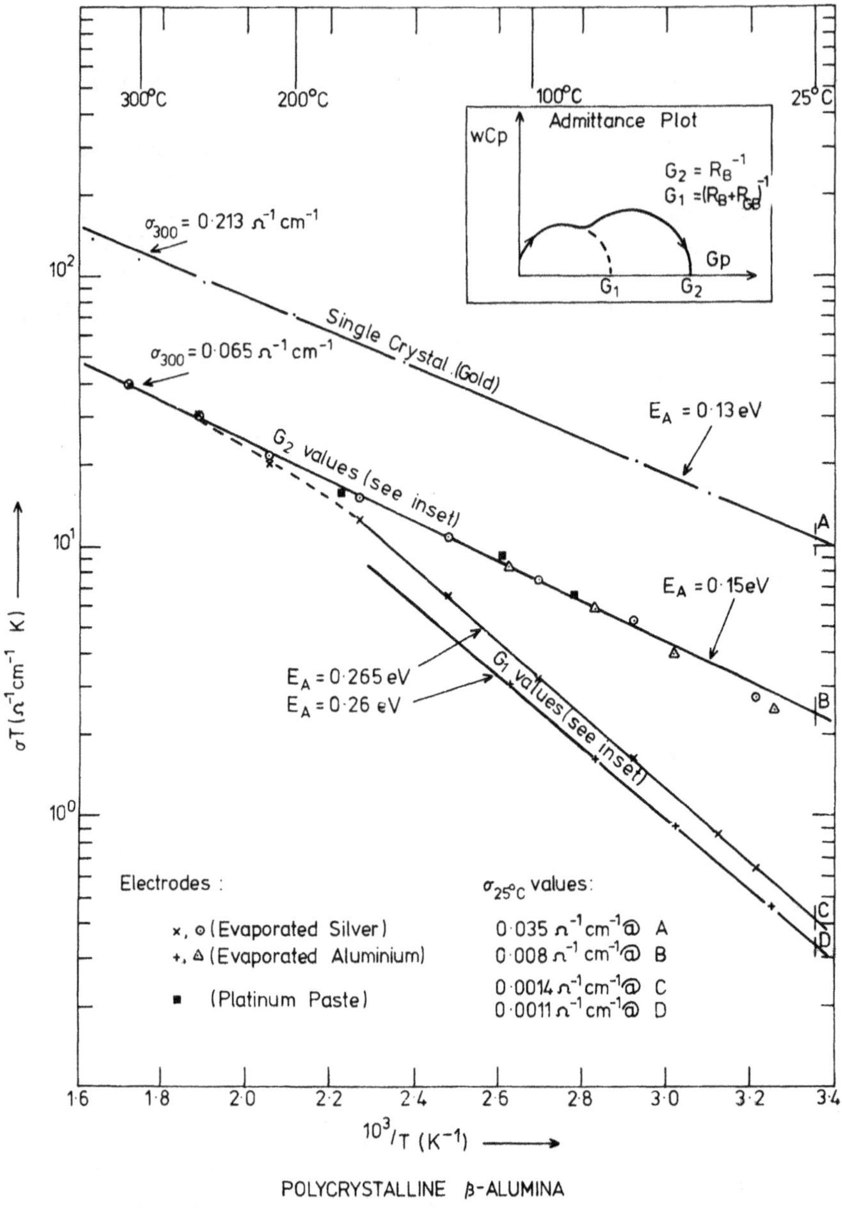

Fig. 2.15 Conductivity of grain boundaries (G_1) and bulk (G_2) of polycrystalline beta alumina (after Hooper, 1977, reproduced with permission from The Institute of Physics).

Table 2.10 Ionic conductivities of beta alumina

Specimen	$\sigma_{300\,°C}$ $(\Omega^{-1}\,cm^{-1})$	*E* *(eV)*	*Reference*
Single crystal β	0.213	0.13	Hooper (1977)
Polycrystalline β	0.08	0.25	May (1978)
Single crystal β''	1.0	0.33 (0–150 °C) 0.10 (> 150 °C)	Engstrom *et al.* (1981)
Polycrystalline β''	0.2–0.4	0.16–0.22	Virkar *et al.* (1978)

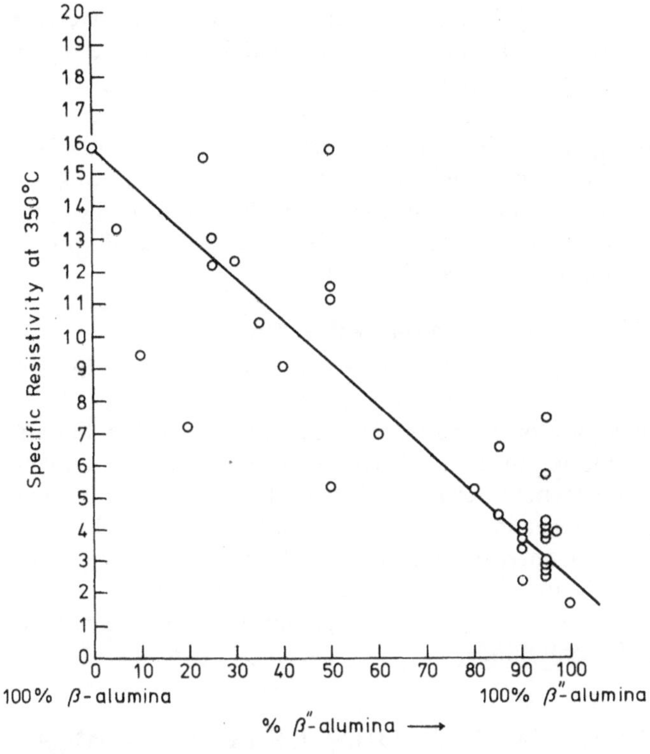

Fig. 2.16 Resistivity of beta alumina as a function of the beta/beta″ ratio. After Sudworth *et al.* (1973).

electrical properties of the polycrystalline material. Statistical calculations (Powers and Mitoff, 1978) indicate that, provided the resistance of the grain boundaries is small, the effect of tortuosity (the factor by which the conduction path is lengthened by the random orientation of individual grains) will also be small. The effect of preferred orientation of the grains constituting the ceramic is potentially quite damaging, since the fabrication procedures used for beta alumina would be expected to cause any texture arising to be in an unfavourable sense, i.e. with the blocking *c* axis oriented in a radial direction of the electrolyte tube. In fact it appears that although such unfavourable texture does occur during the manufacture of both beta (Ohta *et al.*, 1976) and beta'' alumina artefacts (Youngblood and Gordon, 1978) it is small and does not greatly degrade the electrolyte conductivity.

(c) *The effect of pressure*

Modest increases in pressure (up to 4 kbar) cause an increase in the conductivity of Li^+ beta alumina, no change in the conductivity of Na^+ beta alumina and a decrease in the conductivity of K^+ beta alumina (Radzilowski and Kummer, 1971). This is explained in terms of the relative sizes of the three diffusing ions: K^+ is the largest and will be restricted by quite modest compression of the conducting planes. Li^+ is sufficiently small that in conditions of normal pressure it relaxes towards one or other of the neighbouring spinel blocks but at increased pressure it is forced back into a site on the conduction plane where it is more mobile. Na^+, being of intermediate size, suffers no change in conductivity up to 20 kbar of applied pressure. Above 20 kbar, however, the conductivity of Na beta alumina falls progressively with increasing pressure (Itoh *et al.*, 1975).

2.5.2 Sodium ion conducting glass

The ionic conductivity of the sodium containing glass used in the Dow sodium sulfur cell is about $2 \times 10^{-4}\ \Omega^{-1}\ cm^{-1}$ at 300 °C (Levine, 1975). The glass electrolyte/separator must therefore be much thinner than the equivalent beta alumina membrane in order to keep the total resistance low, and this is achieved by using the glass in the form of fine hollow fibres with walls around 10 to 15 μm thick. Many thousands of fibres are assembled in parallel to provide a large electrolyte surface area per unit volume. The fibres are set into a glass tube sheet which forms the lower limit of the liquid sodium reservoir. Liquid sodium thus occupies the inside of the fibres which project down into the cathode chamber of molten sulfur/polysulfide.

2.6 CHEMICAL REACTIONS OF BETA ALUMINA

As a consequence of their high mobility the cations in or near the conducting zones in beta alumina can be replaced without disturbing the rest of the structure. Thus reactions such as

$$(Na_2O)_y 11Al_2O_3 + 2yMX \rightarrow (M_2O)_y 11Al_2O_3 + 2yNaX$$

where MX is the halide of a monovalent metal, represent a highly significant and characteristic behaviour.

Such reactions assume considerable technological importance in the context of high temperature batteries since they represent a mechanism for the introduction into the electrolyte of foreign ions with a potential for cracking the ceramic or increasing resistivity (Yasui and Doremus, 1978). Cationic replacement reactions can also be useful, however. The replacement of sodium by silver is routinely used as a means of chemical analysis (e.g. Bugden *et al.*, 1981; Pitt and Fray, 1982a) and the preparation of substituted beta alumina ceramics offers a potential method of detecting and measuring certain elements by means of electrochemical probes (e.g. Fray, 1981).

The replacement reactions can be accomplished by a variety of procedures as set out in the following sections.

2.6.1 Ion exchange in melts

Beta alumina will undergo cation exchange when subjected to any of a number of molten salts and also with a number of low melting metals. The extent of exchange is governed by the mole fraction of ions in the molten salt (see Fig. 2.17) and by the choice of salt anion (Yao and Kummer, 1967). For a wide range of cations it is possible to ensure virtually complete replacement of sodium in the beta alumina structure by replacing the melt with pure salt a number of times. The rate of exchange depends upon the ionic mobility of the cations concerned (see Table 2.16).

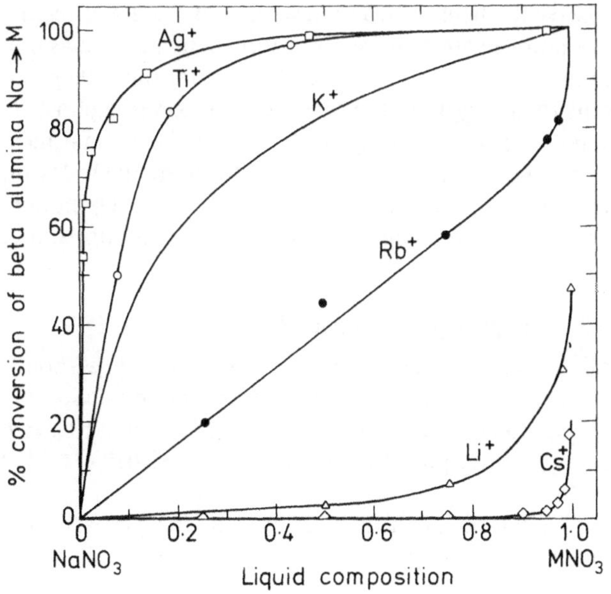

Fig. 2.17 Extent of ion exchange of beta alumina versus composition of molten nitrate with which it is in contact. Reproduced with permission from Yao and Kummer (1967).

Table 2.11 shows the range of substituted beta aluminas which can be prepared with monovalent cations. In some instances it is necessary to proceed via Ag^+ substituted beta alumina in a two stage process in order to achieve 100% conversion. Table 2.12 shows a number of monovalent cation substituted beta″ aluminas.

The range of products of the exchange reactions of beta aluminas with so many different cations brings a wide range of altered properties (bonding, conductivity, colour etc.). The chemistry of beta aluminas is extended by the possibility of replacing the monovalent cations in the plane by divalent cations (one for two). This operation can be achieved for both the beta and the beta″ structures but with differing degrees of facility which reflect the two distinct structures of the conducting zones. Thus beta alumina, which has extra (interstitial) oxygen atoms in the plane as part of a charge compensation scheme (see earlier), accepts divalent cations rather slowly (see Table 2.13) whereas the beta″ structure, which is unencumbered with interstitial oxygen ions, exchanges rapidly with the same cations (Table 2.14).

It will be of interest to explore the limits of the exchange with multivalent ions. The La^{3+} beta alumina structure can be prepared by direct sintering of La_2O_3 and Al_2O_3 (Dexpert-Ghys *et al.*, 1982) and thus it may be possible to ion exchange beta aluminas with rare earth halides.

2.6.2 Electrochemical exchange

As an alternative to chemical reaction with molten salts or metals it is possible to replace the mobile cation by incorporating beta alumina as the electrolyte in an electrochemical cell with metal/metal halide forming the positive electrode. By this means the integrity of the ceramic is preserved against the mechanical damage which takes place during the course of the normal melt exchange.

Electrochemical exchange has been used to insert the following metals into beta alumina: silver and copper (Kutsenok *et al.*, 1981), indium and gallium (Pitt and Fray, 1982a) and copper, silver, thallium, indium and gallium (Fray, 1981).

2.6.3 Vapour phase exchange

An alternative means of exchanging mobile cations without damaging a ceramic is to expose the material to a vapour of the halide of the metal of interest. This method has been successfully used for converting electrolyte tubes of beta″ alumina into potassium analogue by Crosbie and Tennenhouse (1982).

Gas phase exchange has also been used in the preparation of hydrogen (deuterium) beta alumina, details of which are given in Table 2.15.

2.6.4 Reactions with aqueous mineral acids

Concentrated mineral acids exchange with beta alumina to yield a form of hydrogen or hydronium beta alumina (see Table 2.15). There is some debate

Table 2.11 Exchange of beta alumina with monovalent ions in melts

Cation	*Precursor*	*Melt*	*Temperature (°C)*	*Cation radius (Å)*	*Lattice parameters (Å)*		*Reference*
NO^+	$Ag^+\beta$	$NOCl \cdot AlCl_3$	197		5.597	22.71	Radzilowski and Kummer (1969)
NH_4^+	$Na\beta$	NO_3^-	170	1.48	5.596	22.89	Yao and Kummer (1967)
Li^+	$Ag^+\beta$	NO_3^-/Cl^-	350	0.60	5.593	22.64	Yao and Kummer (1967)
Na^+	$-\beta$	—	—	0.97	5.594	22.53	Peters *et al.* (1971)
K^+	$Na\beta$	NO_3^-	300/350	1.33	5.596	22.73	Yao and Kummer (1967)
Rb^+	$Na\beta$	NO_3^-	300/350	1.47	5.597	22.88	Yao and Kummer (1967)
Cu^+	$Na\beta$	Cu/Cu_2Cl_2	527	0.96	5.616	22.43	Little and Fray (1980)
Ag^+	$Na\beta$	NO_3^-	300/350	1.26	5.594	22.49	Yao and Kummer (1967); Roth (1972)
Ga^+	$Ag^+\beta$	Ga/GaI	285	1.13	5.600	22.72	Radzilowski (1969)
In^+	$Ag^+\beta$	In/InI	350/500	1.32	5.599	22.90	Yao and Kummer (1967); Pitt and Fray (1982b)
Tl^+	$Na\beta$	NO_3^-	>206	1.47	5.598	22.93	Kodama and Muto (1976)

Table 2.12 Exchange of beta″ alumina with monovalent ions

Cation	*Precursor*	*Cation radius (Å)*	*Lattice parameters (Å)*		*Reference*
K^+	Naβ″	1.33	5.610	34.03	Crosbie and Tennenhouse (1982)
Ag^+	Naβ″	1.26	5.630	33.42	Bates *et al.* (1979)
Na^+	—	0.95	5.623	33.591	Bates *et al.* (1979)

(Tofield, 1980) as to whether the species involved is truly a hydronium ion or whether it is a proton bound to the spacer oxygen (O(5)) with water accommodated independently in the conduction plane.

2.6.5 Ionic conductivities of substituted beta aluminas

The ionic conductivities (at room temperature) and activation energies for conductivity are given in Table 2.16 for a number of substituted beta aluminas.

2.7 THERMAL PROPERTIES

The need to operate sodium sulfur cells at 300–350 °C and the use of high temperature techniques during cell assembly serve to emphasize the importance of the thermal properties of the material comprising the electrolyte.

Table 2.17 gives values of the thermal expansion coefficients of beta and beta″ alumina obtained after care had been taken to avoid distortion of the results by moisture pick-up or sodium loss (May and Henderson, 1979). The lack of any significant anisotropy in thermal expansion coefficient explains why internal stress is not found to be a problem in the use of beta alumina in the temperature range of normal operation.

A mean value of the expansion coefficient for a polycrystalline beta alumina is compared in Table 2.18 with the values for the sodium conducting glass electrolyte and for alpha alumina. It is clear that no difficulties of thermal mismatch need be anticipated in the construction of bonds between alpha and beta alumina.

Other thermal parameters of beta alumina are compared with the equivalent values for alpha alumina in Table 2.19.

Heat capacity measurements provide some further insight into the behaviour of the mobile ions in the structure of beta alumina. The contribution to the heat capacity made by the atoms in the conducting plane exhibits a maximum value just above 100 K which is thought (Bjorkstam *et al.*, 1980) to signify a two dimensional melting transition.

Table 2.13 Exchange of beta alumina with divalent ions

Cation	*Precursor*	*Melt*	*Temperature (°C)*	*Time (h)*	*Percentage exchange*	*Reference*
Li/Ca^{2+}	$Li\beta$(P)	$LiNO_3/Ca(NO_3)_2$	400	96	4(Li) 96(Ca)	Yao and Kummer (1967)
Sr^{2+}	$Na\beta$(SC)	SrI_2	620	72	92	Yao and Kummer (1967)
Li/Ba^{2+}	$Li\beta$(P)	$LiNO_3/Ba(NO_3)_2$	495	96	21(Ba)	Yao and Kummer (1967)
Zn^{2+}	$Ag^+\beta$(SC)	Zn	720	24	100	Yao and Kummer (1967)
	$Na\beta$(P)	$ZnCl_2$	497	36	20	Rog *et al.* (1981)
Mn^{2+}	$Na\beta$(P)	$MnCl_2$	550	24	30	Yao and Kummer (1967)
Fe^{2+}	$Na\beta$(SC)	$FeCl_2$	700	72	28	Yao and Kummer (1967)
	Na(P)	$FeCl_2$	697	72	74	Rog *et al.* (1981)
Cu^{2+}	$Na\beta$(P)	$CuCl_2$	550	144	10	Yao and Kummer (1967)
	$Na\beta$(P)	$CuCl_2$	697	72	16	Rog *et al.* (1981)
Sn^{2+}	$Na\beta$(P)	$SnCl_2$	300	144	37	Yao and Kummer (1967)
Pb^{2+}	$Na\beta$(SC)	$PbCl_2$	550	70	86	Yao and Kummer (1967)

(P) = powder (SC) = crystals ~ 1 mm
Mercury did not exchange at all.

Table 2.14 Exchange of beta″ alumina with divalent ions

Cation	*Precursor*	*Melt*	*Temperature (°C)*	*Time (h)*	*Reference*
Ca^{2+}	Naβ''(SC)	$Ca(NO_3)_2/CaCl_2$	500/700	1	Farrington (1981)
Sr^{2+}	Naβ''(SC)	$SrCl_2$	550	1	Farrington (1981) Dunn and Farrington (1980)
Ba^{2+}	Naβ''(SC)	$Ba(NO_3)_2/BaCl_2$	550	1	Farrington (1981) Dunn and Farrington (1980)
Cd^{2+}	Naβ''(SC)	$CdCl_2$	600	1	Farrington (1981) Dunn and Farrington (1980)
Pb^{2+}	Naβ''(SC)	—	—	—	Dunn *et al.* (1981)

(SC) = single crystal.

2.8 MECHANICAL PROPERTIES

2.8.1 Polycrystalline samples

The fracture strength σ_F of beta alumina is related to the stress necessary to propagate a microstructural flaw according to the Griffith equation

$$\sigma_F = (2E\gamma_i/\pi C)^{\frac{1}{2}}$$

where E represents Young's modulus, γ_i is the fracture surface energy and C is the half-length (radius) of the strength controlling flaw. A plot of log C against log σ_F is approximately linear with slope -2 (Fig. 2.18) and allows the determination of γ_i (Tan and May, 1977a). Hence values of the fracture toughness or critical stress intensity factor K_{IC} ($=\sqrt{(2E\gamma_i)}$) may be derived.

The strength of beta alumina ceramics is strongly influenced by microstructure, so that while a dense material with a grain size of 5 microns or less (Fig. 2.19) may exhibit values of σ_F in excess of 250 MN m^{-2}, the strength of a coarse grained (>200 microns) specimen may be an order of magnitude lower.

Typical values for the various mechanical parameters of polycrystalline beta

Table 2.15 Hydrogen (deuterium) and hydronium beta aluminas

Cation	*Precursor*	*Preparation*	*Temperature (°C)*	*Lattice parameters (Å)*		*Reference*
D^+	$Ag^+\beta$	D_2 gas	450	5.606	22.62 (4.5 K)	Tofield *et al.* (1979)
D^+	$Ag^+\beta$	D_2 gas	450	5.622	22.61 (550 °C)	Tofield *et al.* (1979)
H_3O^+	$Na\beta$	H_2SO_4 aqueous	100	5.596	22.70	Breiter *et al.* (1977)

Table 2.16 Activation energies and conductivities (at room temperature) of some exchanged beta aluminas (Bates *et al.*, 1982)

	Beta alumina		*Beta″ alumina*	
Cation	*Activation energy E (eV)*	*Conductivity* σ_{RT} $(\Omega^{-1}\ cm^{-1} \times 10^4)$	*Activation energy E (eV)*	*Conductivity* σ_{RT} $(\Omega^{-1}\ cm^{-1} \times 10^6)$
Na^+	0.15	358	0.33	250
Li^+	0.24	30	0.30	50
K^+	0.28	1.6	0.15	1000
Ag^+	0.16	130	0.19	40
Rb^+	0.39	0.12	—	—

Table 2.17 Thermal expansion coefficients of beta and beta″ alumina in the range 293–1178 K (after May and Henderson, 1979)

Beta alumina	*a* axis	$8.0 \pm 0.2 \times 10^{-6}\ K^{-1}$
	c axis	$8.3 \pm 0.3 \times 10^{-6}\ K^{-1}$
Beta″ alumina	*a* axis	$8.1 \pm 0.3 \times 10^{-6}\ K^{-1}$
	c axis	$7.0 \pm 0.4 \times 10^{-6}\ K^{-1}$

Table 2.18 Thermal expansion coefficients of polycrystalline alpha and beta alumina and the conducting glass

Alpha alumina	$8.8 \times 10^{-6}\ K^{-1}$	May (1978)
Beta plus beta″ alumina	$7.2 \times 10^{-6}\ K^{-1}$	May (1978)
Beta″ alumina	$7.8 \times 10^{-6}\ K^{-1}$	Miller *et al.* (1979)
Conducting glass	$10\text{–}12.5 \times 10^{-6}\ K^{-1}$	Levine (1975)

alumina together with the tensile strength of the conducting glass are given in Table 2.20.

Beta alumina ceramic exhibits an abnormally low time dependence of strength under stress both in air and in liquid sodium. For example, the mean strength of specimens tested at two different strain rates in air (Fig. 2.20) are quite similar (Davidge *et al.*, 1979). The lack of significant stress corrosion cracking in beta alumina has implications for the various failure modes which are dealt with later.

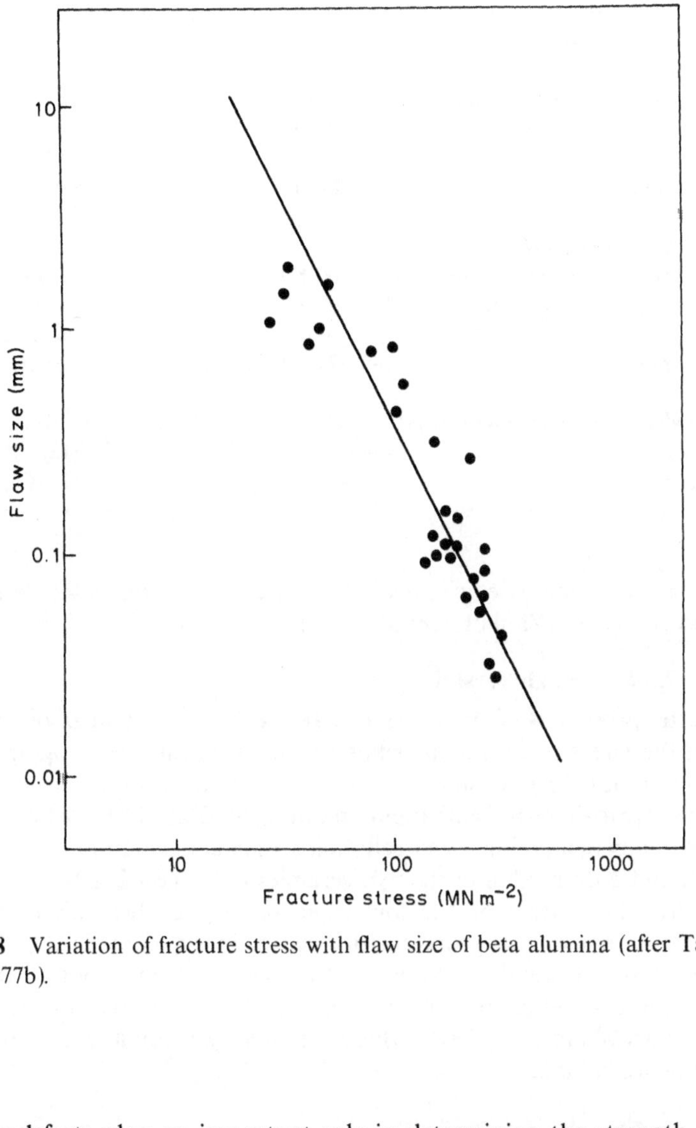

Fig. 2.18 Variation of fracture stress with flaw size of beta alumina (after Tan and May, 1977b).

Macrodefects play an important role in determining the strength of beta alumina ceramics (Tan *et al.*, 1980). The presence of such features as very large grains (possibly containing pores, as shown in Fig. 2.21), regions of intergranular porosity or localized regions of high impurity content results in mean fracture strengths for electrolyte tubes of ~90 MN m^{-2}, falling well below the fracture strengths (~250 MN m^{-2}) of small specimens cut from tubes and free from flaws. There is clearly a substantial incentive for the elimination of macroimperfections. It is possible that acoustic emission methods may assist in monitoring the quality of beta alumina, since under a four point bend test there

Table 2.19 Thermal properties of beta alumina and alpha alumina compared

Specific heat (cal mol^{-1} K^{-1})				
Beta alumina	Polycrystalline	273 (300 K)	316 (680 K)	May (1978)
	Single crystal	100 (300 K)		McWhan *et al.* (1977)
Alpha alumina		20 (294 K)		MCIC (1981)
Thermal conductivity (W m^{-1} K^{-1})				
Beta alumina	Polycrystalline	3.0 (300 K)	2.9 (680 K)	May (1978)
	Single crystal	3.0 (100 K)		Anthony and Anderson (1976)
Alpha alumina		69 (294 K)	6.5 (1366 K)	(MCIC (1981)
Thermal shock resistance (W m^{-1} $\times 10^{-2}$)				
Beta alumina		34.80		May (1978)
Alpha alumina		301.00		MCIC (1981)

is a substantial increase in emission as the load is increased but before breakdown occurs (Yankulov *et al.*, 1981).

2.8.2 Single crystal

The elastic properties of single crystals are defined by a matrix of equations relating the stress/strain relationships to the principal crystallographic axes (Koerber, 1962). The constants from these equations for beta alumina are compared with those for alpha alumina in Table 2.21. The elastic constants indicate that beta alumina is generally softer than alpha alumina (compare the Young's and bulk moduli of the two ceramics in Table 2.20).

Of particular interest is the low value of C_{44} for beta alumina which represents the constant appropriate for shearing in the basal planes. The value of C_{44} is sensibly lower than any of the other elastic constants but not as low as that (around 2) of the equivalent parameter for the transition metal dichalcogenides (see McWhan *et al.*, 1975) where the interlayer bonding is even weaker than in beta alumina.

2.9 INTERACTIONS WITH ELECTRODE MATERIALS

In order to function efficiently the surface of the electrolyte in a sodium sulfur cell must be completely wetted by the electrode materials. The wetting of beta alumina by sodium is facilitated by use of a smooth surface rather than a rough one (Armstrong *et al.*, 1973; McDonough *et al.*, 1978) and by use of increased temperatures; a partially wetted surface at 300 °C becomes totally wetted at 350 °C (Gibson, 1977). Wetting is of great importance for cell longevity, as well as for reasons of cell resistance, because the presence of even small non-wetted

Table 2.20 Mechanical properties of polycrystalline beta alumina and alpha alumina compared, and tensile strength of sodium conducting glass

Fracture strength σ_F (MN m^{-2})		
Beta alumina (fine grain)	230–330	May (1978)
Alpha alumina	240–400	MCIC (1981)
Sodium conducting glass (tensile)	550	Levine (1981)
Fracture surface energy γ_i (J m^{-2})		
Beta alumina	24	Tan and May (1977a)
Alpha alumina	20–40	MCIC (1981)
Critical stress intensity factor K_{IC} (MN $m^{-3/2}$) (fracture toughness)		
Beta alumina	2.7	May (1978)
Alpha alumina	3–6	MCIC (1981)
Hardness (Vickers test; kg mm^{-2})		
Beta alumina	1400	May (1978)
Alpha alumina	3000	MCIC (1966)
Young's modulus E (N m^{-2})		
Beta alumina	2.1×10^{11}	Tan and May (1977b)
Alpha alumina	4.4×10^{11}	MCIC (1981)
Bulk modulus (N m^2)		
Beta alumina	1.5×10^{11}	May (1978)
Alpha alumina	2.2×10^{11}	MCIC (1966)
Poisson's ratio		
Beta alumina	0.25	Trantina (1981)
Alpha alumina	0.25	MCIC (1966)

Table 2.21 Elastic constants (in 10^{11} dyne cm^{-2}) of beta alumina and alpha alumina

	Elastic constants					
Material	C_{11}	C_{12}	C_{33}	C_{44}	C_{66}	*Reference*
Beta alumina*	33.8	14.3	24.5	5.0	—	McWhan *et al.* (1975)
Beta alumina†	35.1	13.6	23.0	4.5	10.8	Hattori *et al.* (1981)
Alpha alumina	46.5	12.4	56.3	23.2	10.1	Huntington (1958)

* Measured by inelastic neutron scattering.
† Measured by optical interferometry.

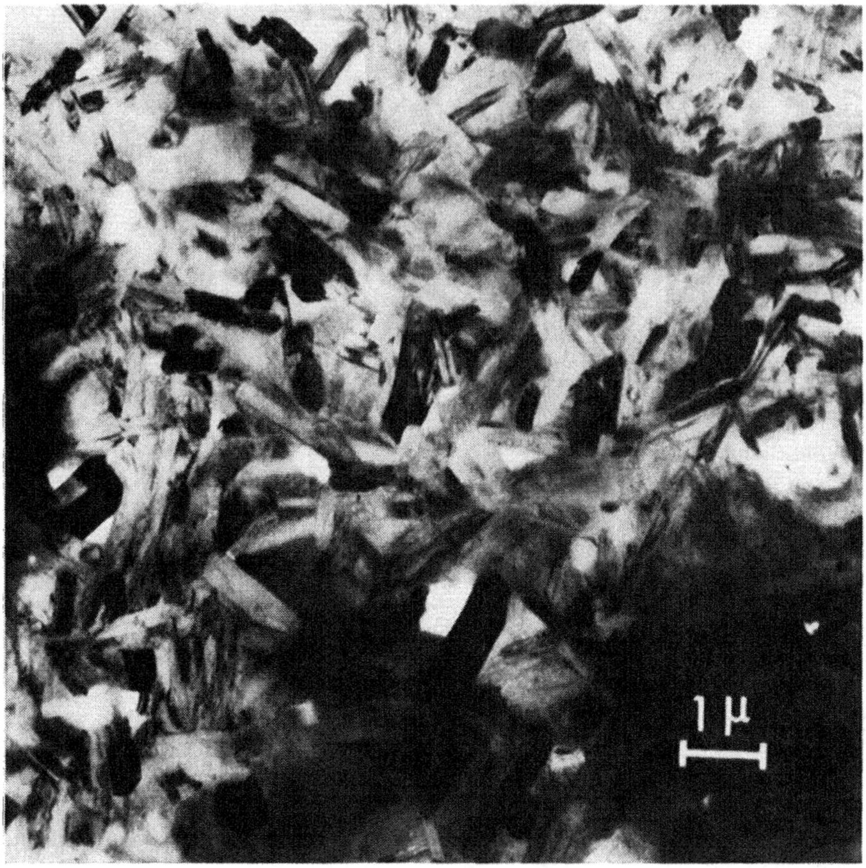

Fig. 2.19 Microstructure of fine grained beta alumina ceramic. Note the sharply faceted grains. The pronounced faulting visible in several grains is normal to the crystallographic *c* axis (Hudson and Moseley, 1973).

areas gives rise to high local current concentrations and premature electrolyte failure (Virkar *et al.*, 1980). In the absence of such features very high critical current densities (up to 4 A cm^{-2}) can be achieved (Viswanathan and Virkar, 1982).

Once wetting is complete it is maintained on cooling, and a properly constructed sodium sulfur cell should show no interfacial polarization so that the measured cell resistance is merely that of the electrolyte and the electrodes together.

Ideally there should be no interaction between the electrolyte and the electrode materials apart from the transfer of sodium ions. Any other interactions taking place at the surface of the electrolyte will probably lead to an

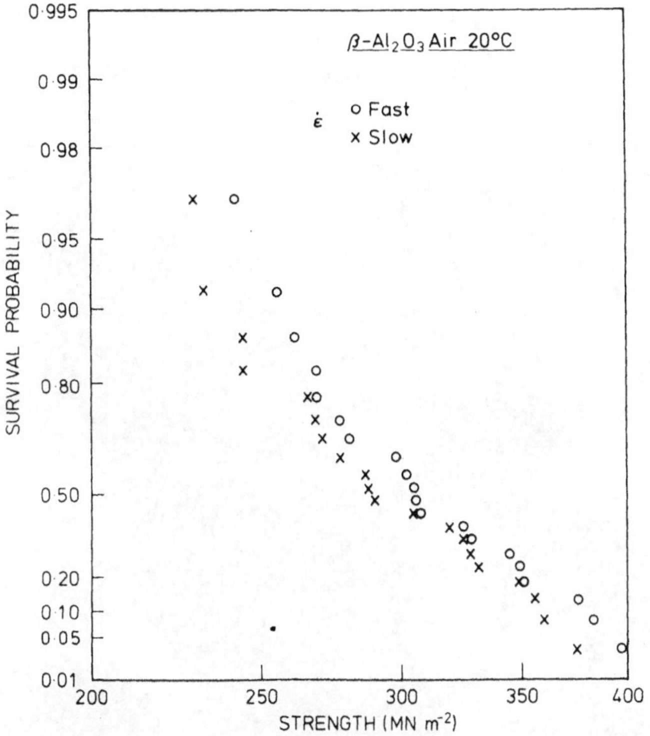

Fig. 2.20 Variation of strength of beta alumina in air with strain rate (after Davidge *et al.*, 1979, reproduced with permission from the American Ceramic Society).

increase in interfacial resistivity or to electrolyte degradation, either of which will be detrimental to cell performance. A number of difficulties may occur in this area.

Resistance increases in sodium sulfur cells cycled for extended periods have been reported from several laboratories (e.g. Park *et al.*, 1981). This phenomenon seems to originate at the beta alumina surface, and experiments in which the sodium electrode is renewed following a rise in cell resistance show that the original cell resistance is restored (Demott, 1980). The rate of resistance rise is influenced by the Li_2O content and, to a lesser extent, by the Na_2O content of the bulk electrolyte (Bugden *et al.*, 1981). It appears that higher levels of alkali ion assist the transport of impurity ions from the bulk to the surface of the electrolyte where they form a resistive barrier to sodium ion transport. The resistance rise problem can be overcome empirically by a combination of a surface treatment on the electrolyte and the use of an oxygen getter in the sodium (Demott *et al.*, 1981).

Beta″ alumina ceramic with high soda compositions can also exhibit an

Fig. 2.21 Large grain of beta alumina containing pores (Hudson and Moseley, 1973).

asymmetric resistance to direct current flow, the voltage measured across the electrolyte being noticeably larger during discharge than during charge (Breiter *et al.*, 1980; Breiter and Dunn, 1981). Such behaviour is attributed to a thin passivating film of Na_2O between the beta″ alumina surface and the liquid sodium. This layer may arise as a result of the ceramic bearing a soda concentration above the equilibrium level (see Section 2.4). Its presence is detrimental to both efficiency and cell life.

Interactions between the electrolyte and the cathodic material are also important in maintaining the uniformity of current density. An analysis of failed beta″ alumina tubes (Gupta and Tennenhouse, 1979) has revealed a strong correlation between the shape of the electrode in contact with the electrolyte and the crack patterns observed. This is likely to be due to high local current densities resulting from non-uniform current distribution as a result of shaped electrodes.

2.10 DEGRADATION AND FAILURE

2.10.1 Beta alumina

One of the principal causes of failure of sodium sulfur cells is electrolyte breakdown. Although individual cells may provide useful lives of several thousand cycles, there are mechanisms of electrical and mechanical degradation which cause progressive or precipitous failure of the ceramic, and the *average* cell life falls far short of the optimum. It is clearly of great importance for the several failure mechanisms to be understood in order that the proportion of cells offering acceptable life spans should approach 100%.

The failure of beta alumina may result from electrical breakdown by internal short circuiting or through the application of an excessive voltage, from external mechanical shock or from thermomechanical effects of heating and cooling. These various modes are closely interactive but it is convenient to treat them individually.

(a) *Dendrite formation*

It is sometimes found that a sodium sulfur cell will, after a period of successful charge/discharge cycling, fail to hold its charge because an internal short circuit has developed, apparently spontaneously, and this worsens rapidly if further cycling is attempted (Armstrong *et al.*, 1974; Tennenhouse *et al.*, 1975; Chu-Kun and Shan-Tin, 1981). This phenomenon is conveniently studied by life testing in a sodium sodium cell, using an applied voltage to pump Na^+ ions through the electrolyte at a controlled current density (Knödler and Baukal, 1978). Such experiments show that under favourable circumstances beta alumina electrolyte can pass several thousand ampere hours per square centimetre at current densities of 1 A cm^{-2} (Fischer *et al.*, 1978). Failure ultimately occurs by penetration of liquid sodium through the electrolyte: on cooling this is seen as a sodium filament or 'dendrite' (see Fig. 2.22) that grows through the electrolyte starting from the exit side. It is also clear that:

(1) This phenomenon is associated with the sodium formation reaction $Na^+ + e^- \rightarrow Na$, and is therefore a function of cell charging and independent of cell discharge;
(2) the higher the current density on charge, the shorter the electrolyte life;
(3) long electrolyte life is favoured by fine grained ceramic;
(4) highly doped material does not generally show good resistance to sodium penetration (there is more sodium available to contribute to the dendrite).

The flaws involved in the breakdown may be decorated electrolytically or chemically to render them more easily visible (De Jonghe and Feldman, 1980).

Dendrites are initiated at microcracks or imperfections at the exit surface and these propagate progressively into the body of the ceramic. Two explanations were originally advanced for this phenomenon. One invoked crack growth in a subcritical manner by stress induced dissolution of the ceramic in the liquid sodium (Richman and Tennenhouse, 1975). The other proposed that sodium was discharged preferentially into the microfissure, raising the stress intensity to exceed the critical stress intensity factor K_{IC}. Both mechanisms depend upon the microcracks acting as current concentrators, thereby causing a Poiseuille pressure in the cracks due to flow of sodium. However, the absence of stress corrosion cracking in beta alumina (Davidge *et al.*, 1979) seems to indicate that the mechanism involving subcritical crack growth is not directly applicable.

The concept of pressure developing due to sodium flowing out through a crack can be extended by evaluating the crack shape and crack opening displacement under the action of the Poiseuille pressure instead of assuming the crack width (Shetty *et al.*, 1977). Hence the critical current density i_{cr}, above which sodium penetration is allowed is given by $i_{cr} = A/\eta l$, where A is a constant containing elastic and fracture properties of the solid electrolyte, η is the viscosity of sodium and l is the crack length. The validity of this expression appears to be affirmed by the observation that sodium penetrates a beta alumina specimen at a current density of 30 mA cm^{-2} below the melting point of sodium but not above (Virkar and Viswanathan, 1979).

The presence of high local current densities arising out of non-uniform operation of the sodium electrodes (non-wetting) is likely to be an important factor in affecting the growth of dendrites. Models of this type of degradation (Brennan, 1980; Virkar *et al.*, 1980) agree that it proceeds via a relatively long induction period, during which time changes may occur at the sodium/electrolyte interface, followed by a rapid (a few mm s^{-1}) crack growth phase and electrolyte failure. Once the difficulties of the interfacial region (dendrite initiation) are overcome then critical current densities of several A cm^{-2} are achievable (Virkar and Miller, 1979). Indeed, when low current densities are used at the electrode contacts bulk beta alumina can support a current of 10 A cm^{-2} (De Jonghe *et al.*, 1979).

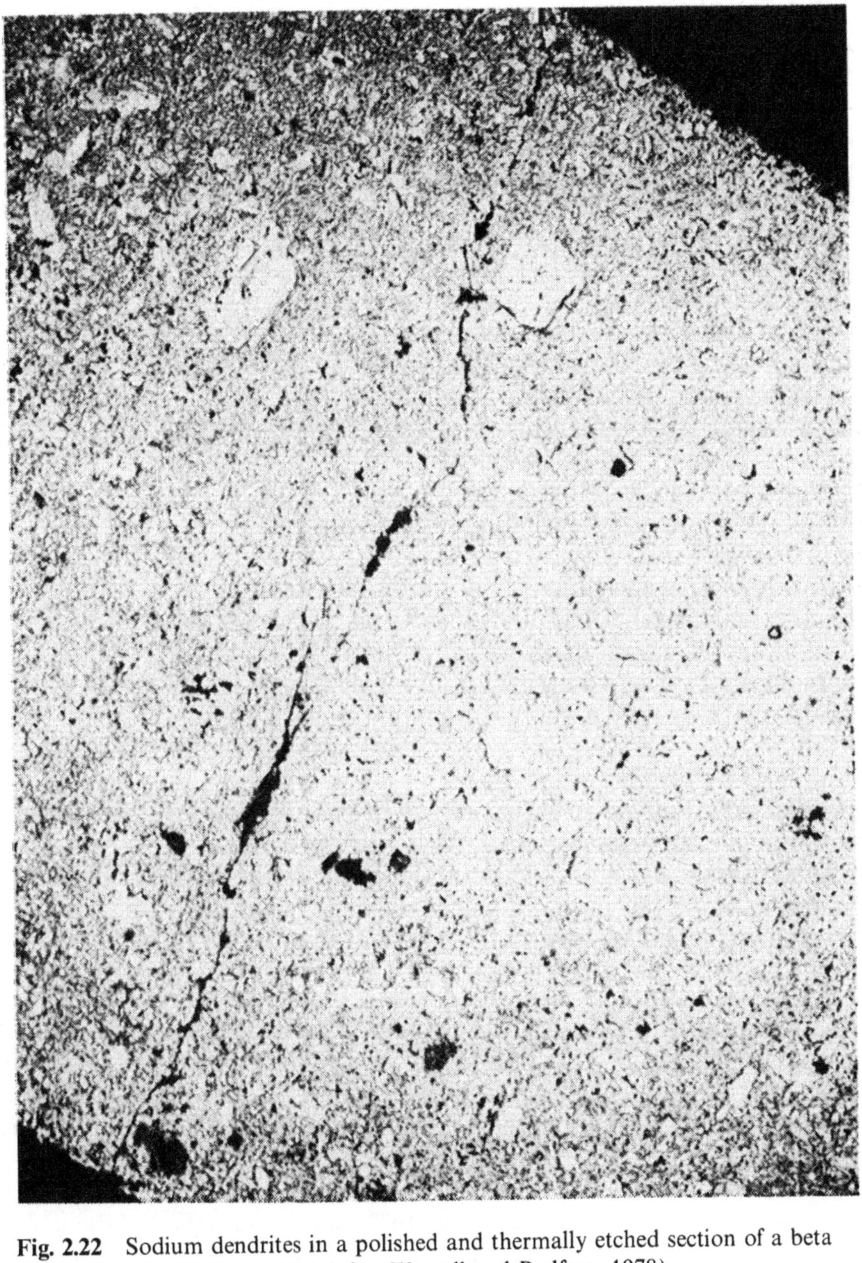

Fig. 2.22 Sodium dendrites in a polished and thermally etched section of a beta alumina tube wall 2 mm thick (after Worrell and Redfern, 1978).

There remain details of the mechanism of dendrite propagation which are not satisfactory. Critical current densities calculated on the basis of measured critical stress intensity factors yield values which are several orders of magnitude higher than average measured critical current densities (Feldman and De Jonghe, 1982). Also, a theoretical model (Virkar, 1981) indicates that the pressure generated in the sodium filled cracks of beta alumina actually decreases with increasing crack length for a given current density. This appears to be at odds with a simple application of the original model.

The electrical breakdown of beta alumina during the transport of sodium ions is conveniently monitored by means of the acoustic emission technique (Worrell and Redfern, 1978, De Jonghe *et al.*, 1981b). In beta alumina compositions which are susceptible to dendrite penetration, many acoustic events occur prior to breakdown. By contrast, compositions which exhibit long life in sodium sulfur cells show no acoustic events until the current density exceeds 0.5 A cm^{-2}. There is adequate evidence that progressive microcracking precedes breakdown. Strength measurements reveal a marked decline in fracture stress during this period prior to electrical breakdown (Tennenhouse *et al.*, 1975).

It follows from the various points made that care should be exercised in charging sodium sulfur cells to ensure that the current density remains below the critical level at which dendrites start to form. This will vary from one ceramic to another but will lie in the range 0.1 – 0.5 A cm^{-2} for beta alumina and surface finishes presently produced on a large scale.

(b) *Coloration of beta alumina*

Beta alumina exposed to liquid sodium is often found to develop a dark coloration (brown, grey or black) which progressively penetrates the ceramic from the surface in contact with the metal. The colour generated varies greatly from one case to another and the effect is sometimes accompanied by a sharp abbreviation in life expectancy (Gourier *et al.*, 1982) or an increase in resistivity (Bourke *et al.*, 1980). In other cases, however, the coloration of beta alumina occurring during normal electrolyte duty is associated with no deleterious changes whatever (Armstrong *et al.*, 1982). The reason for this apparent diversity of behaviour may lie in differences of material from different origins, or there may be more than one type of reaction between sodium and beta alumina depending on, for example, the presence of an electric potential or the presence of oxygen.

The darkening of the ceramic is likely to involve colloidal sodium centres (Armstrong *et al.*, 1982; Gourier *et al.*, 1982) but the formation of these may depend upon the initial generation of point defects such as oxygen vacancies associated with electrons. The effective diffusion coefficient for the coloration process is of the order of 10^{-10} cm^2 s^{-1} (De Jonghe and Buechele, 1982). The production of colloidal sodium in the structure of the ceramic is likely to contribute to either or both the dendrite mechanism (mode I degradation) or the internal deposition of sodium (mode II degradation; De Jonghe *et al.*, 1981a). The important observation that colouring of the electrolyte may take place without any degradation of performance may indicate the presence of point defects below some critical level of concentration and temperature necessary for coalescence.

Other colorations in beta alumina (e.g. blue) can be traced to the presence of transition metal ions trapped near microstructural defects (May, 1978) and merely indicate the need for tighter quality control during fabrication.

(c) *Internal sodium formation*

A mode of beta alumina degradation which is distinct from the dendrite formation described earlier arises from the progressive deposition of sodium metal within the electrolyte, as a layer starting from the sodium side. This mechanism proceeds more slowly than does the coloration of the ceramic (De Jonghe *et al.*, 1981b) but may arise as a consequence of the formation of the point defects causing the coloration, when the electrolyte is not operated within certain bounds of purity and voltage limits.

Following the partial reduction of the electrolyte by interaction with sodium, a gradient in the electronic/ionic transport number ratio is generated. When current is passed through such an electrolyte the possibility exists (De Jonghe, 1982a) for internal deposition of sodium. The resistance overvoltages at which such sodium deposition is expected are of the order of one volt (De Jonghe *et al.*, 1981b), and since such overvoltages are not normally encountered in the charging of a sodium sulfur cell it should be possible to suppress this degradation mechanism by good management of the charging regime. However, if a high resistance layer were to develop on the electrolyte as a consequence of the presence of impurities the critical resistance overvoltage could be substantially lowered (De Jonghe, 1982b) and the internal deposition of sodium would once more become a threat.

(d) *Voltage breakdown*

The modes of electrical breakdown dealt with so far concern the operation of the cell within a rather narrow voltage range. A completely separate behaviour results when high voltages are applied to membranes of beta alumina under blocking conditions (Hooper, 1980). At a given temperature there is a well defined voltage limit (of some tens of volts) above which the electronic insulating properties of the ceramic break down and gross physical damage occurs. The significance of this observation is that although the charging voltage limit might be set at 2–3 V for a single sodium sulfur cell, the connection together of many cells into a battery configuration will generally require a higher total overvoltage and this is likely to be several tens of volts. If, as may happen, one cell in such a chain should become fully charged before the others, then all of the applied charging voltage could be dropped across that one high resistance element, leading to voltage breakdown.

Above the threshold voltage, which is a function of temperature, the electronically insulating properties of the electrolyte are lost intermittently and gross damage takes place at the positive electrode face. Fig. 2.23 shows the variation of breakdown voltage with temperature for fine grained and coarse grained ceramic specimens with both electrodes blocking. Samples with two non-blocking electrodes do not appear to suffer breakdown (Hooper, 1980). A model for this type of degradation is shown in Fig. 2.24. On application of a DC bias, the region of electrolyte close to the positive blocking electrode becomes depleted of sodium ions. The resistivity of this layer increases and, consequently, the majority of the applied voltage is dropped across it so that dielectric breakdown accompanied by melting and possibly charring ensues. The threshold voltage above which breakdown occurs is independent of the lifetime of the electrolyte in a sodium sulfur cell (Hooper, 1980). At room temperature a

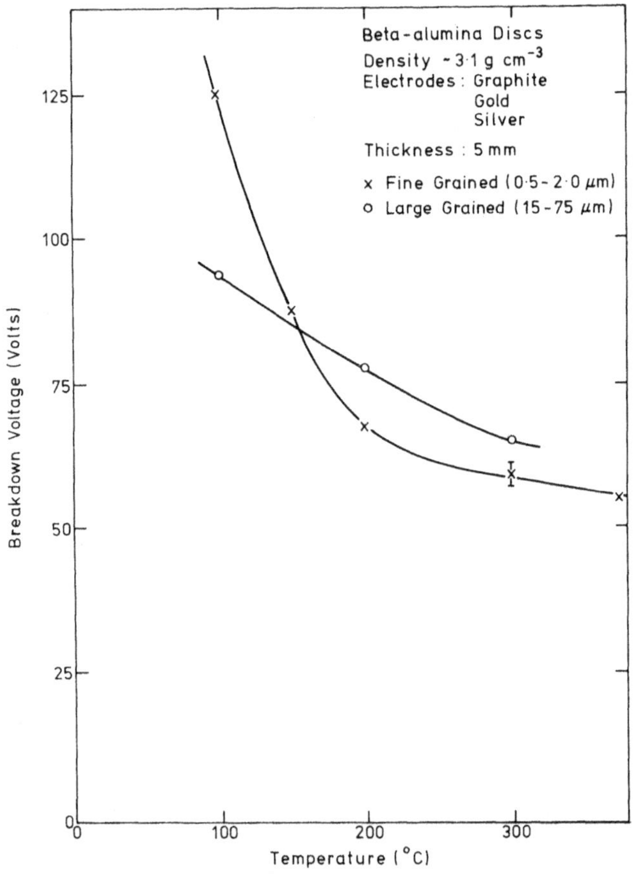

Fig. 2.23 Variation of breakdown voltage with temperature for two grades of beta alumina. Reproduced with permission from Hooper (1980).

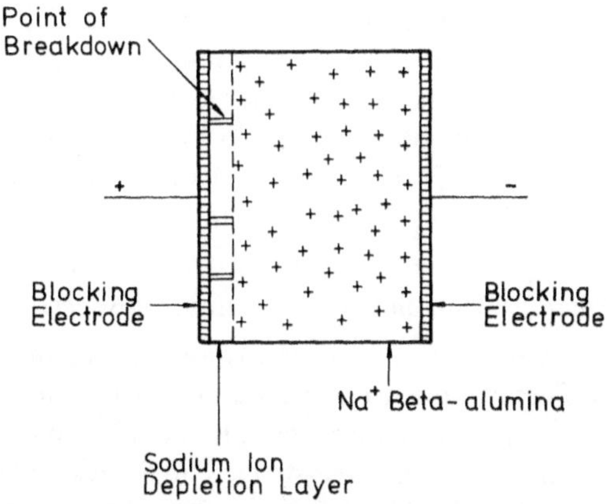

Fig. 2.24 Schematic drawing of the mechanism of voltage breakdown of beta alumina. Reproduced with permission from Hooper (1980).

polycrystalline ceramic breaks down at a much lower voltage (40–80 V) if only the negative electrode is blocking (De Jonghe *et al.*, 1979) than if both electrodes are blocking (270 V; Tang and Chaudhri, 1982).

(e) *Mechanical breakdown*

Mechanical breakdown of the electrolyte may result from inbuilt stresses in the tube, from stresses imposed during cell assembly, or from externally applied shock. Cracking of the electrolyte tube tends to occur preferentially at weak spots, e.g. near the closed end, near the glass seal or at gross microstructural defects. Because there is often a progressive fall in strength of the tubes in use (Tennenhouse *et al.*, 1975) as a result of microcracks and dendrites, it is important that mechanical shock tests should be carried out on cells which have been subjected to a large number of cycles as well as on new cells.

(f) *Thermal breakdown*

Thermal breakdown of beta alumina can arise for any one of several reasons:

(1) A mismatch in the thermal expansion coefficients of the beta alumina and the glass which is used to bond it to the insulating alpha alumina collar. As mentioned earlier, the thermal expansion coefficients of alpha and beta alumina are fairly closely matched, and a glass with a similar response must be used in order to avoid the generation of stresses leading to cracking as the cell is warmed or cooled.
(2) Mixing of sodium and sulfur through a crack initially caused by an electrical or mechanical failure is likely to lead to a rapid localized rise in temperature. The associated thermal shock may result in destruction of the electrolyte tube.

(3) Phase changes of the electrode material (sodium and sulfur plus polysulfides) are accompanied by molar volume changes which must be accommodated within the cell design in order to avoid stresses being applied to the electrolyte tube during heating or cooling. Repeated heating and cooling rates of $>50\,^{\circ}C\,h^{-1}$ are tolerated by known designs of cell (Hattori *et al.*, 1977).

(g) *Interaction with components of the atmosphere*

Beta alumina exposed to air tends to pick up moisture and increase in resistivity. The extent of this effect is influenced by a number of ceramic properties: rough surfaces and a coarse microstructure facilitate the take-up of water as does the presence of high levels of lithium (Armstrong and Sellick, 1980). The soda level of the ceramic is also likely to be a factor influencing water pick-up.

Anhydrous specimens take up water in two processes, one complete in a few hours, the other typically lasting for several days (Flor *et al.*, 1981). The total water uptake is around one molecule per cation in the conduction plane and the process is accompanied by a shift in the *c* lattice parameter of the crystal structure.

Tubes of beta and beta″ alumina respond to exposure to flowing air by exhibiting an increase in resistivity in the radial direction only (Dunn, 1981). This is because only a thin layer at the surface of the tube is hydrated. The layer is of the order of 5 μm deep in high density beta alumina (Heavens, 1982). The change in resistivity is more pronounced for the beta″ tubes than for beta but in each case the resistivity is restored to its original value by heating in dry air to 600 °C. The heat treatment also drives off carbon dioxide and it is likely that exposure to air allows the formation of phases such as Na_2CO_3 or $NaHCO_3$ at the electrolyte surface.

2.10.2 Sodium ion conducting glass

The sodium ion conducting glass represents one of the main sources of failure in the Dow version of the sodium sulfur cell. However, the consequences of its failure are much less serious than is fracture of a beta alumina electrolyte membrane because a fractured glass capillary is generally self-sealing. Failure of one of the glass capillaries then results in some small loss of capacity but not in a catastrophic breakdown.

The glass electrolyte is degraded chemically by the attack of a caustic impurity (Na_2O, NaOH) in the sodium electrode, and this is avoided (Levine, 1978) by the use of an oxygen getter (Zr/Ti alloy).

Fracture of the glass may also occur as a result of formation of solids in the cathode. Cells discharged too rapidly contain local regions of supercooled sodium polysulfide melt which crystallize suddenly to cause fibre breakage. Restriction of the lower limit of the closed circuit potential may obviate this problem (Levine, 1978).

2.11 SUMMARY

The ionic conductivity of beta alumina membranes at the temperature of operation of the sodium sulfur cell is of the same order as the conductivity of the electrolyte in lead acid cells. This remarkable property of the ceramic allows the concept of electrical storage cells based on a solid electrolyte to be realized and gives rise to a rich ion exchange chemistry.

The conductivity of beta alumina and its dependence on composition arise as a result of the unique crystal structure of the material, and a good understanding of the implications of this type of structure has been achieved.

In other respects the properties of beta alumina resemble those of more conventional ceramics, so that electrolyte tubes with acceptable thermal and mechanical properties can be constructed in large numbers.

The passage of high current densities through a solid electrolyte allows the possibility of failure modes unknown in the realm of liquid electrolyte cells, but these modes are well understood and appear to be avoidable.

The conductivity of the glass electrolyte is far less than the spectacular values exhibited by beta alumina, but cells can be constructed with acceptable electrical performance by making the glass extremely thin.

2.12 REFERENCES

Allen, S. J., Cooper, A. S., De Rosa, F., Remeika, J. P. and Olasi, S. K. (1978) *Phys. Rev.*, **B17**, 4031.

Anthony, P. J. and Anderson, A. C. (1976) *Phys. Rev.*, **B14**, 5198.

Armstrong, R. D., Dickenson, T. and Turner, J. (1973) *Electroanal. Chem. Interfacial Electrochem.*, **44**, 157.

Armstrong, R. D., Dickenson, T. and Turner, J. (1974) *Electrochim. Acta*, **19**, 187.

Armstrong, R. D. and Sellick, D. P. (1980) *Electrochim. Acta*, **25**, 1199.

Armstrong, R. D., Sellick, D. P. and Tan, S. R. (1982) *Solid State Ionics*, **6**, 203.

Baffier, N., Badot, J. C. and Colomban, Ph. (1981) *Mater. Res. Bull.*, **16**, 259.

Bailey, W. C., Story, H. S., Ochadlick, A. R. and Farrington, G. C. (1979) in *Fast Ion Transport in Solids* (eds P. Vashishta, J. N. Mundy and G. K. Shenoy), North-Holland, p. 281.

Barklie, R. C., Niklas, J. R. and Spaeth, J. M. (1980) *J. Phys. C.: Solid State Phys.*, **13**, 1745.

Bates, J. B., Brown, G. M., Kaneda, T., Brundage, W. E., Wang, J. C. and Engstrom, H. (1979) in *Fast Ion Transport in Solids* (eds P. Vashishta, J. N. Mundy and G. K. Shenoy), North-Holland, p. 261.

Bates, J. B., Engstrom, H., Wang, J. C., Larson, B. C., Dudney, N. J. and Brundage, W. E. (1981) *Solid State Ionics*, **5**, 159.

Bates. J. B., Wang, J. C. and Dudney, J. C. (1982) *Phys. Today*, July, 46.

Beevers, C. A. and Ross, M. A. S. (1937) *Z. Krist.*, **97**, 59.

Bettman, M. and Peters, C. R. (1969) *J. Phys. Chem.*, **73**, 1774.

Bettman, M. and Terner, L. (1971) *Inorg. Chem.*, **10**, 1442.

Bevan, D. J. M., Hudson, B. and Moseley, P. T. (1974) *Mater. Res. Bull.*, **9**, 1073.

Bevan, D. J. M., Hudson, B. and Moseley, P. T. (1975) *Mater. Res. Bull.*, **10**, 229.
Bjorkstam, J. L., Ferloni, P. and Villa, M. (1980) *J. Chem. Phys.*, **73**, 2932.
Boilot, J. P., Collin, G., Colomban, Ph. and Comès, R. (1980) *Phys. Rev.*, **B22**, 5912.
Boilot, J. P., Collin, G., Colomban, Ph. and Comès, R. (1981) *Solid State Ionics*, **5**, 157.
Boilot, J. P. and Théry, J. (1976) *Mater. Res. Bull.*, **11**, 407.
Bourke, M. A. M., Hooper, A., Moseley, P. T. and Taylor, R. G. (1980) *Solid State Ionics*, **1**, 367.
Bovin, J-O. (1979a) *Acta Cryst.*, **35A**, 572.
Bovin, J-O. (1979b) in *Fast Ion Transport in Solids* (eds P. Vashishta, J. N. Mundy and G. K. Shenoy), North-Holland, p. 315.
Bovin, J-O. and O'Keefe, M. (1980) *J. Solid State Chem.*, **33**, 37.
Bragg, W. L., Gottfried, C. and West, J. (1931) *Z. Krist.*, **77**, 255.
Breiter, M. W. and Dunn, B. (1981) *J. Appl. Electrochem.*, **11**, 685.
Breiter, M. W., Dunn, B. and Powers, R. W. (1980) *Electrochim. Acta*, **25**, 613.
Breiter, M. W., Farrington, G. C., Roth, W. L. and Duffy, J. C. (1977) *Mater. Res. Bull.*, **12**, 895.
Brennan, M. P. J. (1980) *Electrochim. Acta*, **25**, 621.
Brown, G. M., Schwinn, D. A., Bates, J. B. and Brundage, W. E. (1981) *Solid State Ionics*, **5**, 147.
Buechele, A. C. and De Jonghe, L. C. (1979) *Am. Ceram. Soc. Bull.*, **58**, 861.
Bugden, W. G., Barrow, P. and Duncan, J. H. (1981) *Solid State Ionics*, **5**, 275.
Bugden, W. G. and Duncan, J. H. (1977) *Sci. Ceram.*, **9**, 348.
Chandrashekhar, G. V. and Kim, K. K. (1981) *Solid State Commun.*, **37**, 299.
Chu-Kun, K. and Shan-Tin, L. (1981) *Ceram. Int.*, **7**, 98.
Collin, G., Comès, R., Boilot, J. P. and Colomban, Ph. (1980) *Solid State Ionics*, **1**, 59.
Crosbie, G. M. and Tennenhouse, G. J. (1982) *J. Am. Ceram. Soc.*, **65**, 187.
Davidge, R. W., Tappin, G., McLaren, J. R. and May, G. J. (1979) *Am. Ceram. Soc. Bull.*, **58**, 771.
De Jonghe, L. C. (1976) *J. Mater. Sci.*, **11**, 206.
De Jonghe, L. C. (1977a) *J. Mater. Sci.*, **12**, 497.
De Jonghe, L. C. (1977b) *Mater. Res. Bull.*, **12**, 667.
De Jonghe, L. C. (1979) *J. Mater. Sci.*, **14**, 33.
De Jonghe, L. C. (1982a) *J. Electrochem. Soc.*, **129**, 752.
De Jonghe, L. C. (1982b) *Solid State Ionics*, **7**, 61.
De Jonghe, L. C. and Buechele, A. (1982) *J. Mater. Sci.*, **17**, 885.
De Jonghe, L. C. and Feldman, L. (1980) *Mater. Res. Bull.*, **15**, 777.
De Jonghe, L. C., Feldman, L. and Buechele, A. (1981a) *J. Mater. Sci.*, **16**, 780.
De Jonghe, L. C., Feldman, L. and Buechele, A. (1981b) *Solid State Ionics*, **5**, 267.
De Jonghe, L. C., Feldman, L. and Millet, P. (1979) *Mater. Res. Bull.*, **14**, 589.
De Vries, R. C. and Roth, W. L. (1969) *J. Am. Ceram. Soc.*, **52**, 304.
Demott, D. S. (1980) *J. Electrochem. Soc.*, **127**, 2312.
Demott, D. S. and Redfern, B. A. W. (1976) *J. Phys.* (France), **37**, C7 423.
Demott, D. S., Wright, M. L. and Hames, M. D. (1981) Electrochemical Society Extended Abstracts 81–1, 469.
Dernier, P. D. and Remeika, J. P. (1976) *J. Solid State Chem.*, **17**, 245.
Dexpert-Ghys, J., Faucher, M. and Caro, P. (1982) *J. Solid State Chem.*, **41**, 27.
Dubin, R. R., Story, H. S., Powers, R. W. and Bailey, W. C. (1979) *Mater. Res. Bull.*, **14**, 185.

Dunn, B. (1981) *J. Am. Ceram. Soc.*, **64**, 125.
Dunn, B. and Farrington, G. C. (1980) *Mater. Res. Bull.*, **15**, 1773.
Dunn, B., Ostrom, R. M., Seevers, R. and Farrington, G. C. (1981) *Solid State Ionics*, **5**, 203.
Elrefaie, F. A. and Smeltzer, W. W. (1981) *J. Electrochem. Soc.*, **128**, 1443.
Engstrom, H., Bates, J. B., Brundage, W. E. and Wang, J. C. (1981) *Solid State Ionics*, **2**, 265.
Ewart, F. T., Moseley, P. T. and Sharpe, B. M. (1983) *Solid State Ionics*, **8**, 151.
Farrington, G. C. (1981) Extended Abstract 458, Electrochemical Society Fall Meeting, Denver.
Feldman, L. A. and De Jonghe, L. C. (1982) *J. Mater. Sci.*, **17**, 517.
Flor, G., Marini, A., Masarotti, V. and Villa, M. (1981) *Solid State Ionics*, **2**, 195.
Fischer, W., Haar, W., Bartmann, B., Meinhold, H. and Weddigen, G. (1978) *J. Power Sources*, **3**, 299.
Foster, L. M. (1979) in *Fast Ion Transport in Solids* (eds P. Vashishta, J. N. Mundy and G. K. Shenoy), North-Holland, p. 249.
Fray, D. J. (1981) UK Patent 1 602 564.
Gibson, A. (1977) in *Power Sources*, vol. 6 (ed. D. H. Collins), Academic Press, 673.
Gourier, D., Wicker, A. and Vivien, D. (1982) *Mater. Res. Bull.*, **17**, 363.
Gratias, D., Boilot, J. P., Le Cars, Y. and Théry, J. (1976) *Phys. Stat. Sol.*, **A38**, 595.
Gupta, N. K. and Tennenhouse, G. J. (1979) *J. Electrochem. Soc.*, **126**, 1451.
Harata, M., Imai, A., Ohta, T. and Sugaike, S. (1981) *Solid State Ionics*, **3/4**, 409.
Hattori, T., Nakata, H., Imanishi, T. and Mitsuishi, A. (1981) *Solid State Ionics*, **3/4**, 69.
Hattori, S., Yamaura, M., Kimura, S. and Iwabuchi, S. (1977) Proceedings SAE Congress, Detroit, paper 770281.
Hayes, C. E. and Ailion, D. C. (1979) in *Fast Ion Transport in Solids* (eds P. Vashishta, J. N. Mundy and G. K. Shenoy), North-Holland, p. 297.
Hayes, W., Holden, L. and Tofield, B. C. (1980) *Solid State Ionics*, **1**, 373.
Heavens, S. N. (1982) *J. Mater. Sci.*, **17**, 965.
Hooper, A. (1977) *J. Phys. D: Appl. Phys.*, **10**, 1487.
Hooper, A. (1980) *Trans. & J. Br. Ceram. Soc.*, **79**, 134.
Hudson, B. and Moseley, P. T. (1973) UKAEA Harwell, AERE MEMO-2613.
Hull, R., Cherns, D., Humphreys, C. J. and Hutchison, J. L. (1981) Institute of Physics Conference Series EMAG (Cambridge).
Humphreys, C. J., Hutchison, J. L., Wickens, V. J., Bugden, W. G. and Duncan, J. H. (1980) Proceedings 7th European Congress on Electron Microscopy.
Huntington, H. B. (1958) *Solid State Phys.* **7**, 213.
Imai, A. and Harata, M. (1972) *Jpn J. Appl. Phys.*, **11**, 180.
Itoh, K., Kondo, K., Sawaoka, A. and Saito, S. (1975) *Jpn J. Appl. Phys.*, **14**, 1237.
Jorgensen, J. D., Rotella, F. J. and Roth, W. L. (1981) *Solid State Ionics*, **5**, 143.
Kennedy, J. H. and Sammells, A. F. (1972) *J. Electrochem. Soc.*, **119**, 1609.
Kim, K. K., Mundy, J. N. and Chen, W. K. (1979) *J. Phys. Chem. Solids*, **40**, 743.
Knödler, R. and Baukal, W. (1978) *J. Power Sources*, **3**, 23.
Kodama, T. and Muto, G. (1976) *J. Solid State Chem.*, **17**, 61.
Koerber, G. G. (1962) *Properties of Solids*, Prentice-Hall.
Kummer, J. T. (1972) *Progr. Solid State Chem.*, **7**, 141.
Kutsenok, I. B., Kaul', A. R. and Tret'yakov, Yu.D. (1981) *Inorg. Mater.*, **17**, 628.
Le Cars, Y., Théry, J. and Collongues, R. (1972) *Rev. Int. Hautes Temp. Refract.*, **9**, 153.

Levine, C. A. (1975) 10th IECEC Conference Proceedings, 621.
Levine, C. A. (1978) Proceedings 28th Power Sources Symposium, 12.
Levine, C. A. (1981) 16th IECEC Conference Proceedings, 823.
Levine, C. A., Heitz, R. G. and Brown, W. E. (1972) 7th IECEC Conference Proceedings, 50.
Little, J. A. and Fray, D. J. (1980) *Electrochim. Acta*, **25**, 957.
McDonough, W., Flinn, D. R., Stern, K. H. and Rice, R. W. (1978) *J. Mater. Sci.*, **13**, 2403.
MCIC (1966) Metals and Ceramics Information Centre, Battelle, Columbus, Ohio, Engineering Properties of Selected Ceramic Materials.
MCIC (1981) Metals and Ceramics Information Centre, Battelle, Columbus, Ohio, Report MCIC-HB-07.
McWhan, D. B., Shapiro, S. M., Remeika, J. P. and Shirane, G. (1975) *J. Phys. C: Solid State Phys.*, **8**, L487.
McWhan, D. B., Varma, C. M., Hsu, F. L. S. and Remeika, J. P. (1977) *Phys. Rev.*, **B15**, 553.
Matsui, Y. and Horiuchi, S. (1981) *Acta Cryst.*, **A37**, 51.
May, G. J. (1978) *J. Power Sources*, **3**, 1.
May, G. J. and Henderson, C. M. B. (1979) *J. Mater. Sci.*, **14**, 1229.
May, G. J. and Hooper, A. (1978) *J. Mater. Sci.*, **13**, 1480.
Miller, G. R., McEntire, B. J., Hadnagy, J. D., Rasmussen, J. R. and Gordon, R. S. (1979) in *Fast Ion Transport in Solids* (eds P. Vashishta, J. N. Mundy and G. K. Shenoy), North-Holland, p. 83.
Newsam, J. M. (1982) *Solid State Ionics*, **6**, 129.
Ohta, T., Harata, M. and Imai, A. (1976) *Mater. Res. Bull.*, **11**, 1343.
Ormrod, S. E. and Kirk, D. L. (1977) *J. Phys. D.: Appl. Phys.*, **10**, 1497.
Park, D. S., Powers, R. W. and Breiter, M. W. (1981) *Solid State Ionics*, **5**, 271.
Pauling, L. (1963) *The Nature of the Chemical Bond*, 3rd edn, Cornell University Press.
Peters, C. R., Bettman, M., Moore, J. W. and Glick, M. D. (1971) *Acta Cryst.*, **B27**, 1826.
Pitt, M. G. and Fray, D. J. (1982a) *Electrochim. Acta*, **27**, 15.
Pitt, M. G. and Fray, D. J. (1982b) *J. Solid State Chem.*, **43**, 227.
Powers, R. W. and Mitoff, S. P. (1978) in *Solid Electrolytes* (eds P. Hagenmuller and W. Van Gool), Academic Press, p. 123.
Radzilowski, R. H. (1969) *Inorg. Chem.*, **8**, 994.
Radzilowski, R. H. and Kummer, J. T. (1969) *Inorg. Chem.*, **8**, 2531.
Radzilowski, R. H. and Kummer, J. T. (1971) *J. Electrochem. Soc.*, **118**, 714.
Rankin, G. A. and Merwin, H. E. (1916) *J. Am. Chem. Soc.* **38**, 568.
Richman, R. H. and Tennenhouse, G. J. (1975) *J. Am. Ceram. Soc.*, **58**, 63.
Rog, G., Kozinzki, S. and Kozlowska-Rog, A. (1981) *Electrochim. Acta*, **26**, 1819.
Roth, W. L. (1972) *J. Solid State Chem.*, **4**, 60.
Roth, W. L., Benenson, R. E., Tikku, V. K., Briant, J. L. and Dunn, B. (1981) *Solid State Ionics*, **5**, 163.
Roth, W. L. and De Vries, R. C. (1977) *J. Solid State Chem.*, **20**, 111.
Roth, W. L. Reidinger, F. and La Placa, S. (1976) in *Superionic Conductors* (eds G. D. Mahon and W. L. Roth) Plenum Press, p. 223.
Sato, H. and Horotsu, Y. (1976) *Mater. Res. Bull.*, **11**, 1307.
Shetty, D. K., Virkar, A. V. and Gordon, R. S. (1977) in *Fracture Mechanics of Ceramics*, vol. 4 (eds R. G. Bradt, D. P. H. Hasselman and F. F. Lange), Plenum Press, New York, p. 651.

Sieh, M. Y. H. and De Jonghe, L. C. (1978) *J. Am. Ceram. Soc.*, **61**, 185.
Stevens, R. and Miles, L. J. (1976) *J. Mater. Sci.*, **11**, 1911.
Sudworth, J. L., Hames, M. D., Storey, M. A., Azim, M. F. and Tilley, A. R. (1973) in *Power Sources*, vol. 4 (ed. D. H. Collins), Oriel Press, p. 1.
Takabatake, M. and Fukatsu, Y. (1973) *Rep. Res. Lab. Asahi Glass Co.*, **23**, 45.
Tan, S. R. and May, G. J. (1977a) *Sci. Ceram.*, **9**, 103.
Tan, S. R. and May, G. J. (1977b) *J. Mater. Sci.*, **12**, 1058.
Tan, S. R., May, G. J., McLaren, J. R., Tappin, G. and Davidge, R. W. (1980) *Trans. & J. Br. Ceram. Soc.*, **79**, 120.
Tang, T. B. and Chaudhri, M. M. (1982) *J. Mater. Sci.*, **17**, 157.
Tennenhouse, G. J., Ku, R. C., Richman, R. H. and Whalen, T. J. (1975) *Bull. Am. Ceram. Soc.*, **54**, 523.
Tofield, B. C. (1980) AERE-R9901.
Tofield, B. C. (1982) *Intercalation Chemistry*, Academic Press, p. 181.
Tofield, B. C. and Farrington, G. C. (1979) *Nature*, **278**, 438.
Tofield, B. C., Jacobson, A. J., England, W. A., Clarke, P. J. and Thomas, M. W. (1979) *J. Solid State Chem.*, **30**, 1.
Trantina, G. G. (1981) *J. Am. Ceram. Soc.*, **84**, C48–51.
Virkar, A. V. (1981) *J. Mater. Sci.*, **16**, 1142.
Virkar, A. V. and Miller, G. R. (1979) in *Fast Ion Transport in Solids* (eds P. Vashishta, J. N. Mundy and G. K. Shenoy), North-Holland, p. 87.
Virkar, A. V., Miller, G. R. and Gordon, R. S. (1978) *J. Am. Ceram. Soc.*, **61**, 250.
Virkar, A. V. and Viswanathan, L. (1979) *J. Am. Ceram. Soc.*, **62**, 528.
Virkar, A. V., Viswanathan, L. and Biswas, D. R. (1980) *J. Mater. Sci.*, **15**, 302.
Viswanathan, L. and Virkar, A. V. (1982) *J. Mater. Sci.*, **17**, 753.
Walker, J. R. and Catlow, C. R. A. (1980) *Nature*, **286**, 473.
Walker, J. R. and Catlow, C. R. A. (1982) *J. Phys. C: Solid State Phys.*, **15**, 6151.
Wang, J. C., Bates, J. B., Dudney, N. J. and Engstrom, H. (1981) *Solid State Ionics*, **5**, 35.
Wang, J. C., Gaffari, M. and Choi, S. (1975) *J. Chem. Phys.*, **63**, 772.
Weber, N. and Kummer, J. T. (1967) Proceedings 21st Annual Power Sources Conference, p. 37.
Weber, N. and Venero, A. (1970) *Bull. Am. Ceram. Soc.*, **49**, 498.
West, A. R. (1979) *Mater. Res. Bull.*, **14**, 441.
Whittingham, M. S. and Huggins, R. A. (1971) *J. Chem. Phys.*, **54**, 414.
Wolf, D. (1979) *J. Phys. Chem. Solids*, **40**, 757.
Worrell, C. A. and Redfern, B. A. W. (1978) *J. Mater. Sci.*, 13, 1515.
Yankulov, P. D., Staikov, G. and Budevski, E. (1981) *J. Mater. Sci.*, **16**, 3510.
Yao, Y-F. Y. and Kummer, J. T. (1967) *J. Inorg. Nucl. Chem.*, **29**, 2453.
Yasui, I. and Doremus, R. H. (1978) *J. Electrochem. Soc.*, **125**, 1007.
Youngblood, G. E. and Gordon, R. S. (1978) *Ceram. Int.*, **4**, 93.

3 Solid electrolyte production

J. H. DUNCAN, R. S. GORDON, R. W. POWERS and R. J. BONES

3.1 INTRODUCTION

The specification of the beta alumina electrolyte is far more exacting than that of conventional ceramics. The main requirements are long life in cells, high ionic conductivity and low interfacial impedance. These have to be achieved reproducibly with a high manufacturing yield and close dimensional control using a low cost production process. Several of the requirements are conflicting but nevertheless it has been possible to produce satisfactory ceramic electrolyte by several production routes. The key points in the production process are:

(1) Careful control of chemical composition and purity of the starting materials.
(2) Homogeneous mixing of the constituents.
(3) Careful attention to detail in the forming process.

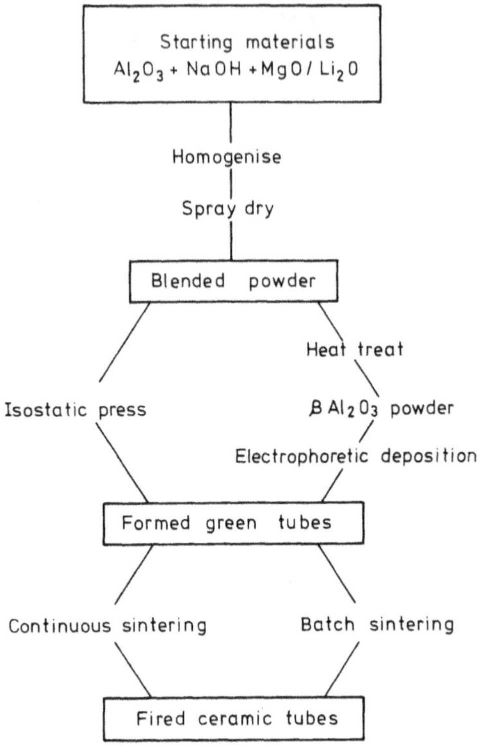

Fig. 3.1 Plan of the possible beta alumina manufacturing routes.

(4) Accurate control of the sintering process.
(5) Development of suitable quality control procedures.

The main production routes used at present are shown in Fig. 3.1, and the various stages of powder preparation, forming and sintering are discussed in detail in the following sections. It will be apparent from the differing views of the contributors to this chapter that there is no general agreement on the preferred production route. Several of the methods described here have produced electrolyte which has shown excellent life in cells, demonstrating that long life material can be produced in different laboratories and by different methods. The process chosen for production of beta alumina on a large scale will need to meet stringent cost goals and to result in a high quality electrolyte.

3.2 POWDER PREPARATION

3.2.1 Introduction

The methods available for powder preparation range from milling beta alumina prepared by fusion (Fally *et al.*, 1973) to reaction sintering of alpha alumina and a sodium compound, a method which developed from the observation by Duncan and Creyke (1969) that beta alumina can form as an impurity in high purity alpha alumina ceramics.

For those fabrication routes where conversion to beta" alumina occurs during sintering, selection of the alpha alumina raw material is of particular importance. Typical examples of aluminas having the necessary reactivity are superground Alcoa A-16 and Reynolds calcined alumina RC-HPS-DBM. These aluminas have a median particle size of approximately 0.5 μm and are calcined to >95% alpha alumina.

A systematic assessment of various alpha alumina raw materials including those made by Alcoa, Linde and Reynolds was reported by Gordon (1978) and by Miller *et al.* (1979) (Table 3.1). These workers preferred the more pure aluminas, but in Europe a less pure (and lower cost) material, Alcoa A16, has been widely used and has been found to produce beta alumina with acceptable properties. All the aluminas fit the description of 'reactive', and the choice of

Table 3.1 Emission spectrographic analysis and ultimate particle size of alpha aluminas. Reproduced with permission from Gordon (1978)

		Element (ppm)				*Particle size*
Material	*Source*	*Si*	*Ca*	*Fe*	*Ni*	*(μm)*
A	Linde C	142–179	12–16	25–48	<2	1
B	Meller (deagglomerated)	59	<9	12	<2	0.3
C	Reynolds RC-HP-DBM	106–111	42–52	28–29	<2	0.5
D	Reynolds RC-HPS-DBM*	130–146†	8–10	21–25	<2	0.5
E	Alcoa XA-139SG	164	121	164	3	<1
F	Alcoa A-15, 325 mesh‡	112	133	53	<2	2.5
G	Alcoa A-15SG	239	103	114	3	2.5
H	Reynolds RC-172-DBM	469	296	169	27	0.65
K	Reynolds RC-24‡	82	279	121	<2	2.8
L	Alcoa XB-2SG §	930	420	310	<10	4–8

* D is a less expensive variant of C, containing about 0.12 wt% Na_2O.
† ~75 ppm according to vendor analysis.
‡ Material supplied unmilled by manufacturer. All other materials are milled.
§ Ground beta alumina.

material is principally determined by the level of impurity. The two common impurities are silica and calcium. Silica, being a glass former, is likely to block ionic transport at grain boundaries, and calcium, which is mobile in the beta alumina structure, could be oxidized in preference to sodium to form an insulating interface (Hames and Duncan, 1975).

3.2.2 Sol gel processes

Both the mechanical and electrical properties of beta alumina electrolyte are directly related to the quality of the ceramic. It is known that physical flaws cause a decrease in strength of the material (Davidge *et al.*, 1979); impurity or non-equilibrium phases increase the electrical resistance. Many such defects have their origin in inadequate mixing or classification at the powder preparation stage.

As early as 1971 Markin *et al.* (1975) explored sol gel routes in which all of the constituents were mixed in a liquid form as either colloidal sols or aqueous solution. In the process, the components are mixed thoroughly on the ionic or microcrystallite (< 100 Å) scale and subsequently converted into a solid gel by either chemical or physical means without segregation. By suitable choice of the gelation process, it is possible to predetermine the particle shape and size fraction, giving a very high yield. The sol gel process had been used prior to this study to fabricate other oxide solid solutions, e.g. yttria stabilized zirconia (Woodhead, 1968), where it was demonstrated that much lower sintering temperatures were required to convert homogeneous powder to a dense single phase product compared with conventional powder mixing routes.

The early work utilized two types of alumina sol. One was formed by thermally denitrating $Al(NO_3)_3 9H_2 0$ to yield a sol having an Al/NO_3 ratio of 1:1; the other was formed by exchanging the chloride ion with the nitrate ion in commercially available aluminium chlorohydrate. The sodium and magnesium dopants were added to the stable sols in the form of aqueous solutions, preferably as acetate. Gelation was effected either by air drying, or by chemical denitration using an organic amine in a support solvent of Genclene.

Transparent gel particles were obtained which sintered to give 90% theoretical density, with a developed beta alumina structure, at 1500 °C. The free flowing nature of these powders rendered them ideal for arc plasma spraying. Attempts were made to form beta alumina tubes by this process. Considerable soda loss was found due to the high temperatures (in excess of 2000 °C) which the powder experiences for a short time during the spraying process. By chemical analysis, the system was calibrated such that the sodium oxide content of the sprayed product could be preselected by using soda rich feed powders. The soda loss was found to be dependent on particle size.

The resultant formed tube was not very crystalline, requiring subsequent sintering at 1500 °C to restore the beta alumina phase. The tube wall has a laminated structure and it was difficult to achieve adequate densities.

The use of dispersible boehmite alumina monohydrate sols was not pursued in the early work as the subsequent doping with aqueous solutions of magnesium and sodium salts led to considerable coagulation and the formation of viscous, thixotropic sols. This method, in which a viscous gel was formed by the addition of the dopants, was, however, reported by Eddy *et al.* (1977). The thixotropic gel was spray dried to form a powdered beta alumina precursor material. Sintering of a green body formed from this powder resulted in densification to >95% of theoretical at 1550–1600°C.

Recent phase equilibria studies on beta alumina materials have reported a new metastable mullite type structure designated M–Al_2O_3 (Perrotta and Young, 1974). Takahashi and Kuwabara (1980) reported that this M–Al_2O_3 material is a promising precursor from which to synthesize beta alumina. A very fine M–Al_2O_3 powder is formed by the thermal decomposition of a complex compound containing both sodium and aluminium:

$$Na_x(NH_4)_{3-x}[Al(C_2O_4)_3]yH_2O$$

$$(0.09 \leqslant x \geqslant 0.333) \qquad (y=3)$$

This process has been studied in great detail; tablets of M–Al_2O_3 have been converted to dense (>95% theoretical) beta alumina when sintered to 1600 °C for 30 minutes.

Workers at Batelle (Baukal *et al.*, 1977) have reported the preparation of beta alumina precursor powder by direct spray hydrolysis of aqueous mixed solutions of aluminium and sodium salts (and dopants if necessary). There is a tendency for this process to yield an initial low tap density powder which requires careful treatment to enable dense green bodies to be fabricated.

The preparation of thin films of beta alumina on suitable substrates has been reported by Yoldas and Partlow (1980). In this process, a clear precursor liquid containing sodium and aluminium ions in a chemically polymerized form is prepared from metal organic compounds. A sodium alkoxide is dissolved in alcohol and hydrolysed using approximately one mole of water to one mole of alkoxide, at room temperature:

$$Na(OR) + (1-x)H_2O \xrightarrow{\text{alcohol}} Na(OH)_{1-x}(OR)x + (1-x)R(OH)$$

where R is an alkyl group and x is a small fraction, e.g. 1/20. The solution is then reacted with an aluminium alkoxide such as aluminium sec-butoxide, $Al(OC_4H_9)_3$, which has been diluted in butanol:

$$Na(OH) + Al(OR')_3 \rightarrow Na\text{–}O\text{–}Al(OR')_2 + R'OH$$

These equations are only approximate. The important requirement is that by the reaction of the hydroxyl groups of the sodium complex with the alkyl groups of the aluminium alkoxide, an oxide bond is formed chemically between the aluminium and sodium ions via the oxygen (i.e. Al–O–Na) at room temperature. Excess water chemically removes the remaining alkyl groups.

The resultant hydroxide complex, if heated to ~500 °C, yields a particulate oxide material containing aluminium and sodium in a chemically polymerized form. When this oxide is heated to 1200 °C, beta alumina is formed.

To form thin films from the transparent complex a peptizing agent, e.g. acetic acid, is added to prevent the agglomeration of crystallites, thus stabilizing the alumina based sol which can be used for producing thin coatings by spray drying or dipping. Careful drying and heat treatment to 1200 °C forms beta alumina films without crazing.

These sol/solution routes to beta alumina are all capable of yielding very homogeneous materials which sinter to give the beta alumina structure at lower temperatures than the powders obtained by more conventional routes. At the present stage of development they are not as readily adapted to mass production as the slurry solution spray drying process described in the next section. However, such novel routes should not be overlooked in the long term, especially if durability of more conventional beta alumina proves to be a problem. Even at present these routes are capable of yielding 'purer' samples of the solid electrolyte for fundamental studies.

3.2.3 Preparation from alpha alumina powder

(a) *Direct mixing of raw materials*

The simplest and most direct powder preparation route is dry milling. Jones and Miles (1971) vibro milled a reactive alumina with sodium aluminate and the resulting powder could be isostatically pressed to form closed ended tubes without the addition of binder. The 'green' tubes (in this context green means unfired) were prefired to remove volatiles, then fired by the zone sintering process which will be described in Section 3.4.4. Magnesia and/or lithia were used as dopants to stabilize the beta″ phase, and a range of compositions was prepared including 8% Na_2O, 2% MgO and 9% Na_2O, 0.7% Li_2O.

Perhaps the most straightforward of the preparation routes for beta alumina is slurry solution spray drying where the alpha alumina powder is mixed with a solution of sodium and lithium hydroxides and the resulting slurry spray dried (Duncan, 1980). This powder, as we will see in Section 3.4.2, is fired in a single operation to produce beta alumina ceramics possessing low resistivity and high strength.

A slip with a solids content of around 50% by weight is necessary for efficient spray drying, and the hydroxides are the only suitable compounds of sodium and lithium which are soluble enough to dissolve completely in such a slip. The spray dried powder should not only have flow characteristics compatible with the next stage (powder compaction), but also produce a well sintered artefact. For good flow properties coarse powders, or those with a narrow particle size range, are required. Successful sintering requires the particles to deform fully during compaction. With the strong particles produced by slurry solution spray drying these requirements are conflicting, and it is necessary to compromise between flowability and sinterability.

With manufacturing rates at the level of pilot plant or less, the size of spray dryers employed is small. The particle size too is small, since the larger the droplet, the longer the path length needed to dry it and the larger the dimensions of the dryer. Finer powder is favoured for high fired density and this is made by rotary atomization where the slip enters at the top of the dryer.

The chemical composition is consistent across the particle diameter and hollow particles are rare, in contrast with solution spray drying, where all the particles that dry out from the atomized wet mix are hollow.

For a powder to deform under load, eliminating large interparticle voids, the particles must not be too strong. When the drier is operated with a low throughput of slip, the dry powder obtained has good flow properties but is hard to compact. On the other hand, when the throughput is high the powder is softer but the flow may be very poor. Success lies somewhere between these extremes. Duncan (1980) reported that, using a Niro Atomizer Production Minor spray dryer, outputs of 20 kg h^{-1} could be obtained. The manner in which output can vary with drying conditions is shown in Table 3.2.

Table 3.2 Spray drying trial with a Niro Production Minor spray drier. Slip solids content 50%. Reproduced with permission from Duncan (1980)

	Temperature (°C)				
Batch	*Inlet*	*Outlet*	*Product (kg)*	*Time (min)*	*Rate (kg h⁻¹)*
1	300	125	5.70	20	17.1
2	300	150	3.00	13	13.5
3	250	145	5.14	20	15.4
4	200	100	1.92	13	10.4
5	200	130	1.29	19	4.0
6	180	125	1.50	24	3.8
7	175	80	4.00	25	9.6

Failure to achieve high density during sintering can arise from several causes such as composition, firing atmosphere and powder properties. The nature of the inside surface of a tube compact provides a convenient criterion for assessing the contribution of the powder properties to any porosity remaining after firing. Having been formed against a polished steel mandrel, this tube surface can be examined using the scanning electron microscope to determine the relative deformation of the spray dried spheres. The sinterability of powders is also reflected in the dependence of fired density on isostatic forming pressure. This is

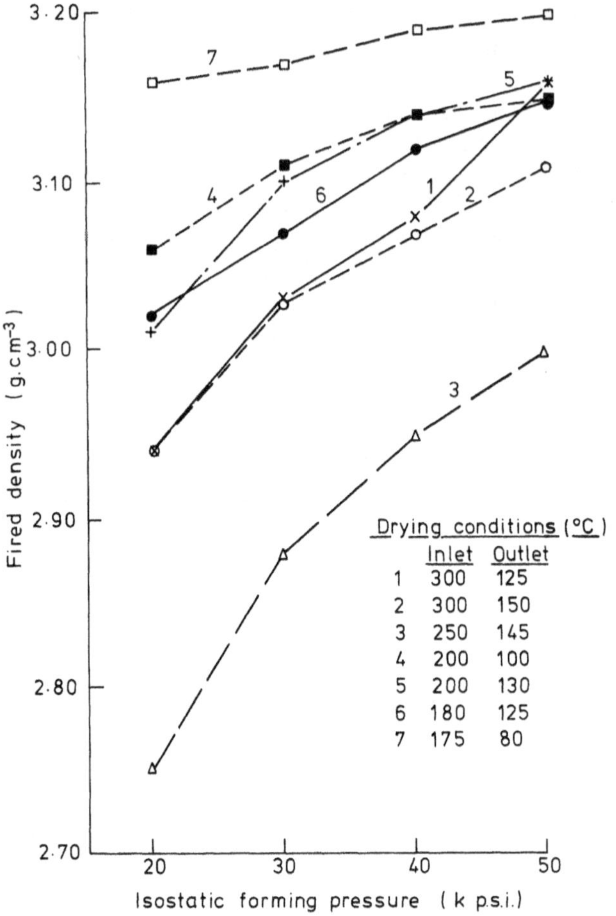

Fig. 3.2 Fired density versus forming pressure for spray dried powder. Duncan (1980). Reproduced with permission from British Rail Research.

illustrated by Fig. 3.2 (Duncan 1980), where the parameters can be compared with the spray drying conditions.

In these experiments the particle size range of the powder derived from rotary atomized slip was 1–50 μm. Within the dryer the swirling mist causes particles to agglomerate before these are dry. These lumps of four or five globules are grossly oversize and were removed by sieving through 45 μm aperture mesh.

An effect that is somewhat surprising is the high green strength of isostatically pressed shapes of powders containing the alkali metal compounds. The relative softness and high volume content of these compounds is an aid to pressing and the resulting compacts are strong enough to handle easily. Speronello (1980) postulated that the strength of the spray dried granules and the green density of the compacts was controlled by the amount of soluble salts deposited in the pore space of the alumina agglomerate.

(b) *Calcination, prereaction of raw materials*

The original preparation of beta″ alumina at the Ford Motor Company (1968) involved prereaction (calcination) of the raw materials – high purity alumina (Linde B), magnesium oxide and sodium oxide – in two stages.

Haar *et al.* (1976; also Haar, 1980) described a calcination method for preparing beta alumina powder from gamma alumina, sodium carbonate and magnesium oxide. The gamma alumina was converted to alpha alumina before the calcination stage to avoid low fired densities. The flow properties of the powder were dependent upon the final moisture content. The density of the sintered ceramic was found to be affected by the chemical composition. In particular, at the higher magnesia levels required to produce a high beta″ alumina content, a higher level of soda was necessary to produce a density near to theoretical.

The zeta process developed at the University of Utah by Virkar *et al.* (1978) and further described by Youngblood *et al.* (1977) and Gordon (1978) produces high quality beta alumina. The process is named after zeta $LiAl_5O_8$ spinel which is formed by reaction of lithium nitrate or lithium oxalate with alpha alumina (Reynolds RC–HPS–DBM) at 1250 °C. Prereaction of the small lithium addition with some of the alumina is expected to lead to better distribution of this element in the green compact. Low temperature eutectic liquids (< 1400 °C) have been reported (Gordon, 1978) in the Al_2O_3–Na_2 O–Li_2O system. These could also enhance sintering.

There are two routes described for the zeta process. Each route starts with the preparation of $LiAl_5O_8$ by calcination. In the first route the $LiAl_5O_8$ is dry milled with sodium carbonate and alpha alumina to give the composition 8.85% Na_2O, 0.75% Li_2O, balance Al_2O_3. The resulting powder is then calcined. In the second route, a mixture of sodium carbonate and alumina is calcined, and the product is mixed with $LiAl_5O_8$ to give the required composition.

Spray drying of calcined powder from the zeta process has been reported by Gordon (1978). He used a small stainless steel dryer which produced powder at the rate of 2 kg h^{-1} at an efficiency of 75%. Production of a flowable powder by spray drying requires preparation of a slip which is stable, i.e. settles out only slowly. The powder when formed and sintered must result in a beta alumina ceramic with high density, low resistivity and long life in a cell environment.

When calcined powder is suspended in water, the pH and the viscosity both rise. The workers at the University of Utah (Gordon 1978; Ford Motor Co. 1978) added citric or tartaric acids to reduce the pH, and thus the viscosity, allowing the solids content of the slips to be raised, so raising the rate of powder production. The spray dried powders had excellent flow properties, and when sintered the properties of the ceramic were acceptable. In general, the best flowing powder was obtained by drying with a high pressure on the two fluid nozzle, i.e. good atomization coupled with a low slip feed rate. With any spray drying operation it is necessary to find the optimum combination of slip solids content, slip feed rate, size distribution of the atomization and spray dryer inlet and outlet temperatures.

3.2.4 Preparation from beta alumina powder

(a) *Commercial beta alumina powder*

There are some advantages in using commercially available beta alumina powder. One such powder is the Alcoa XB2 beta alumina, whose level of impurities is similar to that of the Alcoa A-16 alpha alumina. Table 3.3 gives the properties of the original calcined material (Francis *et al.*, 1971). Processes where the XB2 powder has proved useful have been extrusion (Gordon, 1978) and slip casting (Byckalo *et al.*, 1976; Rivier and Pelton, 1978); no chemical alterations were made to the beta alumina for the slip casting.

Table 3.3 Properties of calcined beta alumina. Reproduced with permission from Francis *et al.* (1971)

Constituent	*Amount (%)*
Chemical	
Na_2O	7.30
Fe_2O_3	0.03
CaO	0.03
SiO_2	0.02
K_2O	0.01
Li_2O	0.01
Ga_2O_3	0.01
Al_2O_3 (by difference)	>92.5
Physical	
Apparent specific gravity	3.24
Microscopic crystal size	
Median	5–15 μm
Range	1–50 μm
Surface area	1.2 $m^2\ g^{-1}$
Particle size	
+100 mesh	13%
+200 mesh	91%
+325 mesh	99%
Bulk density	
Loose	1.8 $g \cdot cm^{-3}$ (44 lb ft^{-3})
Packed	2.3 $g \cdot cm^{-3}$ (57 lb ft^{-3})

The powder for the extrusion experiments was supplied as a superground material. It is interesting to note that in spite of the platey nature of beta alumina crystals, under the scanning electron microscope the powder particles were seen to be equant in all directions. By doping the extrusion mix with additional soda and lithia in the form of oxalates, the standard composition 8.7% Na_2O, 0.75% Li_2O could be fabricated and readily sintered.

Results (Francis *et al.*, 1971) of dry milling trials on the calcined beta alumina using various milling aids are given in Table 3.4. Using the criteria of green density after pressing to 35 MPa (5000 psi) and fired density after sintering to 1700 °C for one hour, the best additives were monoethanolamine, methanol, ethylene glycol and 1-amino 2-propanol.

Table 3.4 Grinding aids for dry milling calcined beta alumina. Reproduced with permission from Francis *et al.* (1971)

	Bulk density (g cm^{-3})		*Diametric linear shrinkage*
Grinding aid (0.5%)	*Fired* 1700 °C, 1 h*	*Green 35 MPa (5000 psig)*	*(%)*
None	3.139	1.85	16.1
Monoethanolamine	3.200	1.93	15.5
Methanol	3.197	1.86	16.5
Ethylene glycol	3.196	1.94	15.4
1-amino 2-propanol	3.196	1.92	15.6
Capric acid	3.188	1.83	16.1
Triethylene glycol	3.187	1.91	15.7
Ethylene glycol monoethyl ether	3.186	1.91	15.7
Oleic acid	3.184	1.93	15.4
n-propanol	3.184	1.90	15.8
Ethyl acetate	3.183	1.94	15.2
Myristic acid	3.182	1.91	15.6
Triethanolamine	3.181	1.87	16.2
Isopropanol	3.180	1.87	16.2
Ethanol	3.179	1.88	16.1
Octanol	3.172	1.90	15.6
Isostearic acid	3.166	1.87	16.1

* Alkaline atmosphere provided.

(b) *Fusion of raw materials*

Fally *et al.* (1973) were singularly different in fusing their own beta alumina and milling this to a powder to form tubes by electrophoretic deposition. The compositions prepared varied between $Na_2O \cdot 5.33Al_2O_3$ and $Na_2O \cdot 11Al_2O_3$ with, in some cases, additions of magnesia and zirconia incorporated before milling. The fused material was first crushed then milled in a stainless steel ball mill. Metallic impurities were removed by leaching with hydrochloric acid which also removed any sodium aluminate. The surface area of powders prepared in this manner was 2 $m^2 g^{-1}$. Subsequently, on the grounds of cost, the fusion stage was replaced (Desplanches, 1979) by dry milling with alumina media and calcination at 1050 °C, giving alpha alumina and sodium aluminate, or at 1280 °C to give a mixture of beta and beta" aluminas. Isostatic pressing gave a final compaction to tubes shaped by electrophoretic deposition. High purity alpha alumina raw material was selected in line with the following limits to ensure long cell life: potassium and magnesium each less than 10 ppm, silicon and calcium each less than 100 ppm and iron less than 300 ppm.

Desplanches (1980) reported that calcined powder prepared to the composition 9.2% Na_2O, 90.8% Al_2O_3 could be formed into tubes by isostatic pressing alone as well as by electrophoretic deposition, and powders giving the beta" alumina composition 9.9% Na_2O, 0.77% Li_2O, remainder Al_2O_3 were calcined at 1200 °C and shaped only by isostatic pressing.

(c) *Commercial fusion cast Monofrax H brick*

Crushed Monofrax H fusion cast brick was dry milled to 10 μm then calcined at 1400 °C with sodium carbonate and magnesium oxide to give beta and beta" alumina compositions (Sugaike, 1979) and finally milled to 1 μm particle size in acetone. This preparation method produced fired tubes with excellent physical properties.

3.3 FORMING

3.3.1 Slip casting

Two reports of slip casting of beta alumina (Byckalo *et al.*, 1976; Rivier and Pelton, 1978) describe procedures that result in satisfactory products. The laboratory route of Rivier and Pelton could be useful for slip casting a calcined powder which has had only intermediate heat treatment and which would be particularly susceptible to leaching of the sodium. No problems were encountered when slip casting a suspension of XB2 alumina in normal plaster moulds.

Particle orientation parallel to the tube wall has long been regarded as a disadvantage of slip cast beta alumina tubes where the low resistivity direction is then perpendicular to the direction of current flow. When tubes from the two routes were fired, there was strong particle orientation in the product cast from the coarse 3.5 μm powder in the deflocculated aqueous suspension. However, it was reported that no texture was observed in the tubes rapidly cast in a finer, 0.5 μm powder from additive-free methanol.

3.3.2 Extrusion

Extrusion has been reported in detail by Gordon (1978). Four binder systems were studied: polyvinyl pyrrolidone/ethylene glycol (PVP/EG), polyvinyl alcohol, beeswax for hot extrusion, and vaseline/mineral oil.

(a) *Polyvinyl pyrrolidone/ethylene glycol*

In this system the powder, based on calcining Alcoa C-33 ($Al(OH)_3$), was aimed at the composition 8.7% Na_2O, 0.7% Li_2O, remainder Al_2O_3.

The principal drawback of the PVP/EG system with alumina derived from Alcoa C-33 was the low green density which limited the fired density. Green densities of at least 55% are needed to reach acceptable sintered densities of 96%, and that would require a higher ratio of powder to binder. The consequence of this was a requirement for a better degree of mixing of the powder and the binder and a need to employ de-airing to reduce entrapped air and consequent large voids.

Improved green densities with the binder system were possible where the powder was the Alcoa XB2 beta alumina. The binder was PVP/EG in the ratio 30:70. With a powder:binder ratio of 5.3:1 extrusions with a green density of 68% were obtained. After sintering at 1650 °C the beta alumina, which was 97% dense, was produced. Unfortunately it did not prove possible to amend the composition to one that would result in the three-block beta″ alumina after firing.

Producing straight, crack-free extruded tubing requires a combination of good design of the extruder and the correct binder system. Successful extrusion of tubing required extrusion rates as high as 500 cm min^{-1}. These were achieved with the PVP/EG system.

(b) *Polyvinyl alcohol*

In an early report of extrusion (Charles *et al.*, 1971), sodium meta-aluminate $Na_2O \cdot Al_2O_3 \cdot 3H_2O$ (19 wt%) was employed as the source of sodium together with Linde A alumina, magnesia and yttria to stabilize the beta alumina phase. The binder system was polyvinyl alcohol added as a 10% solution in water. The fired density was 3.08 g cm^{-3} after sintering at 1825 °C for two hours in dry oxygen and the resistivity was 5–10 Ω cm.

(c) *Hot extrusion with beeswax binder*

This method was used in an attempt to obtain a rigid extrusion after rapid cooling, whilst using low extrusion forces of 2.0–2.5 tonnes for the comparatively high extrusion rate of 0.5 m min^{-1}. Good dimensional control was possible.

Extrusion was performed at 96–122 °C and green densities were as high as 55% of theoretical, but the fired density at 92–93% was not satisfactory, owing in part to inadequate mixing. There were also problems with damage arising during the binder removal stage of the firing cycle.

(d) *Vaseline/mineral oil*

Experience with this system highlights the pitfalls of the extrusion process. Success in obtaining the desired phase composition depends on the selection of the raw materials. Unfortunately, the material that suited extrusion did not finally convert to the low resistivity beta″ alumina.

The successful extrusion of the Alcoa XB2 beta alumina in a previous system was repeated here. High ratios of powder to binder at low pressures gave green densities as high as 60% of theoretical when the binder was fired off at 760 °C. Subsequent sintering at 1650 °C resulted in a density 99% of theoretical.

When sodium oxalate and lithium oxalate were added to produce beta″ alumina as the final product, extrusion was moderately successful but the phase conversion from beta to beta″ alumina was not.

Where closed ended tubes are to be made, the extrusion may be closed at one end with a high green density disc of the same material. Differential firing shrinkage during sintering encourages the two parts to seal. This method is preferable to simply pinching off the closed end.

3.3.3 Isostatic pressing

(a) *The isostatic pressing operation*

This consists of pressurizing with a liquid a tubular polymer bag, containing the pressing powder, on to a concentric mandrel. On depressurization the bag must instantly separate from the powder compact which in turn elastically relaxes from the mandrel, allowing the tube to be slid off even where the mandrel has parallel sides. There is a minimum pressure to achieve the release; this is approximately 35 MPa (5000 psi) with unreacted spray dried powder. In the axial direction the inside length of the compact is 1–2 mm longer than the mandrel on to which it was pressed, owing to the elastic relaxation.

There is absolutely no doubt that the first requirement for success with isostatic pressing is the choice of a polymer which does not interact with the powder. Polymer/compact interaction results in the two materials adhering when the pressure is removed. Adherence is thus a sign that the two materials are not compatible.

Experiences vary with the same elastomer. Whereas silicone rubber was found to be unsuitable for use with spray dried unreacted powder (Street and Duncan, 1978), Haar *et al.* (1976) found it satisfactory with calcined beta alumina powder.

Desplanches (1979) reports first forming a tube by electrophoretic deposition, then compacting it by isostatic pressing. The tube on a mandrel was encased in a latex bag, sealed by an O ring, then isostatically pressed to 200 MPa (29 000 psi). Latex bags are made by a dipping process which can result in an uneven wall thickness. This would produce a compact with an uneven wall thickness if the bag were incorporated in metal tooling. These difficulties were presumably not encountered by Desplanches, as the isostatic pressing was used after the tube had been shaped. Sugaike (1979) reported isostatic compacting of powder prepared from Monofrax beta alumina. The compacts were machined to size before firing. Gordon (1978) did not describe his tooling but reported that tooling was made to match each powder.

The next requirement when producing compacts in quantity is to have good dimensional control of tooling in both the polymer and metal components to produce fired tubes to within given tolerances. The principal dimensions of the tooling are the diameter of the powder cavity and the diameter of the central mandrel within that cavity. These are specific for any one powder. The extent to which the powder compacts on to the mandrel which forms the inner diameter of the tube will determine the tube wall thickness. The firing shrinkage behaviour of the compact will fix the final inner diameter of the tube.

The above points are well illustrated by the tooling described by Street and Duncan (1978). They described steel and polyurethane items of a tooling set for hemispherical ended tubes 37 mm in diameter and 360 mm long. The polyurethane, which was cast in machined metal moulds, was a precise fit in the metal part of the tooling assembly (see Fig. 3.3). In such a system the sizes of the polyurethane bag and the mandrel are readily varied to suit different powders. The outer metal tube or cage is concentric with the mandrel and retains the polyurethane on centre for close control of wall thickness.

Street and Duncan also reported that PVC tubing can be of use; it has better compact release and mould life than silicone rubber and latex. If available at the appropriate diameter it can be incorporated in simple tooling with a cage to keep it concentric.

In the same report it is shown how the appearance of the inner surface of a powder compact that has been in contact with the mandrel can give an indication of the suitability of the powder for isostatic pressing. Where the spray dried powder granules are relatively hard they will be flattened but not broken up by compaction against the surface of the mandrel. In the subsequent sintering stage this coarse intergranular porosity must be eliminated. When the powder is relatively soft, little remains of the spherical shape of the granules after they have been pressed against the mandrel and the intergranular porosity is minimized. The more easily compressed powder is thus more likely to sinter to a high density.

(b) *Assessment of isostatic pressing*

The advantages of isostatic pressing include the following:

(1) The amount of binder that has to be added to the powder to impart strength to the compact is minimal and binder may not even be necessary in some cases. Compared with die pressing, where 5% binder is a common level of addition, there may only be 1% binder in powder for isostatic pressing. Powders produced by calcination would be expected to need the addition of binder as a pressing aid. For consistent distribution this means introducing the binder in solution, followed by a drying stage, comminution and sieving.

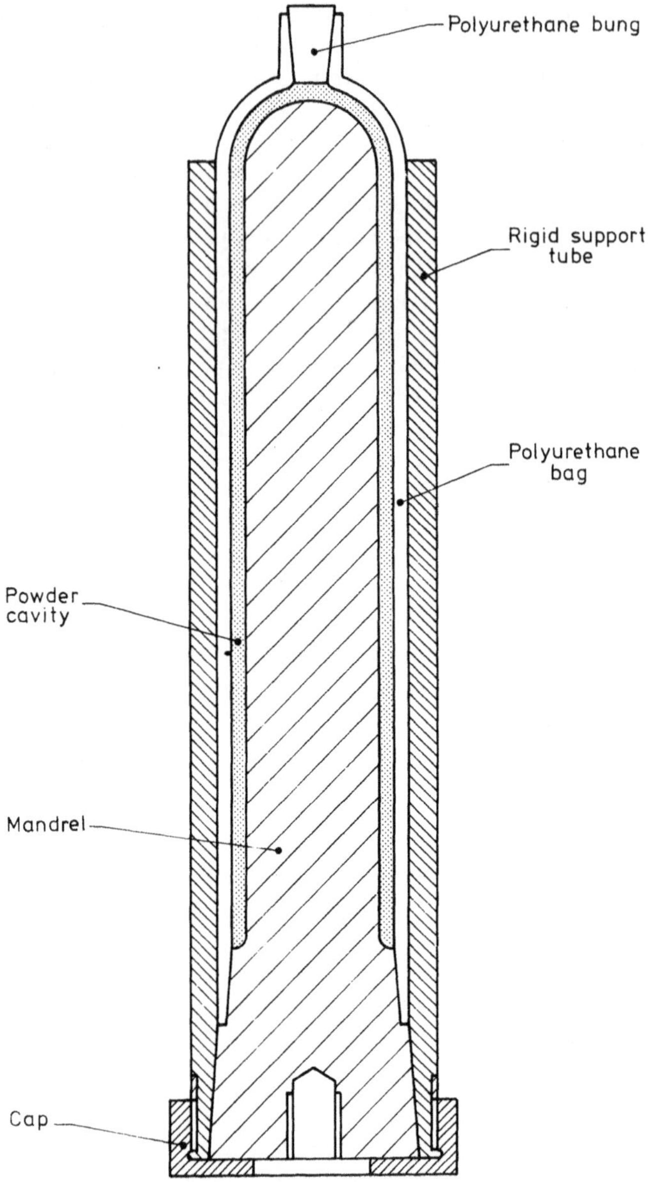

Fig. 3.3 Isostatic pressing tooling for thin walled tubes. Street and Duncan (1978). Reproduced with permission from Metal Powder Report Publishing Services Ltd.

There are instances, e.g. with spray dried powders (Duncan, 1980), where a raw material such as sodium hydroxide will itself act as a binder.

(2) The compact has a high strength and can be handled easily immediately it is released from the mould, unlike shapes produced by slip casting, extrusion or electrophoretic deposition (EPD). After isopressing the shape may only need minor trimming before firing and the closed ended shape is complete, with just the overall length to be adjusted by cutting off the open end after firing. This reduces the risk of damaging the compact before firing when it is at its weakest, and enables a more accurate length to be obtained on the final fired article.

(3) As a cold forming technique isostatic pressing is known to be more efficient than die pressing; it avoids the variation in density throughout the compact that arises through friction against the die walls and poor transmission of pressure to the centre of the bulk. The compaction forces are not limited by surface tension as in slip casting. The applied isostatic pressure can be altered (within the limits of the equipment) to suit the compaction behaviour of a given powder having a given particle size distribution. For thin walled beta″ alumina tubes with a domed end, isostatic pressing offers the combination of flexibility in choice of powder type with consistency of green density throughout the shape. In extrusion and deposition routes the selection of raw materials has to be tailored to the process. Isostatic pressing could compact all the powders used for these routes, although there may be some particular reason for selecting a given route, e.g. the good surface finish associated with electrophoretic deposition. However, the converse is less likely to be the case.

(4) Rates of production can readily be raised with dry bag pressing and even with wet bag pressing. In a dry bag press the tooling is inside the pressure vessel where it is an integral part of the vessel and vessel closure mechanism. In wet bag pressing separate individual tooling sets of different sizes and shapes, if desired, are placed in the hydraulic fluid in the pressure vessel which is then closed and pressurized. Mass production of beta alumina by spray drying followed by automatic isostatic pressing to shape appears to be economically viable. Mass production of beta alumina tubes has been likened to present day manufacture of articles such as spark plug insulators, but these are ground into final shape from an isopressed tubular blank before firing. A better comparison is with the thin walled transparent alumina ceramic tubes used in sodium vapour lamps.

3.3.4 Electrophoretic forming

(a) *Introduction*

Satisfactory electrophoretic forming of beta alumina ceramics is more than just an electrophoretic deposition operation. It also involves removal of the greenware from mandrels without cracking as well as consideration of the compatibility of the starting powder and of subsequent sintering with this kind of forming. Problems specific to certain types of compositions need to be taken into account. These matters are treated briefly in this section. In addition, some problems associated with scale-up are considered. Finally, the outstanding features and limitations of this technique are assessed.

This section is mainly concerned with the method of electrophoretic forming practised at the General Electric Company. A discussion of other methods, as well as background information on electrophoretic deposition, appears in Powers (1975) and in references cited therein. For reasons of brevity, statements in this section will not be supported by documentation. Interested readers can find more detailed information in the references.

Electrophoretic forming involves deposition of particles of diameter $\sim 1\ \mu m$ from a suspension with an electric field. According to the theory of Hamaker and Verwey (1940), the main role of the electric field is to press depositing particles against a mandrel, i.e. the forming electrode. The amount of a deposit is given by Hamaker's equation (1940)

$$Y = \int_0^t aACF\mu \, dt \tag{3.1}$$

Here Y is the deposit mass (g) produced in a time t (s). C is the particle concentration (g cm^{-3}) in a suspension of organic liquid or vehicle; F is the electric field (V cm^{-1}; μ is the particle mobility ($cm^2\ s^{-1}\ V^{-1}$); and A is the electrode area (cm^2). The coefficient a indicates the fraction of particles moving into the neighbourhood of the electrode that get deposited. A, C, F, and t are directly controllable external variables while μ depends on the electrical double layer formed at the interface between a beta alumina particle and the vehicle in which the particle is suspended. Primary amyl alcohol has been the vehicle used at the General Electric laboratory mainly because of its dielectric properties. However, practical considerations such as relative safety and commercial availability also played a role in its selection. One usually wishes to minimize deposition times by using concentrations and electric fields as large as is feasible. Concentrations of 0.5 g beta alumina per cm^3 of amyl alcohol are normally used since problems with viscous flow of deposits occur at concentrations much larger. Forming voltages are usually limited to 500 V as undesirable electrochemical reactions occur at increasingly rapid rates at higher voltages. Gaseous electrolytic decomposition products generated at the forming mandrel can introduce large pores, tens of microns in size, into greenware. Such large pores are usually not removed in subsequent sintering.

While electrophoretic deposition has been used as a technique for applying coatings, about 0.1 mm thick, for the last forty years (DeBoer *et al.*, 1939), its use as a ceramic forming method is a much more recent development (Andrews *et al.*, 1969). Deposit thicknesses are about ten times greater for the latter application. Rates of depositions can be rapid, approximately 10^4 times faster than with conventional electroplating. Consequently, plating times are of the order of minutes. Various shapes are formable, including closed ended tubes, within the limitation that the deposit thickness does not exceed ~3 mm. For ceramic electrolytes where low resistances are usually required, this limitation is not very restrictive. Since throwing power is rather good, deposition into corners and other recesses is usually not a major problem. However, some care is required to obtain uniform thicknesses with plates and discs. With the beta aluminas, where the crystallites tend to be very platey, deposits none the less exhibit only weak preferred orientations. However, the major advantage of electrophoretic deposition is that it can produce very homogeneous greenware of excellent sinterability. These features will be discussed in more detail later.

(b) *Description of the forming operations*

The operations carried out at General Electric are essentially as follows. A suitably processed powder is put into suspension in primary amyl alcohol and vibratory milled to permit acquisition of a negative charge on the particles. The particles are then deposited from the suspension on to a mandrel with the suspension container also serving as counter electrode. The forming voltage is typically 500 V. The shaped greenware are put into a low humidity chamber immediately following deposition and the amyl alcohol is removed using a current of warm air. Following complete evaporation of the vehicle, the formed articles are slid off the mandrels.

Satisfactory electrophoretic forming has been carried out with a wide variety of powders of both phase pure beta alumina as well as of mainly beta″ alumina. They have included both commercially available material and that synthesized at the General Electric laboratory. The alumina source has been of both commercial purity (Bayer process) and higher purity (alum derived). Powders have been synthesized using fusion as well as calcination. An essential feature is that the synthesis be carried out at sufficiently high temperature, ~1100 °C or higher, to yield a well crystallized product. Otherwise, particles cannot be properly charged. Consequently precursor powders, i.e. mixtures of starting reactants, which are converted to beta or beta″ alumina in the sintering operation are not suitable for the forming procedure used at General Electric. Another important powder property is related to the tendency of certain powders to undergo excessive shrinkage on the mandrels during evaporation of the vehicle. This situation can lead to cracking of greenware. The alumina source is one important variable affecting this property. In addition, the surface area of the powders needs to be sufficiently great to permit sintering to high density following forming. Beta alumina surface areas in the range 2–4 $m^2\ g^{-1}$ are usually adequate and these are readily available by dry grinding.

A number of experiments support the hypothesis that there are at least two very different charging mechanisms for the beta aluminas (Powers, 1975). One mechanism involves dissociation of the relatively mobile sodium ions to produce a negatively charged particle. This is the *dissociative mode*:

$$\text{beta alumina} \rightarrow (\text{beta alumina})^- + Na^+$$

This has been the predominant charging mode used at General Electric to form closed ended ceramic electrolyte tubes for the beta (sodium sulfur) battery. Particles are charged merely by ball milling or vibratory milling in amyl alcohol for several hours. Presumably the grinding operation exposes fresh surfaces from which some sodium ions dissociate.

Another charging mechanism involves adsorption of a proton from an acid or possibly some other cation and leads to a positively charged particle. This is the *adsorption or acid charging mode*:

$$\text{beta alumina} + \text{HA} \rightarrow (\text{H/beta alumina})^{+} + \text{A}^{-}$$

This mode has been used by a group at Compagnie Générale d'Électricité at Marcoussis, France (Fally *et al.*, 1973) to form both zirconia and beta alumina tubes as well as by Kennedy and Foissy (1975) to form beta alumina tubes.

The two charging modes are competitive. Since water can serve as a proton donor, the presence of water in a suspension can greatly affect particle charging using the dissociative mode. A sufficient concentration of water can lead to charge neutralization and even to charge inversion. The major precaution for satisfactory electrophoretic deposition is control of exposure of suspensions to moisture. The electrophoretic mobility and consequently the rate of deposition can be greatly affected. In addition to powder and amyl alcohol, suspension containers and grinding media need to be thoroughly dried.

The deposition step *per se* in electrophoretic forming is extremely simple and barely warrants comment. However, a very brief discussion of the mandrels used in electrophoretic forming of closed ended tubes would seem worthwhile since there is a widespread misunderstanding that they require an expensive special polish. Mandrels used at General Electric are constructed of stainless steel to avoid complications from rusting. The taper and surface finish are very nearly the same as for mandrels used in isostatic pressing. Tapers in the range 0.01–0.1 degrees are sufficient to expedite greenware removal. A good quality lathe turned surface suffices. While surface undulations of 0.1 mm amplitude should be avoided, small scratches can be tolerated.

Following deposition, tubes are placed in a chamber of low relative humidity, preferably less than 10%, to avoid shrinkage problems (Powers *et al.*, 1978; Chatterji and Mitoff, 1978). After completion of a forming series, tubes are then exposed to a stream of warm air to evaporate the amyl alcohol residing in the pores of the deposits. This operation is usually carried out overnight. Green tubes can then be pushed off mandrels, usually with little force, and are ready for sintering. Green densities are close to 55% of full value.

(c) *Removal of greenware from forming mandrels*

The operations following deposition are designed to facilitate removal of ware from mandrels. Problems related to deposit removal generally are among the most subtle encountered in electrophoretic forming. If any substantial force is required to push greenware off mandrels, it is likely that the ware is irreversibly damaged by the presence of flaws. These flaws are not always obvious in greenware but can lead to cracking during subsequent sintering.

Deposit removal difficulties arise from two different sources. One set comes about from adherence of ware to the mandrels. The other arises from greenware shrinkage during drying. The latter is the more common difficulty and can be indicated by a simple experiment. If a green tube that is difficult to remove is slit lengthwise, often one can observe that the slit width is greater than the width of the cutting edge. More importantly the tube, after slitting, is readily removed leaving only a thin film on the mandrel.

Two causes of adhesion are known. One is related to incomplete evaporation of the vehicle after deposition. The other is caused by the presence of binders, usually polymeric materials. Binders are not incorporated into suspensions at the General Electric laboratory because, while they may strengthen a much wanted cohesion between the green particles, they also can bind particles to mandrels.

Shrinkage problems seem to be caused mainly by introduction of water into a suspension or exposure of greenware to moisture immediately after deposition but before substantial vehicle evaporation. These facts suggest a water/double layer interaction. As mentioned earlier, shrinkage is more severe in powders synthesized with certain alumina sources. It is also greatly affected by the ceramic composition. Early development work at General Electric on electrophoretic forming was carried out on nearly phase pure beta alumina containing ~7.2% Na_2O and 1.0% MgO as well as some other additives. No shrinkage problems were encountered with these compositions. Green tubes were readily taken from mandrels following vehicle evaporation accomplished merely by allowing tubes to remain overnight in a fume hood. However, when magnesia was deleted from the formulation, difficulties with tube removal arose. This problem was greatly accentuated during periods of high ambient humidity. At one time, cracking was observed in nearly 90% of tubes processed, either in the green state or after sintering. The drying practices described stemmed from these experiences.

(d) *The lithia problem*

Another electrophoretic forming problem related to composition is associated with the presence of Li_2O in the ceramics. This additive is used in phase pure beta alumina ceramics to reduce the intragranular resistivity component at the operating temperature of the beta battery (Powers and Mitoff, 1978a). In mainly beta″ alumina ceramics, it stabilizes the beta″ alumina phase. The lithia problem involves a much reduced electrophoretic mobility (deposition rate) compared with that of similar powder without lithia. At high lithia concentrations, e.g. 0.75 wt%, often no deposition occurs at all with conventionally processed powder. While the origin of this problem is not really understood, it is believed to be another moisture effect since the lithia problem is at its worst in periods of high ambient humidity.

A solution to this difficulty was found in the course of some studies on liquid phase sintering (Chatterji and Mitoff, 1978). It consists in the simultaneous electrophoretic deposition of two separate powders of different soda concentrations (Powers, 1978). The lower soda powder component typically contains 8% Na_2O, while in the higher soda component the soda:alumina mole ratio is usually maintained at 3:7 or 3:11. The proportions of the high to the low soda component in suspension are adjusted to yield the final composition required. Typically, the higher soda component constitutes 10 to 20% of the total powder mass. While additives such as lithia or magnesia can be incorporated in either component, there is indirect evidence that their dispersion is more uniform in the final ceramic if incorporated into the higher soda component.

There are two other aspects of the two powder route which are very important, in fact, the first is crucial to the success of this procedure. The component powders codeposit in the same ratio as present in suspension. This implies that they have the same electrophoretic mobilities. Were this not so, in a series of depositions from a suspension, the first article formed would be relatively richer in the powder with the higher mobility than the last article deposited. It has not been possible to establish a significant difference in composition by chemical analysis between the first and last tubes deposited in a series. Probably a better indication of equal component mobilities during deposition is related to the fact that the densification of beta alumina ceramics during sintering depends sensitively on the soda concentration. No significant differences in the densification/sintering temperature relationship have been found between tubes formed near the beginning and near the end of a series.

The other important fact about electrophoretic codeposition is that the densification rate during sintering of ceramics formed using this route is the same as that of ware formed by the more conventional and straightforward single powder route provided other aspects of powder synthesis, forming and sintering are maintained the same. This fact has been established for both mainly beta″ alumina as well as for phase pure beta alumina ceramics of different compositions.

(e) *Scale-up problems*

Green beta alumina tubes as large as 3 cm in diameter and 45 cm in length have been formed by electrophoretic deposition. The major problems involved wall thickness variations not found in the forming of smaller tubes, 1 cm in diameter by 10 cm in length. Variations between top and bottom, as well as circumferentially, as large as a factor of 1.5 were encountered in early efforts. The two kinds of variations have very different origins.

Longitudinal variations arise because of particle settling in the course of deposition. As a result, the local particle concentration can increase substantially at the closed end located near the bottom of the suspension container. The deposition rate is proportional to the local particle concentration, as indicated by Equation 3.1. Particle settling is a greater problem with larger tubes because thicker deposits and consequently longer deposition times are required. The simple solution to this problem is to circulate the suspension during a deposition. The materials of construction for the pump and circulation lines must be carefully considered however. Otherwise charge on the particles can be adversely affected, or the suspension, and consequently the deposit, can be badly contaminated by wear products from the pumping system. Beta alumina suspensions can be very abrasive.

On the other hand, circumferential variations in wall thickness come about because of non-uniformities in the electric field. At General Electric field non-uniformities have been a much greater problem in the forming of larger diameter tubes because of a different geometric relationship in the deposition cell. During forming, it is useful to deplete a suspension to about 30% of the initial particle concentration. Larger depletions are not practical as deposition times become excessive. In the forming of small tubes of 1 cm nominal diameter, the practice of forming five to seven tubes in succession from the same suspension has been very helpful in obtaining a better understanding of the forming process. The practice is possible with a suspension container of internal diameter about four times the mandrel diameter. This arrangement forces some sacrifice in controlling deposit mass (average wall thickness) as successive deposition times must be increased in inverse proportion to the particle concentration (Equation 3.1). However, in the manufacture of larger diameter tubes, control of average wall thickness is a more important consideration and only a single tube is deposited from a suspension. With tubes of 2.8 cm nominal diameter, about 70% depletion is accomplished in a suspension container with an inner diameter only 1.5 times the mandrel diameter.

In a concentric deposition cell, the field F at the mandrel surface is given by the expression

$$F = V/[r_m \ln(r_c/r_m)] \tag{3.2}$$

where V is the applied voltage, r_m is the mandrel radius, and r_c is the inner radius of the counter electrode (suspension container). Consequently, a small displacement of a mandrel from the centre of the container leads to a field variation almost three times larger for a 2.8 cm tube as for a 1.0 cm tube. The solution to the problem of circumferential variations in wall thickness has been greater care in the fabrication of the forming cell. Specifically, the design and construction of jigging for holding mandrels concentric within the suspension container must be given thoughtful consideration. Moreover, mandrels and suspension containers must be very straight to permit a concentric arrangement along the whole mandrel length.

(f) *Assessment of electrophoretic forming*

The electrophoretic forming procedure described in this section was developed initially to solve the problem of forming a closed ended beta alumina tube. Perspectives change with time, however, and this procedure is now viewed mainly as a way of forming very homogeneous ware with excellent sinterability. Dozens of beta and beta″ alumina compositions have been investigated in the course of optimizing the properties of the ceramic electrolyte to be used in the beta battery. All but a couple have been sintered to densities in excess of 3.25 g cm^{-3}. All have shown that translucency characteristic of ware with less than 1% porosity.

This excellent sinterability is attributed to very fine pores in the greenware. The 'building block' of the green structure produced by electrophoretic forming is believed to be the ultimate crystallite or, at most, very small aggregates of such crystallites. This stands in contrast to the use of large aggregates, tens of microns in size, as the building block for greenware formed by mechanical pressing methods. The size of pores in a green structure is scaled according to the size of the 'block' used to build the structure. The operation of sintering is largely one of pore removal, and small pores are easier to remove than bigger ones (Burke and Rosolowski, 1976). The pores observed in sintered ware formed electrophoretically are 1 to 1.5 μm in size. The 5 to 10 μm or even larger sized pores frequently seen in mechanically pressed ware are not observed.

Deflocculation, the break-up of particle agglomerates, arises from particle charging, a prerequisite for forming by electrophoretic deposition (Hamaker and Verwey, 1940). The electrical force between the charged particles is modified by the dielectric properties of the vehicle in which the particles are suspended. The use of a combustible, somewhat malodorous, organic liquid can be considered the cost to be paid to obtain excellent particle deflocculation and the subsequent benefits resulting from it. The organic liquid must be acquired, stored, dried and, after use, either recovered or in the USA disposed of in accordance with strict OSHA regulations. The ultimate assessment of electrophoretic forming depends on whether the benefits cited justify the cost. A few years ago, the comparative costs of forming beta alumina tubes by electrophoretic deposition and by isostatic pressing were analysed by experienced manufacturing engineers in the General Electric Company (Chatterji and Mitoff, 1978). The financial costs by these two methods appeared to be nearly equal.

3.4 SINTERING

3.4.1 Introduction

Sintering is the most difficult process to control and potentially the most expensive operation involved in the fabrication of beta and beta″ alumina electrolyte tubing. Several important processes or reactions proceed in series or parallel in the firing of this important ceramic material. At temperatures below

1000 °C, volatile constituents (e.g. organic binders, water, salt decomposition products) must be removed in a manner which is non-destructive to the tubing. This portion of the firing cycle, usually referred to as bisquing, can be conducted in the initial stages of the sintering operation (Duncan, 1980) or in a separate batch process (Miller *et al.*, 1979). At temperatures over about 1450 °C, reactive phase sintering, usually with the assistance of liquid phases, takes place in which densification, phase conversion to the beta or beta″ alumina phase, and grain growth all proceed in parallel (Duncan, 1980; Miller *et al.*, 1979). The ultimate objective at this stage is to maximize densification and phase conversion while keeping grain growth under control. In some rapid sintering schemes at temperatures over 1550 °C, conversion to the beta″ alumina phase lags significantly behind densification. In this situation, a separate postsintering anneal at a lower temperature (1300–1500 °C) is required to promote conversion (and hence reduce the ionic resistivity) while at the same time inhibiting any undesired exaggerated grain growth (Youngblood *et al.*, 1977). Annealing can be conducted at the end of a continuous (pass through) or batch sintering cycle or in a separate batch firing operation. Another technique which can be used to control grain growth in beta″ alumina compositions is to subject the electrolyte to a presintering anneal at 1400–1475 °C, prior to normal sintering at temperatures over 1600 °C (Bugden and Duncan, 1978; Jatkar *et al.*, 1978). Postsinter annealing can still be employed to control the final ionic resistivity. All of these firing steps can, in principle, be incorporated into both continuous and batch firing operations. Finally, the technique of seeding catalysis (Gordon *et al.*, 1978; Jatkar *et al.*, 1976) can be employed to minimize or even eliminate the need for any pre- or postsinter annealing. In this technique, seed crystals (<5 μm) of beta″ alumina are incorporated in the precursor powder. During firing, they catalyse the chemical conversion to the beta″ phase. Thus a high conductivity electrolyte can be produced in a short firing cycle with an intermediate grain size microstructure.

In addition to the control of density, resistivity and microstructure, two other important factors must be considered in the sintering of beta and beta″ alumina electrolyte tubing. First, at temperatures over about 1400 °C, sodium oxide is a volatile constituent and techniques must be used which either inhibit volatilization or minimize its effect. Second, large diameter (up to 55 mm) and long (up to 650 mm) tubes with thin walls (1.6–2.5 mm) must be produced at economical production rates, with a minimum of energy consumption and within stringent dimensional tolerances with respect to absolute dimensions, ovality and warpage. Finally, the sintering process which is selected must yield electrolyte tubing with acceptable performance and endurance characteristics in sodium sulfur cells operating at temperatures between 300 and 400 °C.

In this section various aspects of sintering will be treated, including continuous processing and batch firing processes, encapsulation techniques including the use of ceramic setters and saggers, and vertical versus horizontal modes of firing. Some consideration will also be given to economic factors, energy consumption and dimensional control, all of which are critically important in a viable high volume sintering process.

3.4.2 Batch sintering techniques

(a) *Introduction*

Batch sintering of beta alumina tubing has been investigated in several laboratories including British Rail Technical Centre, Brown Boveri & Cie, Ceramatec Inc. and Compagnie Générale d'Électricité (CGE). In order to enhance the furnace capacity per firing cycle, vertical sintering employing some form of encapsulation has been adopted in most cases. A vertical mode of firing also improves the dimensional quality of electrolyte tubing over certain ranges of diameters, lengths and wall thicknesses.

In vertical sintering, several configurations are utilized: free standing, inclined vertical, and hang firing. In the free standing method the bisqued or green tube is placed on a beta alumina support disc. Stability can be increased by leaving the 'elephant's foot' (a diametrical protrusion which is common at the open end of the tube after pressing) on the open end of the tube until firing is complete, at which time it is cut off. The free standing encapsulation tube (alumina or magnesia) is then placed over the electrolyte. In the inclined vertical technique, the electrolyte tubes are placed inside the ceramic setter which is inclined $\sim 5°$ from the vertical. The electrolyte is always in contact with the setter, minimizing any warpage along the length, while at the same time the near vertical configuration minimizes any ovality. Hang firing of tubes, apparently from their open ends, is employed at CGE inside beta alumina setting concretes.

Horizontal firing is sometimes utilized, particularly for electrolytes with very long lengths ($\sim$500–600 mm). Reasonable control of ovality can be achieved in thin walled ($\sim$1–2 mm) tubing (OD $\sim$15–35 mm) by the use of V setter encapsulation saggers (Gordon *et al.*, 1978) which provide two line support over the entire length of the tube. Owing to the extra quantities of saggers and setters, the batch horizontal mode of firing is very inefficient with respect to both product throughput per firing cycle and energy consumption per unit. Furnace capacities can be increased three and a half times by vertically sintering electrolytes ($\sim$35 mm OD) in the same firing space.

In most batch operations, the sintering and annealing (including any presinter or postsintering treatments) are conducted in the same firing cycle which is usually programmed automatically. In some cases, even the bisquing operation is incorporated into the firing cycle. Since transient liquids are important in achieving densification, particularly in the beta″ aluminas, rapid heating rates at temperatures over 1300–1400 °C are sometimes required to maximize their effect. Of course, rapid heating leads to higher power factors in furnace operation which will have an impact on cost.

The characteristics of various techniques which are currently employed in batch sintering processes are given in Table 3.5. This summary will serve as a

Table 3.5 Vertical batch sintering of beta and beta″ alumina electrolyte tubing

Laboratory	*Composition*	*Electrolyte dimensions (mm)*	*Capacity (number/cycle)*	*Cycles/week*	*Yield (%)*	*Encapsulation technique*
British Rail	β'' (Li_2O)	Lengths up to 530 OD up to 50	24–32	4	70–80	MgO
Brown, Boveri & Cie	β-β'' (MgO)	Lengths ~210 OD ~25	28	5–6	~70	β-β'' powder in alpha alumina or Pt lined alpha alumina
Ceramatec Inc.	β''(Li_2O)	Lengths up to 300 OD up to 34	48	5–6	60–70	Pt lined alpha alumina
CGE	β*	Lengths up to 380 OD up to 27	small (?)	5–6	?	Beta alumina setting concretes in alpha alumina

* Gas fired, all others electric.

good reference when batch and continuous sintering are compared in Section 3.4.5.

(b) *Maintenance of sodium rich atmosphere*

The prime consideration in the firing of beta alumina ceramics is control of the surrounding sodium enriched atmosphere. The influence of the firing conditions on this atmosphere was observed by Duncan and Creyke (1969) when investigating beta alumina impurities in alpha alumina ceramics. This is still relevant to firing of pure beta alumina. In those original investigations the presence of sodium from a stearate binder combined with low levels of magnesia was sufficient to form a few per cent of a beta alumina phase, resulting in high dielectric loss in the fired samples. Convection currents in a vertical electric furnace were sufficient to remove the sodium atmosphere and inhibit the formation of beta alumina at the periphery of sample pellets. In a horizontal furnace the relatively stable atmosphere permitted beta alumina crystals to form at the fired surface. Passage of air over samples at temperatures as low as 1300 °C could remove sodium. Even when the air movement was initiated at 1600 °C after densification and elimination of open porosity in the alpha alumina, breakdown of the beta alumina impurity at a depth of a few millimetres was still observed. Similarly, Francis *et al.* (1971) showed that beta alumina refractory materials can partially break down to alpha alumina at 1500 °C in air with the loss of 0.3% Na_2O.

The sodium to provide the atmosphere may come from the object being fired or from a buffer source such as loose powder of the same composition, or from sodium aluminate. As an expedient the container may be alpha alumina refractory, but unless it is protected by, for example, lining with platinum foil the alumina is soon likely to be corroded by the sodium atmosphere, resulting in formation of beta alumina in the crucible. The expansion accompanying this phase change can result in cracking, and the removal of soda from the atmosphere within the crucible could affect the properties of the beta alumina sample. In Section 3.4.4, R. S. Gordon considers a number of containment systems including beta alumina and platinum crucibles.

Diffusion between the article and a source of sodium (often referred to as a buffer) via the vapour phase is such that the sodium content of the fired article can be raised or lowered in direct relation to the level of sodium in the buffer. It is not essential to have a buffer material present, and Duncan (1980) found it sufficient to stand a tube on an unfired disc of beta alumina in a close fitting crucible as shown in Fig. 3.4. As the density of the sodium vapour generated by the tube and the disc is greater than air, all that is required is a close fitting cover on the crucible. Thus when the refractory container is stable with respect to sodium at the sintering temperature it is unnecessary to introduce a buffer, and indeed it would be undesirable if that were to leave loose powder on the tube surface. Sodium analyses of fired samples of beta alumina encapsulated in these

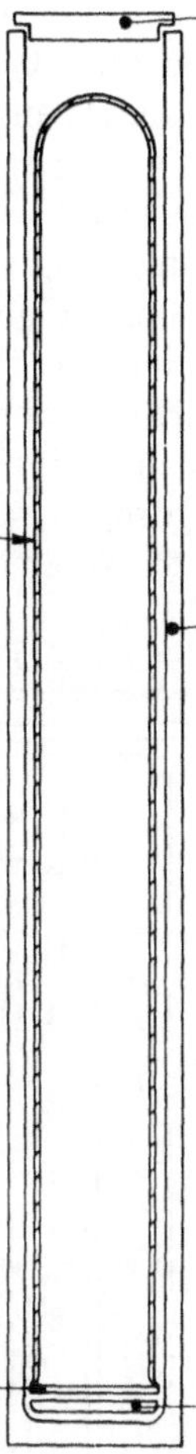

Fig. 3.4 Magnesia system for firing 33 mm × 360 mm beta alumina tubes. Duncan (1980). Reproduced with permission from British Rail Research.

magnesia crucibles show that, within experimental error, their composition remains unchanged.

Surrounding an article with beta alumina powder does not necessarily ensure prevention of sodium loss during firing. Rivier and Pelton (1978) report a loss of 0.7% sodium when beta alumina was fired at 1530 °C for two hours surrounded by 100 mesh beta alumina powder. Also, Byckalo *et al.* (1976) not surprisingly found the soda content of beta alumina made from Alcoa XB2 powder had reduced from 7.3 to 3.1% after sintering embedded in beta alumina powder at 1850 °C for three hours in a gas fired furnace. Obviously in these examples equilibrium between the article and the surrounding vapour phase was not established during firing.

Desplanches (1979) described a muffle (Fig. 3.5) to hold fourteen beta alumina tubes for firings to around 1700 °C in a gas furnace. The muffle was built up of ring sections each composed of an outer layer of an aluminous cement and an inner layer of a beta alumina cement to ensure a sodium rich atmosphere during firing. The fired density was high, indicating little if any loss of soda, but the tubes sagged during firing.

For sintering alpha alumina ceramics, Coble (1962) showed that an oxygen or hydrogen atmosphere was essential to reach theoretical density, as otherwise other gases such as nitrogen that are insoluble in the ceramic could not diffuse out of the closed porosity. Powers and Mitoff (1978b) used this approach and fired beta alumina in oxygen (a hydrogen atmosphere would remove sodium oxide above 680 °C: Trostel, 1965). When sintering different beta alumina

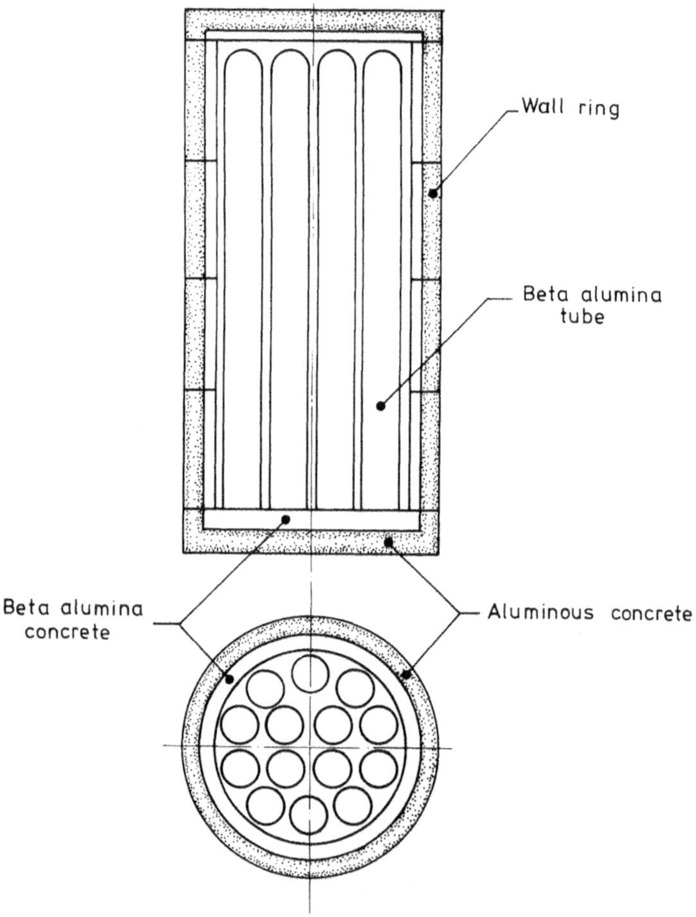

Fig. 3.5 Tubes and a sintering muffle made from beta alumina (Desplanches, 1979).

compositions, they found the bulk density was consistently (0.02–0.04 g cm^{-3}) higher than for firings in air.

Gordon (1978) reports soda loss of 0.2–0.3% from beta″ alumina during continuous sintering in air at 1600 °C. In spite of this, densities were as high as 3.24 g cm^{-3}.

Magnesia was one of the first refractories employed in sintering beta″ alumina and is mentioned by Ford Motor Company (1968) in an early patent. A problem may be encountered with the purity of the magnesia; being more refractory than alumina, high purity magnesia is expensive to process. Considerable levels of silica impurity may be present where it has been manufactured from low cost fused magnesia. At British Rail (Duncan, 1977) beta″ alumina fired in a commercial magnesia crucible that had impurity levels of 0.5% lime (CaO) and 1.0% silica was consistently porous, apparently due to soda loss. The presence of silica in magnesia with such a low level of lime leads to the formation of phases such as $CaO \cdot MgO \cdot SiO_2$ which is liquid at beta alumina sintering temperatures; contamination with iron also leads to the presence of low melting point liquids (Table 3.6). Surprisingly, it appears that such interstitial liquids in a magnesia crucible upset the equilibrium between the article being fired and the surrounding vapour phase to the extent that full densification of the beta alumina does not take place. A low level of silica is an advantage; as lime is soluble in magnesia during sintering (Hatfield *et al.*, 1970) the level of lime should take this into account in order to prevent the formation of liquids with a low lime:silica ratio and low melting point.

In the process described by Duncan (1977) the magnesia crucibles were sintered at approximately 1600 °C, the same temperature as the beta″ alumina.

Table 3.6 Chemical analyses of milled magnesia (Duncan, 1977). Batches as follows: (1) 1.2 kg milled for 16 hours (2) 20 kg milled for 6 hours (3) 20 kg milled for 16 hours (4) 2.0 kg milled for 4 hours in a steel vibro mill

	1	*2*	*3*	*4*
SiO_2	0.20	0.28	1.50	0.14
TiO_2	0.01	0.02	0.08	0.01
Al_2O_3	0.36	0.43	3.20	2.32
Fe_2O_3	0.05	0.06	0.10	0.28
Mn_3O_4	0.02	0.02	ND	ND
CaO	1.00	1.00	1.04	1.00
MgO	98.1	98.0	93.8	96.1
K_2O	0.02	0.03	0.06	0.01
Na_2O	0.22	0.20	0.23	0.18
Li_2O	<0.01	ND	<0.01	<0.01

ND: not determined.

At British Rail these magnesia crucibles were used routinely for firing beta″ alumina in a batch process in an electric furnace; their life of a few hundred cycles was more likely to be terminated by accidental damage than the aggregate time at the electrolyte's sintering temperature. Allowance may have to be made for the poor thermal shock properties of magnesia when selecting a beta alumina firing schedule. This is particularly important during rapid cooldown with modern fibre lined furnaces, where the heat retention is principally in the fired ware and supports rather than in the furnace walls.

(c) *Firing schedules*

The breakdown of precursors during the initial stages of firing has been investigated by Bugden and Duncan (1979), Johnson *et al.* (1979) and Vogel *et al.* (1981). A firing schedule should make allowance for those stages of rapid weight loss and phase decomposition to retain the integrity of the compact. Expansion is seen above 700 °C when sodium aluminate forms, and then above 800 °C there is limited shrinkage as the aluminate and/or the alumina sinters. Between 1150 and 1200 °C a final expansion takes place as a mixture of beta and beta″ aluminas develops rapidly, the proportion of beta″ alumina depending on the original powder route. In the batch firing process it is this mixture that is being considered.

Bugden and Duncan (1978) showed how the densification of a beta/beta″ alumina followed a typical S shaped curve (Fig. 3.6) with the central portion of the curve representing rapid densification at a heating rate of only 70 °C h^{-1}. This pattern closely followed that reported by Huckabee and Palmour (1972) for alumina, where it is regarded as typifying densification proceeding in an uncontrolled manner. The microstructure that develops in a beta″ alumina composition from such a schedule, with a top temperature of around 1600 °C, is characterized by a duplex grain size distribution (Bugden and Duncan, 1977; Youngblood *et al.*, 1977). The grain size is strongly dependent on the maximum temperature and the hold time. This contrasts with pure beta alumina, where very high firing temperatures with long hold times at maximum temperature can produce a relatively fine grained structure with a normal distribution; Fally *et al.* (1973) reported an average crystal size of 15 μm after firing beta alumina at 1850 °C.

A beta″ alumina fired in the manner described has the disadvantage not only of low strength from the duplex microstructure (Duncan and Bugden, 1981), but also of a relatively low degree of conversion of beta alumina to beta″ alumina. The phase change proceeds less rapidly than the densification. The firing conditions for beta″ alumina could thus be on a knife edge, needing a higher temperature and longer time to increase the proportion of beta″ alumina and electrical conductivity but a lower temperature to restrict crystal growth and encourage high strength.

The problem has been solved by at least two routes which may or may not be related (Virkar *et al.*, 1978; Duncan, 1979). The solution lies with the

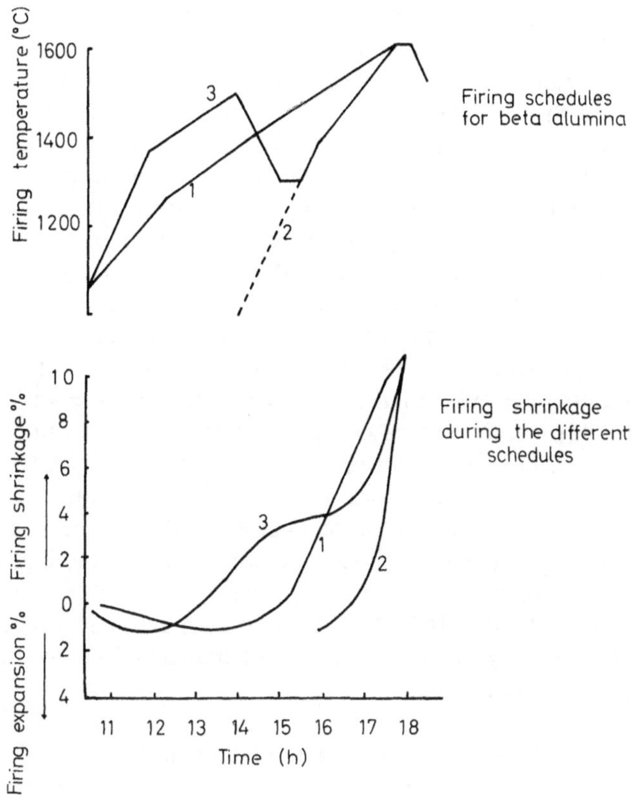

Fig. 3.6 Firing shrinkage during different schedules (Bugden and Duncan, 1978).

observation that the large grains in the duplex microstructure are of beta″ alumina. Denham and Harding, as reported in Duncan and Bugden (1981), noticed on a non-rotation X-ray powder diffraction pattern that the lines specific only to beta alumina did not have any spots due to large grains. From this it may be deduced that the distribution of lithia is poor. Virkar *et al.* (1978) reasoned that prereaction of the small lithia addition with alumina to give the zeta spinel $LiAl_5O_8$ would mean that the lithia could then be better distributed because of its larger volume. The zeta component was equally effective whether added to unreacted soda and alumina or to materials which had been pre-reacted. The proportion of beta″ alumina formed was increased and the final crystal size well controlled.

The microstructural control method of Duncan (1979) was quite different. It originated when 90 mm diameter discs which had been part fired to 1510 °C were interspaced with unfired discs for a firing to 1610 °C on a schedule of 200 °C h^{-1} to 1400 °C followed by 120 °C h^{-1} to 1610 °C. The fracture surface of the once fired discs had the characteristic glint of crystals larger than 150 μm that are visible to the naked eye. In an apparent contradiction there were no visible large crystals in the twice fired material. This was in 1972, and the two peak firing which was developed from this observation became the standard route for production of beta alumina. The grain size anomaly was explained (Duncan and Bugden, 1981) when it was pointed out that the greater amount of heat work from the two peak firings produced a coarser matrix than that in the once fired material (Fig. 3.7). The two peak firing was applicable to a range of compositions doped with magnesia and lithia or with lithia alone, and prepared from slurry solutions of Alcoa A-16SG or Reynolds RC-HPS-DBM with hydroxide precursors of soda and lithia. The firing variable which depended most upon composition was the temperature of the first peak; it was shown (Duncan and Bugden, 1981) that for maximum strength there was an optimum value (Fig. 3.8) to either side of which the duplex microstructure reappeared. The resistivity decreased in firings where the first peak temperature was increased. Of some significance for production of large quantities of electrolyte was the fact that the firing temperature range was widened. The value of this will be clear to anyone who has tried to achieve low resistivity and high strength in beta″ alumina ceramics. Using two peak firing it was possible to raise the final temperature from 1610 to 1625 °C with a minor reduction in strength while at the same time lowering the resistivity. The advantages of the two peak process are illustrated in Table 3.7.

Table 3.7 Strength of beta″ alumina (MN m^{-2})

	Final sintering temperature	
	1610 °C	*1625 °C*
Single peak firing	185	127
Two peak firing	244	204

The mechanism by which this novel firing operates is open to discussion; it is dealt with in detail elsewhere (Duncan, 1979). In brief there are two possible explanations. The first may be that which has been applied to the zeta process and involves increasing the proportion of beta″ alumina at the critical stage before crystal growth rates dominate. However, the cooldown stage is essential in the two peak firing, causing the crystal growth to stop but allowing the phase change of beta to beta″ alumina to continue. Thus when the first peak temperature is reached for a second time there is more beta″ alumina present, offering more nuclei for competitive growth of this faster growing phase. The likelihood of isolated crystals is thus diminished.

The other possibility relates to sintering rate and the findings of Palmour and

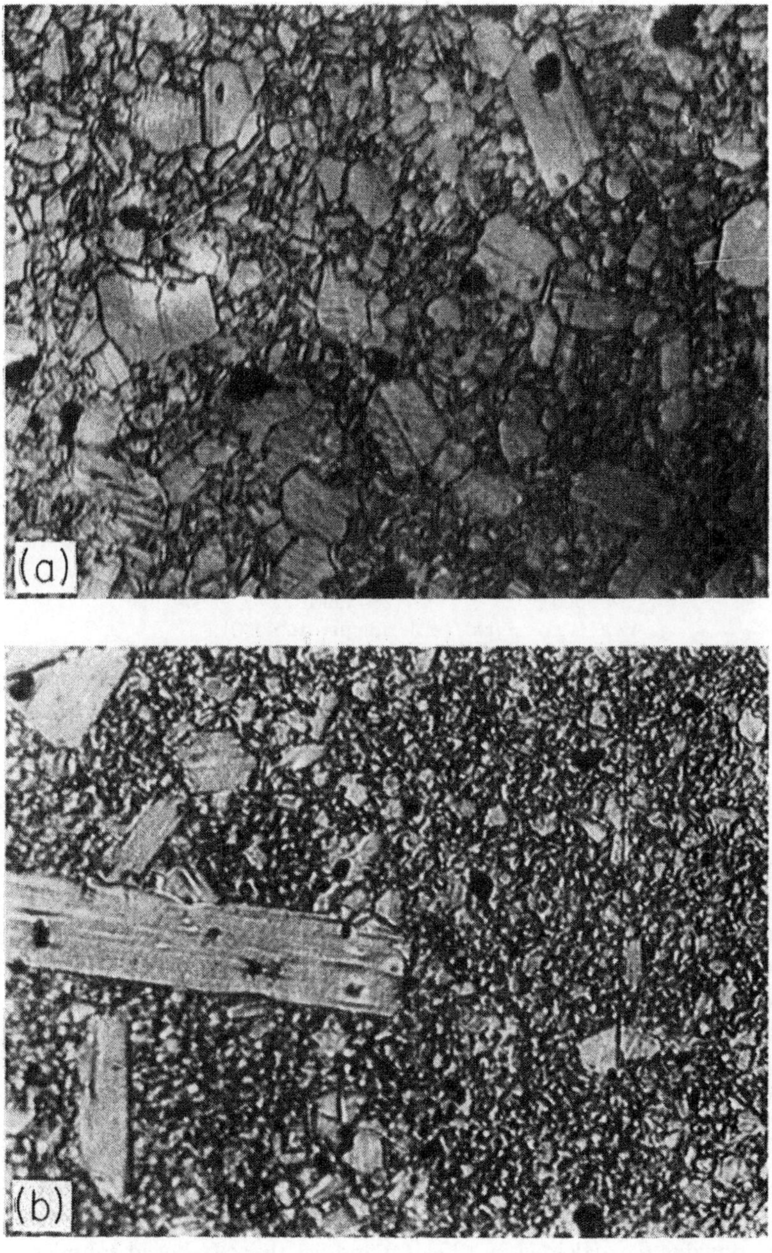

Fig. 3.7 Comparison of the microstructure of beta alumina from (a) an optimum two peak firing schedule and (b) a straight through firing. Magnification ×1000. Duncan and Bugden (1981). Reproduced with permission from the British Ceramic Society.

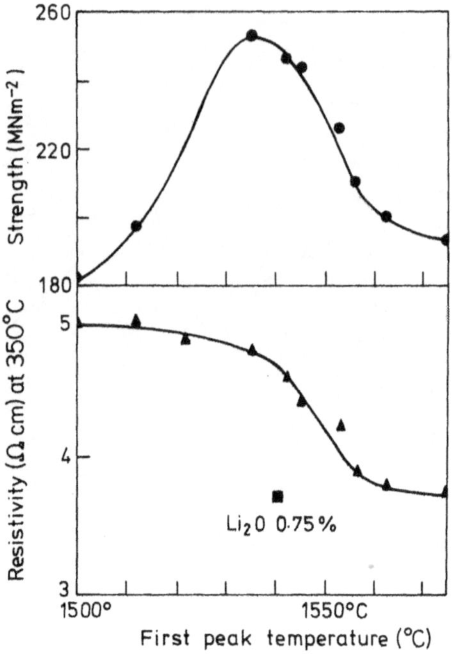

Fig. 3.8 Resistivity and strength of beta″ alumina against first peak temperature showing how the strength is optimized by the first peak in a two peak firing while the resistivity is progressively reduced. Duncan and Bugden (1981). Reproduced with permission from the British Ceramic Society.

his co-workers that rate controlled sintering can control crystal size (Huckabee and Palmour, 1972). In Fig. 3.6 it is evident that the sintering rate is reduced overall in the two peak firing. On the basis that porosity controls crystal growth in the early stages of sintering, they reason that rapid densification in the vertical portion of the S shaped curve can result in uneven distribution of porosity and hence of crystal growth. A critical stage in their rate controlled sintering, 0.85 fractional density, is close to that achieved at the optimum first peak temperature which happens to work out at 0.8 fractional density or 50% of the total firing shrinkage.

Jatkar *et al.* (1978) went a stage further in microstructural control when they combined both the zeta process of powder preparation and a pretreatment at 1480 °C. The fired grain size was reduced for material prepared from partially converted powder (i.e. calcined at 1250 °C) and for the zeta based samples.

Where a beta″ alumina composition has been sintered rapidly to full density the conversion to beta″ alumina may not be complete. Postsinter annealing was used to complete the conversion but the long hold times (8 hours at 1480 °C) were not considered feasible for large scale manufacture. The influence of time and composition on beta″ alumina development at 1500 °C has recently been reported by Duncan and West (1983) (Fig. 3.9). In the investigation of compositions on the $LiAl_5O_8$–$NaAl_5O_8$ join, there was a dramatic drop in the time to reach equilibrium for beta″ alumina formation for only a moderate change in composition. At levels of 9.29 wt% soda, 0.79 wt% lithia (3.5 mol% lithia), the time was 11 hours compared with less than one hour for 8.67 wt%

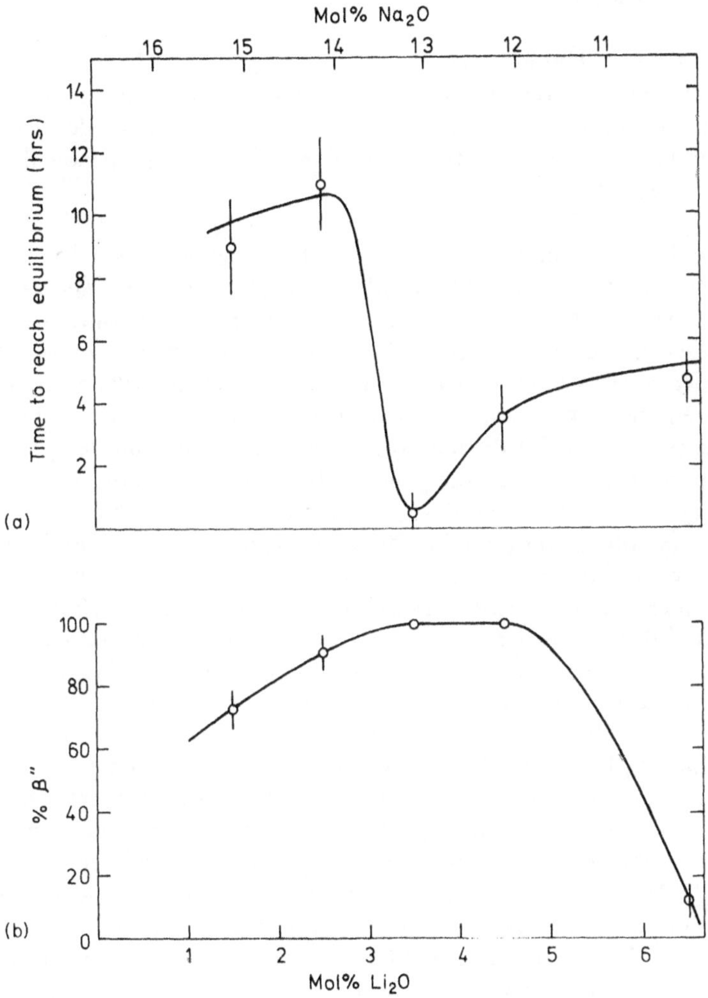

Fig. 3.9 The effect of soda and lithia content on (a) the time taken to achieve equilibrium and (b) the percentage of beta alumina relative to a total beta + beta″ content at equilibrium at 1500 °C for a series of compositions on the $LiAlO_8$–$NaAl_5O_8$ spinel join. After Duncan and West (1983).

soda and 1.11 wt% lithia (4.5 mol% lithia). This result could be associated with the development of a low melting point eutectic.

Jatkar *et al.* (1976) abbreviated the sintering process by seeding partially converted powder with crystals of fully converted beta″ alumina. This increased the number of nuclei so that conversion was completed during a fast firing. By selecting 2% of seeds 5 μm separated by sedimentation, the sintered samples produced had a resistivity (300 °C) of 3.5–4.5 Ω cm after firing at 1600 °C for 5 minutes. They also successfully combined zeta processing, pretreatment at 1480 °C and seeding.

3.4.3 Hot pressing

Hot pressing is a convenient laboratory method of sintering beta alumina powders or precursors which results in full theoretical density and in preferential orientation of the beta alumina crystals perpendicular to the applied pressure.

The dies are fabricated in graphite with graphite plungers. The preferred method has been induction heating with the sintering carried out under vacuum. Babini *et al.* (1977) and McDonough *et al.* (1978) successfully hot pressed Alcoa beta alumina powder. In studying the densification kinetics Babini *et al.* achieved densities of 3.24 g cm^{-3} by hot pressing. In common with Virkar *et al.* (1974) and Green and Hutchinson (1980a, 1980b), Babini *et al.* outgassed the powder before applying the pressure. There seems to be a preference for applying the full pressure after the maximum temperature has been reached; this temperature, 1400–1500 °C, is low compared with normal sintering. The edges of the samples may be converted to alpha alumina or they may be contaminated by graphite. In order to produce white samples Virkar *et al.* used beta alumina discs between the powder and the graphite plungers; Green and Hutchinson used thin sheets of platinum coated with boron nitride in contact with the powder and backed up with graphite spacers between the platinum and the plungers. They also found during preliminary work that inadequate calcination of the powder derived from saturated solutions led to some sulfate remaining. This was reduced in the graphite die, and resultant sulfur imparted a deep red colour to the beta alumina sinter.

Another means of increasing density is to hot isostatically press beta alumina that has been sintered normally, i.e. which contains only closed pores. Hot isostatic pressing at temperatures of 1330–1450 °C at pressures in the range 69–172 MN m^{-2} will eliminate the majority of the closed porosity (May *et al.*, 1980), thus increasing the strength. May *et al.* reported that the resistivity was reduced but this was shown to be related to thermal effects alone. The tubes had initially been sintered rapidly in a pass through furnace, and subsequent heating to 1300–1450 °C in the hot isotatic pressing step was equivalent to the postsinter anneal normally used in conjunction with fast firing. It was concluded that the capital and operating costs would prove prohibitive except for the special applications.

3.4.4 Continuous sintering

(a) *Introduction*

Three basic techniques have been employed to date in the continuous sintering of beta or beta″ alumina electrolyte tubing: unencapsulated zone sintering, continuous encapsulated horizontal sintering, and continuous encapsulated vertical sintering.

In zone sintering, which has been developed by Jones and co-workers (Jones and Miles, 1971; Tan and May, 1977) the electrolyte tube is passed rapidly (3–12 cm min^{-1}) through a hot zone (~1700 °C) which is small (~60 mm) by comparison with the length of the tube (400–600 mm). This type of firing can be conducted vertically, horizontally or at any angle in between. Furthermore no encapsulation devices are employed. The sodium oxide vapour pressure in the furnace is apparently controlled by the tubes themselves. By minimization of the residence time in the hot zone, the problem of soda loss or evaporation can apparently be circumvented. Electrolyte tubing with excellent dimensional characteristics (straight and round) can be produced using this process. After zone sintering, the tubes possess acceptable densities and very fine grained microstructures; resistivities are too high because of incomplete conversion to the beta″ phase. Consequently the tubes must be subjected to a postsintering anneal (1300–1450 °C) in a separate batch operation to lower the ionic resistivity to within acceptable limits. Incorporation of the seeding catalysis technique (Jatkar *et al.*, 1976; Gordon *et al.*, 1978) could possibly promote phase conversion directly during zone sintering and hence eliminate the need for any postsintering annealing. In summary, a high reject rate combined with a need for a long term anneal subsequent to sintering appear to be the primary disadvantages of zone sintering. The advantages are excellent dimensional characteristics in electrolyte tubing with a uniform, fine grained microstructure.

It should be noted that it has not yet been proved that a fine grained microstructure is a necessary requirement for acceptable performance and life in a sodium sulfur cell.

The technique of encapsulated continuous sintering in a horizontal mode has been studied by workers at Ceramatec and General Electric. In contrast to zone sintering, the hot zone is usually longer than or comparable with the length of the tube which is to be fired. Normally some type of encapsulation is utilized to control the volatilization of sodium oxide. The pass through velocity should be set at values between 4 and 8 cm min^{-1} to ensure reasonable production rates (e.g. over 4 to 12 tubes per hour for lengths between 600 and 200 mm, respectively). The length of the furnace will be dictated by the pass through velocity, the time required for densification at the maximum temperature and any annealing steps at lower temperatures. For combined sintering and annealing times no longer than about 1.25 hours, the total length of the furnace can be kept under about 15 metres. To avoid jamming of ware as it passes through the furnace, a 'walking beam' mechanism can be employed instead of a standard pusher operation.

Continuous sintering in a horizontal mode has been developed successfully for beta″ alumina tubing with a length of 200 mm and a diameter of 15 mm. In these studies, sintering and postsinter annealing were conducted in line in a single firing operation. The electrolytes were encapsulated in alumina tubes pretreated with sodium oxide on the inside surface. Further development is required to scale up the operation to larger electrolyte sizes (e.g. lengths up to 500–600 mm and diameters up to 30–50 mm). Also the development of furnaces sufficiently large to permit multitube passes (up to six) needs to be undertaken to obtain optimum furnace capacities.

Some laboratories have studied encapsulated vertical sintering in a continuous mode. Owing to vertical height constraints, the pass through velocities have been very small (0.2–2.2 cm min^{-1}). At these low velocities, the economics of the process are doubtful because of very small product throughput. Under these conditions, batch sintering may be more economical. At higher velocities, postsinter annealing operations may have to be conducted in separate batch furnaces, otherwise the length of the vertical furnace may be too long for practical operation. In a typical continuous sintering operation conducted in the vertical mode, the ware is passed downward through the furnace. The ware is encapsulated in some form of ceramic refractory.

In Table 3.8 the basic operating characteristics are summarized for the different pass through sintering techniques which have been investigated. It is clear that both zone and continuous (encapsulated) horizontal sintering processes are amenable to high heating rates and throughputs (i.e. high pass through velocities) which are essential for low cost firing operations.

(b) *Encapsulation techniques for continuous sintering*

Several methods have been used to inhibit the loss of sodium oxide in the sintering of beta″ alumina electrolytes. These include:

(1) Rapid zone sintering at elevated temperatures (1600–1900 °C) such that soda loss is minimal in the short period the tubes reside in the sintering zone (Jones and Miles, 1971; Tan and May, 1977);

(2) Encapsulation of the ceramic:

(i) in a sealed platinum envelope (Miller *et al.*, 1979; Youngblood *et al.*, 1977)

(ii) in a chemically inert magnesia or zirconia refractory container (Duncan, 1980)

(iii) in an alpha alumina ceramic tube, either pretreated on the inside surface with sodium oxide or lined with a permanent, reusable platinum liner, or

(iv) in a larger beta″ alumina tube of similar sodium oxide content (Miller *et al.*, 1979), and

(v) packing soda rich powders around the green ceramic to inhibit volatilization (Haar, 1980).

Some of the encapsulation techniques are shown schematically in Fig. 3.10.

Table 3.8 Continuous sintering of beta and beta″ alumina electrolyte tubing

Laboratory	*Composition*	*Mode*	*Heating rate* (°C min^{-1})	*Pass through velocity* (cm min^{-1})	*Max. temp.* (°C)	*Sintering time* (min)	*Annealing temp.* (°C)	*Annealing time* (min)	*Encapsulation technique*
Brown, Boveri & Cie	β-β″	Vertical	50–80	1.2–2.2	1610–1630	10–25	1450–1470	300–600*	β-β″ powder in alpha alumina
Chloride Silent Power	β-β″	Zone: Vertical Horizontal Inclined	~100	3–12	1700	1–10	Batch	Annealing	None
Ceramatec Inc.	β″	Horizontal	20–80	2–8	1550	5–15	1350–1525	20–40	Pt lined alpha alumina Pretreated alpha alumina
CGE	β″	Vertical	~20	0.2	1630	Long	?	?	β-β″ alumina
	β	Vertical	~20	0.2	1630	Long	?	?	Setting concretes & powders

* Too long for a process in line with sintering (separate batch annealing or tube is stopped in the anneal zone).

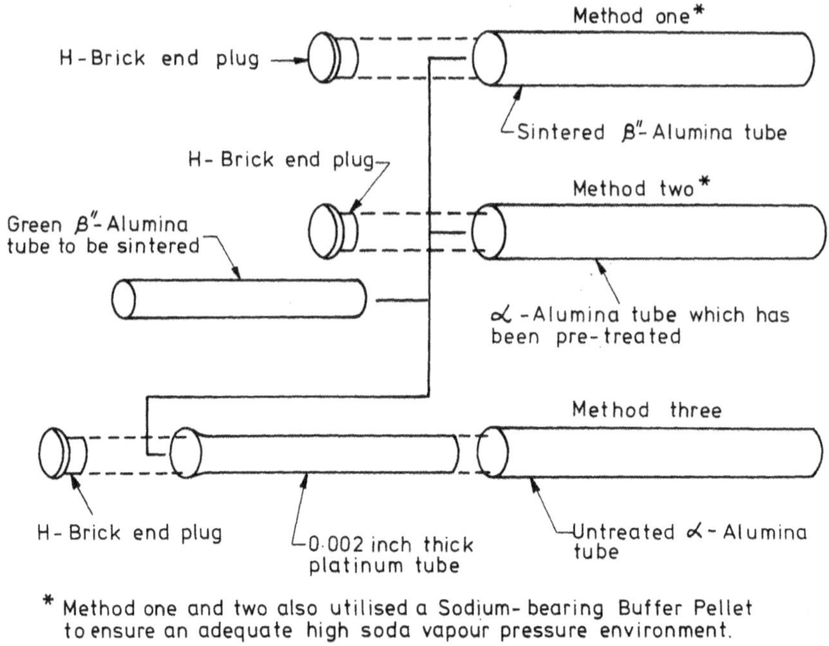

Fig. 3.10 Encapsulation techniques for sintering beta alumina.

The use of platinum as an encapsulation device has two major disadvantages: a high initial cost, and the mechanical deterioration of the envelope owing to repeated sealing and unsealing which leads to frequent reprocessing and refabrication of the platinum. The packing of soda rich powders about the preformed ceramic results in high labour and firing costs in addition to the production of ceramic surfaces of rather poor quality. Chemically inert refractory carriers such as magnesia and zirconia, which are currently available in the United States, have rather poor thermal shock resistance leading to limited refractory life in continuous sintering or rapid firing schedules.

The pretreated alpha alumina encapsulation tubes closed at one end are prepared by filling them with either a prefired powder mixture of alpha alumina and sodium oxide (at the desired sodium oxide composition) or sodium aluminate, or by coating the inside of the tubes with a saturated solution of sodium carbonate. These filled or coated tubes are then fired overnight at 1400–1475 °C prior to use as encapsulation devices in batch or continuous sintering.

Lifetime data are available for beta″ alumina, pretreated alpha alumina and platinum lined alpha alumina carriers. In the case of beta or beta″ alumina carriers or saggers, the useful lifetime is relatively short (5–10 firing cycles). Encapsulation tubes with sodium oxide contents of >9.25% by weight were subject to severe blistering on both the inside and outside surfaces. For tubes with lower sodium oxide contents (<8.8%) no blistering was observed; however, the tubes lost their translucent character and developed a chalky surface depleted of sodium oxide. After several firings, the tubes would shatter catastrophically into several pieces.

With proper design, the useful lifetime of pretreated alpha alumina encapsulation tubes will be over 100 sintering cycles. Essentially an indefinite lifetime has been observed for the platinum lined alpha alumina carrier tubes. More than 300 batch firing cycles have been endured by several dozen carrier tubes with no intrinsic failure of the ceramic. It is expected that this mode of encapsulation will lead to refractory and platinum life in excess of 1000 cycles.

(c) *Projected sintering encapsulation costs*

Since several encapsulation techniques have been developed successfully in various programmes, it is useful to compare the relative costs of each technique. A summary of projected costs for an annual production in excess of 100 000 electrolytes is given in Table 3.9. Of the four encapsulation methods analysed, the use of permanent platinum liners in alpha alumina refractory tubes has the lowest potential cost in large scale ($\geqslant 10^5$ per annum) production. These costs are expected to be less than 5% of the total tube cost. The only possible exception to this case would be perhaps the use of chemically inert refractory carriers, which possess a lifetime in excess of approximately 700 sintering cycles. Magnesia and zirconia refractory tubes, which are currently available in the United States, do not possess the requisite thermal shock properties required for such lifetimes in continuous sintering, where some rapid heating and sharp thermal gradients will be encountered. Magnesia refractories developed by workers at British Rail may possess the required thermal shock resistance for multiple temperature cycling (Duncan, 1980).

The principal disadvantage in the use of platinum lined alpha alumina carriers is the relatively large capital investment in platinum. However, as an appreciating asset, the increase in the value of platinum could be expected to more than compensate for the cost of money involved in its investment. Furthermore, as will be discussed later, substantially fewer carriers (factor of ten) are required in continuous sintering than those needed for multiple batch firing in large scale production. Thus substantially more expensive carriers can, in principle, be employed with little or no cost penalty in continuous firing.

The use of pretreated alpha alumina carriers, although somewhat more expensive (<10% of the total tube cost), is considered a viable alternative provided that reasonable refractory lifetimes (200–300 cycles) can be achieved. The use of *sealed* platinum containers (a very destructive and expensive way to use platinum) is too expensive (> 30% of the tube cost) to be considered a viable option. It is expected that the use of beta″ alumina encapsulation tubes would also be too expensive because of the limited life of the carrier, although reject tubes from beta″ alumina production may find possible application in sintering

Table 3.9 Beta″ alumina sintering encapsulation costs. Over 10^5 units produced annually 250 Wh size electrolyte (33 mm OD × 260 mm length)

Method	*Ultimate costs (1980 dollars)* $ per kWh	$ per MKII electrolyte	¢ per cm²	*Current costs (1980 dollars)* ¢ per cm²
Platinum encapsulation				
Pt life 100 cycles	5.00	1.25	0.46	0.46
Alpha carrier life 1200 cycles				
Platinum lined alpha Al_2O_3 carriers				
Pt life 2400 cycles	0.35	0.09	0.03	~0.10
Alpha carrier life 1200 cycles				
Pretreated alpha Al_2O_3 carrier				
Carrier life 240 cycles	0.72	0.18	0.07	0.16
Inert ceramic refractories				
US (MgO, ZrO_2)				
Carrier life 240 cycles	4.56	1.14	0.419	>1.00
British (MgO)				
Carrier life 360 cycles	0.78	0.20	0.07	<0.20 (?)

Pt costs (1980 dollars): initial metal + fabrication = $909, refabrication = $77.
Refinement = $44 (3% loss), alpha carrier cost ~$43; MgO or ZrO carrier cost ~$114 (US) and ~$70 (British).

electrolytes of smaller size. Finally, the cost data in Table 3.9 do indicate, however, that while some reasonably economical encapsulation methods can be projected for large scale production, a strong incentive still exists to eliminate the need for encapsulation providing rejects in the sintering process can be held to less than about 10%.

3.4.5 Comparison between continuous and batch sintering in large scale production

An appropriate conclusion to this chapter is to discuss the relative advantages of continuous (pass through) and batch sintering when production levels in a single plant approach 10 million electrolyte assemblies per year. This volume of production is reasonable for the expected load levelling and electric vehicle markets. For the purpose of illustration, the electrolyte size is assumed to be 33 mm in diameter and 460 mm long.

For the case of batch sintering, the fabrication of beta″ alumina electrolyte tubing will probably consist of the following basic steps:

(1) Slurry solution spray drying of low cost alpha alumina slurries containing alkalis in solution (no calcination and milling operations are contemplated);
(2) automatic isostatic pressing;
(3) bisque firing, which will be part of the batch sintering process;
(4) sintering in a vertical mode with encapsulation using furnaces with a capacity of approximately 100 electrolytes.

Higher capacity batch sintering furnaces, although possible in principle, do not appear realistic when constraints of uniform heating, size of heating elements and reasonable turnaround times are taken into account. While the actual heating and sintering/annealing times in batch sintering are expected to vary between 4 and 14 hours, depending on the process utilized (e.g. zeta process, spray drying, seeding, double peak firing; Bugden and Duncan, 1978), at least 24 hours will be required to accomplish heating and cooling, loading and unloading, bisquing, sintering and annealing. Finally it is expected that some type of encapsulation device would be required to control the sodium oxide composition.

In a continuous sintering process, the following unit operations are contemplated:

(1) Powder preparation prior to isostatic pressing is expected to be similar to that used in a batch sintering process (i.e. slurry solution spray drying).
(2) Continuous bisque firing will be in line with sintering and annealing. The bisquing furnace would be about 6 metres long and situated on the front end of the continuous sintering furnace.
(3) Sintering and annealing are sequential and continuous operations conducted in the same furnace expected to be approximately 15 metres in

length. The use of seeding (Jatkar *et al.*, 1978; Gordon *et al.*, 1978) to catalyse the phase conversion to beta″ alumina is expected to eliminate or substantially reduce the time required for annealing and thus decrease the length of the sintering furnace.

(4) In continuous bisquing and firing, six electrolyte tubes (in two lines, side by side, three carriers tall) are passed together at a pass through velocity of at least 8 cm min^{-1}.

(5) The sintering mode is expected to be horizontal with the use of alpha alumina refractories in the form of V setter encapsulation tubes.

(6) The daily capacity (24 hours) of each furnace will be at least 1400 tubes.

The comparison between large volume batch and continuous firing is summarized in Table 3.10. While batch firing has been demonstrated to be an excellent technique for producing high quality electrolyte tubing at moderate

Table 3.10 Comparison of continuous versus batch sintering, large scale ($\sim 9 \times 10^6$) production, electrolyte size (34 mm OD × 460 mm length)

	Continuous	*Batch*
Energy consumption		
(1) kWh/tube	4.4	13†
(2) Energy cost ($ per tube)	0.143*	0.452*
Labour (man-hour per tube)	0.003	0.02
Sintering furnaces		
(1) Number	29	380
(2) Capital ($10^6$$)	~9	~45
(3) Floor space (ft^2)	40 500	170 000
(4) Annual cost of space ($10^3$$)	203	760
(5) Daily capacity per furnace‡	~1150	~100
(6) Number encapsulation devices	~5200	~57 000
Uniformity of properties	Expected to be better than batch processing; some kiln to kiln variation	Some variation tube to tube, batch to batch, kiln to kiln
Process yield	>90% (?)	80% maximum (?)

* Based on electric utility rates from Utah Power & Light Company effective 3 January 1980.
† 5.3–7.0 kWh/tube estimated by workers at British Rail. This number will be very sensitive to furnace design. Well insulated furnaces will use less energy but the cycle time will be longer, affecting throughput and capital. Less well insulated furnaces can be heated and cooled more rapidly but will consume more energy.
‡ Based on a furnace length of 15 m.
All costs are 1980 dollars.

production levels (several thousand tubes), continuous firing appears to be economically more attractive in mass production at levels approaching 10 million units per annum. Energy consumption (based on electric resistance heating) is significantly less, power requirements are lower, labour is substantially lower (continuous sintering is expected to be nearly fully automated), capital requirements are about a factor of five smaller, plant space requirements are at least a factor of four smaller, and over an order of magnitude fewer encapsulation carriers are required in sintering. Furthermore, since each electrolyte tube undergoes the same thermal history in firing, process yields and uniformity of properties and dimensions are expected to be higher in continuous sintering. Taking all factors into consideration, economics of manufacture are clearly in favour of continuous firing operations.

Little attention has been paid to the method of heating furnaces. Electrical resistance heating appears to be preferred by most manufacturers because of the control it affords. Some groups have investigated radiofrequency induction (intrinsically inefficient) and gas (difficult to control?) heating as alternatives. It is clear that energy costs play a significant role in determining the economics of the fabrication process. High efficiency heating methods must be researched to include recirculated air preheat and the minimization of the charge to be heated. Other potentially efficient techniques such as microwave heating should be given importance in research planning.

3.5 REFERENCES

Andrews, J. M., Collins, A. H., Cornish, D. C. and Dracass, J. (1969) *Proc. Br. Ceram. Soc.*, no. 12, 211.

Babini, G. N., Bellosi, A. and Vincanzini, P. (1977) *Ceram. Int.*, **3**, 147.

Baukal, W., Beck, H., Kuhn, W. and Sieglen, R. (1977) in *Power Sources*, vol. 6 (ed. D. H. Collins), Academic Press, p. 655.

Bugden, W. G. and Duncan, J. H. (1977) *Sci. Ceram.*, **9**, 348.

Bugden, W. G. and Duncan, J. H. (1978) Second International Meeting on Solid Electrolytes, St Andrews, Scotland.

Bugden, W. G. and Duncan, J. H. (1979) Institution of Metallurgists Conference on Phase Transformations, York, Series 3, no. 11, vol. 2.

Burke, J. E. and Rosolowski, J. H. (1976) in *Treatise on Solid State Chemistry*, vol. 4 (ed. N. B. Hannay), Plenum Press, New York, pp. 621.

Byckalo, W., Robenblatt, J. L. and Nicholson, P. S. (1976) *Am. Ceram. Soc. Bull.*, **55**, 286.

Charles, R. J., Mitoff, S. P. and Morris, W. G. (1971) US Patent 3 607 435.

Chatterji, D. and Mitoff, S. P. (1978) EPRI Report EM-683.

Coble, R. L. (1962) *J. Am. Ceram. Soc.*, **45**, 123.

Davidge, R., McLaren, J., May, J. G. and Tappin, G. (1979) *Am. Ceram. Soc. Bull.*, **58**, 771.

DeBoer, J. H., Hamaker, H. C. and Verwey, E. J. W. (1939) *Rec. Trav. Chim.*, **58**, 662.

Desplanches, G. (1979) Energy Procs Meeting on Prospects for Battery Applications and Subsequent R & D Requirements, Brussels, p. 259.

Desplanches, G. (1980) 4th DoE/EPRI Sodium Sulfur Battery Workshop, Marcoussis.

Duncan, G. K. and West, A. R. (1983) *Solid State Ionics*, **9/10**, 259.
Duncan, J. H. (1977) UK Patent 1 478 317.
Duncan, J. H. (1979) US Patent 4 167 550.
Duncan, J. H. (1980) 4th DoE/EPRI Sodium Sulfur Battery Workshop, Derby.
Duncan, J. H. and Bugden, W. G. (1981) *Proc. Br. Ceram. Soc.*, no. 31, 221.
Duncan, J. H. and Creyke, W. E. C. (1969) *Trans. & J. Br. Ceram. Soc.*, **68**, 137.
Eddy, D. S. and Rhodes, J. F. (1977) US Patent 4 052 538.
Fally, J., Lasne, C., Lazennec, Y., Cars, Y. and Margotin, P. (1973) *J. Electrochem. Soc.*, **120**, 1296.
Ford Motor Company (1968) UK Patent 1 137 079.
Ford Motor Company (1978) UK Patent 1 529 771.
Francis, T. L., Phelps, F. E. and MacZura, G. (1971) *Am. Ceram. Soc. Bull.*, **50**, 615.
Gordon, R. S. (1978) Final Technical Report, Ford Sodium/Sulfur Battery Program, National Science Foundation, Washington DC, Contract NSF-C-805.
Gordon, R. S., McEntire, B. J., Miller, M. L. and Virkar, A. V. (1978) in *Processing of Crystalline Ceramics* (eds H. Palmour, R. F. Davis and T. M. Hare), Plenum, New York, p. 405.
Green, D. J. and Hutchinson, S. T. (1980a) in *Energy and Ceramics* (ed. P. Vincenzini), Elsevier, New York, p. 964.
Green, D. J. and Hutchinson, S. T. (1980b) *J. Can. Ceram. Soc.*, **49**, 7.
Haar, W. (1980) 4th DoE/EPRI Sodium Sulfur Battery Workshop, Heidelberg/Mannheim.
Haar, W., Fischer, W., Kleinschmager, H. and Weddigen, G. (1976) Battery Workshop, Argonne National Laboratories.
Hamaker, H. C. (1940) *Trans. Faraday Soc.*, **36**, 279.
Hamaker, H. C. and Verwey, E. J. W. (1940) *Trans. Faraday Soc.*, **36**, 180.
Hames, M. D. and Duncan, J. H. (1975) Proceedings SAE Congress paper 750375.
Hatfield, T., Richmond, C., Ford, W. F. and White, J. (1970) *Trans. & J. Brit. Ceram. Soc.*, **69**, 205.
Huckabee, M. L. and Palmour III, H. (1972) *Am. Ceram. Soc. Bull.*, **51**, 574.
Jatkar, A. D., Cutler, I. B. and Gordon, R. S. (1976) in *Ceramic Microstructures* (eds R. M. Fulrath and J. A. Pask), Westview Press, Boulder, Colorado, p. 412.
Jatkar, A. D., Cutler, I. B., Virkar, A. V. and Gordon, R. S. (1978) in *Processing of Crystalline Ceramics* (eds H. Palmour, R. F. Davis and T. M. Hare), Plenum, New York, p. 421.
Johnson, D. W., Granstaff, S. M. and Rhodes, W. W. (1979). *Am. Ceram. Soc. Bull.*, **58**, 849.
Jones, I. W. and Miles, L. J. (1971) *Proc. Br. Ceram. Soc.*, no. 19, 161.
Kennedy, J. H. and Foissy, A. (1975) *J. Electrochem. Soc.*, **122**, 482.
McDonough, W. J., Flinn, D. R., Stern, K. H. and Rice, R. W. (1978) *J. Mater. Sci.*, **13**, 2403.
Markin, T. L., Bones, R. J., Woodhead, J. L. and Wassel, L. L. (1975) UK Patent 1 386 244.
May, G. J., Tan, S. T. and Jones, I. W. (1980) *J. Mater. Sci.*, **15**, 2311.
Miller, M. L., McEntire, B. J., Miller, G. R. and Gordon, R. S. (1979) *Am. Ceram. Soc. Bull.*, **58**, 522.
Perrotta, A. J. and Young, J. E. (1974) *J. Am. Ceram. Soc.*, **57**, 405.
Powers, R. W. (1975) *J. Electrochem. Soc.*, **122**, 490.

Powers, R. W. (1978) US Patent 4073711.
Powers, R. W. and Mitoff, S. P. (1978a) in *Solid Electrolytes* (eds P. Hagenmuller and W. Van Gool), Academic Press, New York, p. 123.
Powers, R. W. and Mitoff, S. P. (1978b) *Am. Ceram. Soc. Bull.*, **57**, 456.
Rivier, M. and Pelton, A. D. (1978) *Am. Ceram. Soc. Bull.*, **57**, 183.
Speronello, B. K. (1980) US Patent 4193954.
Street, R. J. and Duncan, J. H. (1978) First International Conference on Isostatic Pressing, Loughborough.
Sugaike, S. (1979) *Interceram.*, **28**, 70.
Takahashi, T. and Kuwabara, K. (1980) *J. Appl. Electrochem.*, **10**, 291.
Tan, S. R. and May, G. J. (1977) *Sci. Ceram.*, **9**, 103.
Trostel, L. J. (1965) *Am. Ceram. Soc. Bull.*, **44**, 950.
Virkar, A. V., Miller, M. L., Cutler, I. B. and Gordon, R. S. (1978) US Patent 4113928.
Virkar, A. V., Tennenhouse, G. J. and Gordon, R. S. (1974) *J. Am. Ceram. Soc.*, **57**, 508.
Vogel, E. M., Johnson, D. W. and Yan, M. F. (1981) *Am. Ceram. Soc. Bull.*, **60**, 494.
Woodhead, J. L. (1968) *Sci. Ceram.*, **4**, 105.
Yoldas, B. E. and Partlow, D. P. (1980) *Am. Ceram. Soc. Bull.*, **59**, 640.
Youngblood, G. E., Virkar, A. V., Cannon, W. R. and Gordon, R. S. (1977) *Am. Ceram. Soc. Bull.*, **56**, 206.

4 Cell configuration

A. R. TILLEY

4.1 INTRODUCTION

Sodium ion conducting borate glass and beta alumina ceramic are the two types of solid electrolyte that have been used in cells which have liquid sodium and liquid sulfur electrodes. The resistivity of the glass electrolyte is approximately $2 \times 10^4\ \Omega$ cm at 300 °C, which is several orders of magnitude greater than that of beta alumina (approximately 6 Ω cm at 300 °C). The cells which use these two electrolytes are completely different in design because of this large difference in resistivity. The glass is incorporated into a cell as thousands of hollow glass fibres to give a high surface area, thin wall ($\sim 10\ \mu m$) electrolyte. With the glass in this form, even at high discharge rates the current density on the electrolyte can be as low as 2 mA cm^{-2}. The lower resistivity beta alumina can be incorporated into cells as plates or tubes in more conventional cell designs.

The features required for any design, such as high energy and power density, low manufacturing cost, high reliability and acceptable safety, are all interrelated and the final cell design has to be a compromise of these features. The requirements for the various applications vary considerably, and these requirements strongly influence design. For vehicle traction applications the sodium sulfur battery will most probably be power limited, so the cell design with the maximum electrolyte area for a given cell volume should be preferred. Peak power output is not so critical in applications such as load levelling, and so designs having larger capacity (thicker sulfur electrode and smaller electrolyte area) can be considered. The orientation of cells in a battery is also determined by the choice of cell design and the application. For instance, in some road traction applications long cells would have to be packed horizontally to maintain a low vehicle floor height, whereas with stationary load levelling or railway traction applications vertical orientation is possible even for long cells.

Each of the various cell configurations is briefly introduced in this chapter, and the advantages and disadvantages discussed. More detailed information on cell performance and the incorporation of the various electrodes, current collector designs and safety features is given in later chapters.

4.2 THE DOW CELL

This cell was first described by Levine and his co-workers at the Dow Chemical Company (Levine *et al.*, 1968). The company had expertise in glass and glass fibre technology, and they were primarily interested in this cell for in-house load levelling applications. A typical design is shown in Fig. 4.1. The hollow glass fibres are typically 70 μm ID × 90 μm OD and approximately 100 mm long. They are sealed at one end, and at the open end the fibres are held together by a glass header which is connected to a sodium reservoir. The liquid sodium extends down the inside of each of the fibres and the level of sodium falls in the sodium reservoir on cell discharge. A large cathode area is provided to match

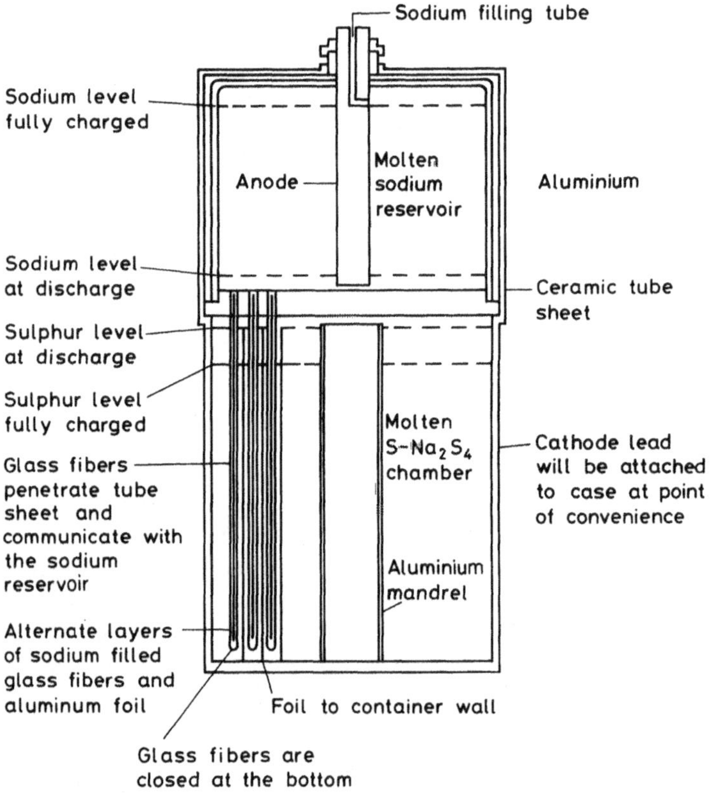

Fig. 4.1 The Dow sodium sulfur cell (Levine, 1983). Reproduced by permission of Academic Press Inc., New York.

this large anode and electrolyte area by using layers of coated aluminium foil between the fibres as the cathode current collector, with the sulfur/sodium polysulfide filling the space between the glass fibres and the aluminium foil. The spacing between the fibres and between the aluminium foil layers is an important design parameter. For a 70 μm ID $\times$ 90 μm OD fibre, the sodium discharge equals 0.1 mAh per centimetre of fibre length. This requires a sulfur chamber volume of 1.8×10^{-4} cm^3 per centimetre of fibre length, and typically this requires a foil to foil spacing of about 125 μm and a fibre to fibre spacing of 300–400 μm (Levine, 1983). These spacings are set during the cell assembly process.

Most of the initial cell development work at Dow was undertaken with cells of 6 Ah capacity ($\sim$2000 fibre bundle); larger cells of 40 Ah ($\sim$14 000 fibres) have been constructed and 400 Ah cells have been proposed (Levine, 1980). The various electrode and materials problems of this design are discussed in the relevant chapters, but particular problems specific to the Dow cell are the effects of impurities in the sodium such as calcium, potassium and sodium oxide which can cause resistance rise and premature failure of the glass fibres. The glass tube sheet header has also been the cause of cell failures and poor cell production pass rates. The glass of the tube sheet must melt at a lower temperature than the fibre glass and has to have a similar coefficient of thermal expansion. The tube sheet is made by extruding a paste of glass powder; the liquid is then dried and the assembly is heated to fuse the powder together to form a non-porous glass disc. To make a good seal between the fibres and the tube sheet header, close control of the glass powder particle size is required and moisture must be excluded during the milling and sieving operations.

The main disadvantages of this design compared with the beta alumina based sodium sulfur cells are the problems of cell reliability and cell safety. To obtain a high level of cell reliability when using such large numbers of glass fibres, there must be rigorous control of all stages of cell construction to ensure high cell assembly pass rates. In addition, if fibres do break in use, then they must be self-healing to allow the cell to continue to operate. Cell safety is slightly different in this type of cell, owing to the presence of the aluminium foil. Bulk mixing of the sodium and sulfur has to be prevented for, if the temperature on failure exceeds the melting point of aluminium, the sulfur/aluminium reaction can occur, liberating even more heat (Levine, 1980). The main advantage of this cell is its potentially low cost compared with beta alumina electrolyte cells, owing to the lower raw material and production cost of the glass electrolyte.

4.3 SODIUM SULFUR CELLS USING BETA ALUMINA ELECTROLYTE

Plate and tubular cell designs have been investigated. The tubular designs can be further subdivided into three design types: central sodium, central sulfur and multitubular cells.

4.3.1 Flat plate cells

The majority of flat plate designs have used thin discs of beta alumina glassed to an outer alpha alumina ring. This alumina ring acts as the electrically insulating component to which the sodium and sulfur compartment containers are sealed. The cells can be stacked in series to form a bipolar battery with no external intercell connections. A typical plate cell design is shown in Fig. 4.2. Cells of this type, which use glass to metal seals, have been constructed and tested by TRW Inc. in the USA (Silverman *et al.*, 1976), and plate cells using compression seals have been tested by British Rail and TRW (Sudworth *et al.*, 1973; Silverman, 1974, 1975, 1976).

The advantages and disadvantages of the plate cell design have been extensively covered by these workers. In the evaluation by Silverman in 1975 there were only two areas where plate cells were considered to be inferior to tube cell designs. These two areas were sealing and the mechanical strength of the beta alumina electrolyte. In the other areas considered – cell packaging, thermal control, mass production of the electrolyte, ease of assembly, safety, cell orientation and quality control – Silverman concluded that the plate cell was far superior to tube cells. This is still the case, but unfortunately the two

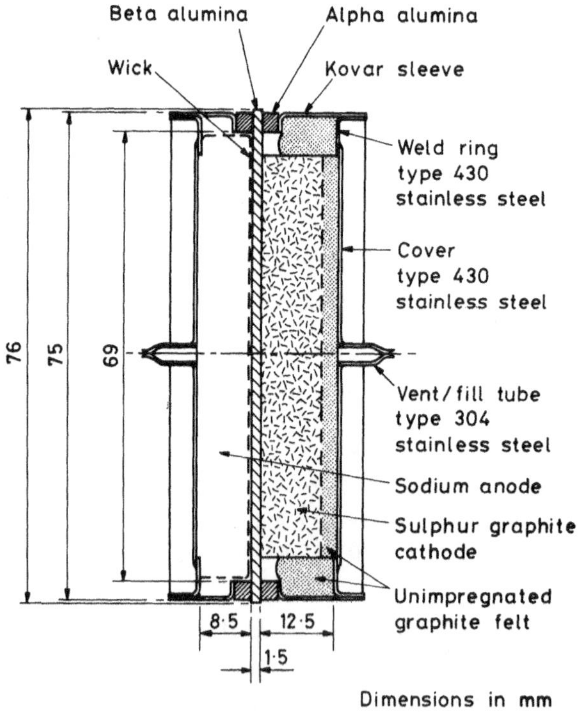

Fig. 4.2 A plate cell design (TRW) (Silverman, 1976). Reproduced by permission of the Electric Power Research Institute.

disadvantages are major and have yet to be totally overcome. For a given sulfur electrode thickness, the capacity of the plate cell is proportional to the square of the diameter. TRW and BR have shown that for a practical battery a disc size of at least 80 mm diameter is required to give a reasonable specific energy, and the optimum diameter is as high as 250 mm. Reducing the beta alumina thickness gives the greatest improvement in performance, owing to the reduction in cell internal resistance. Unfortunately, large diameter, thin section beta alumina discs are unable to withstand quite small pressure differentials, and sealing techniques such as compression seals or thermocompression seals cannot be used with large diameter plate cells, owing to the inherent weakness of the thin section alpha alumina sealing rings. A successful large diameter plate cell requires the development of evacuable hermetic seals such as glass or braze seals, thus eliminating the need to apply a mechanical load to the alpha alumina rings. These problems have halted the development of the plate cell, but if suitable seals could be developed it could compete with tubular cells, particularly in high rate applications. The plate cell has found a useful role in the development of the sulfur electrode and in testing current collectors. The operating area can be accurately defined, orientation effects can be easily eliminated, reference electrodes can be inserted into the beta alumina to enable the various electrode and electrolyte effects to be evaluated, and with small diameter compression sealed plate cells post mortem work is relatively easy. The plate cell was particularly useful for development work on the sulfur electrode, as a metal-free sulfur compartment could be achieved by using vitreous carbon plates as the conducting bipoles (Piercy and Sudworth, 1977).

4.3.2 Tubular cells

Tubular cells can have the sodium inside the beta alumina tube (central sodium) or outside the tube (central sulfur). The central sodium design can use smaller diameter beta alumina tubes than those required for the central sulfur design, and for this reason all the early development work was confined to central sodium cells with beta alumina tube diameters of approximately 15–20 mm. As the ceramic production processes were developed, suitable larger diameter tubes were produced, and workers in the United Kingdom then concentrated on the development of the central sulfur design. Multitubular designs have been considered, but these proved difficult to construct and all the practical cells have been single tube designs.

(a) *Central sodium tube cells*

A typical central sodium cell design is shown in Fig. 4.3. The cell shown is operated vertically and the sodium is gravity fed from a reservoir. The lower sulfur container consists of steel with a protective layer of corrosion resistant material. Virtually all the central sodium designs use a glassed-on alpha alumina collar at the open end of the beta alumina tube. As in the plate cell, this collar acts as the electrically insulating member to which the sodium and sulfur

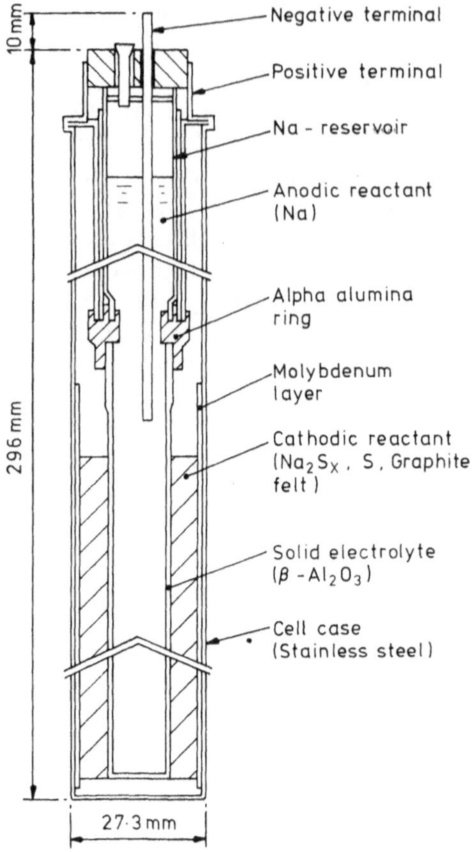

Fig. 4.3 A central sodium tubular cell (Yuasa Battery Co.) (Hattori *et al.*, 1977). Reproduced by permission of the Society of Automotive Engineers.

compartments are attached. Unlike the plate cell design, however, with maximum tube sizes of approximately 25 mm OD the alpha alumina collars are small enough to withstand the sealing loads for compression seals or thermocompression bonds. Central sodium cell designs have been constructed where all the sodium is stored in the beta alumina tube; a sodium wick is used to keep the surface of the electrolyte covered with sodium metal. This type of cell is shown in Fig. 4.4

As shown in Chapter 9, the central sodium cell has the highest specific energy of the two tubular designs but it has two main disadvantages. The outer cell case forms the sulfur electrode current collector, and protecting the cell case from corrosion by sodium polysulfides is difficult not only because of the materials problem but also because it is difficult either to coat the inside of a tube or to form a suitable corrosion resistant material into a cell case. Large capacity cells

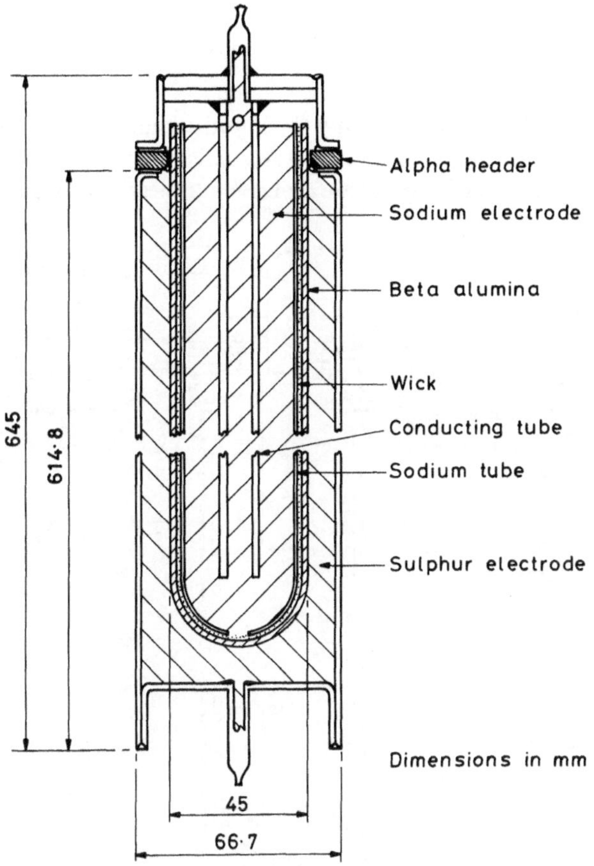

Fig. 4.4 A central sodium tubular cell utilizing a sodium wick (General Electric Co.) (J. Bast, 1982). Reproduced by permission of the Electric Power Research Institute.

in which all the sodium is stored inside the beta alumina tube have been found to be unsafe, and the maximum capacity of this type of cell is at present limited to 40 Ah.

(b) *Central sulfur tube cells*

This cell design was proposed by B. Partridge *et al.* (1976) to circumvent the central sodium cell case corrosion problem. In the central sulfur design, the liquid sulfur/sodium polysulfide is only in contact with the beta alumina electrolyte and a central current collector. This current collector is a rod or tube and is more easily protectively coated than a cell case. It can be fabricated out of materials such as carbon or conducting ceramics from which it would be difficult to fabricate cell cases for the central sodium cell. The range of materials suitable for a corrosion resistant current collector is considerably enlarged. A typical central sulfur cell is shown in Fig. 4.5. Other important advantages of the central sulfur cell are:

(1) The outer cell case contains only the liquid sodium, so it can be fabricated from mild steel. This can give a large cost saving over equivalent central sodium designs which use more expensive stainless steel or superalloy outer cases.

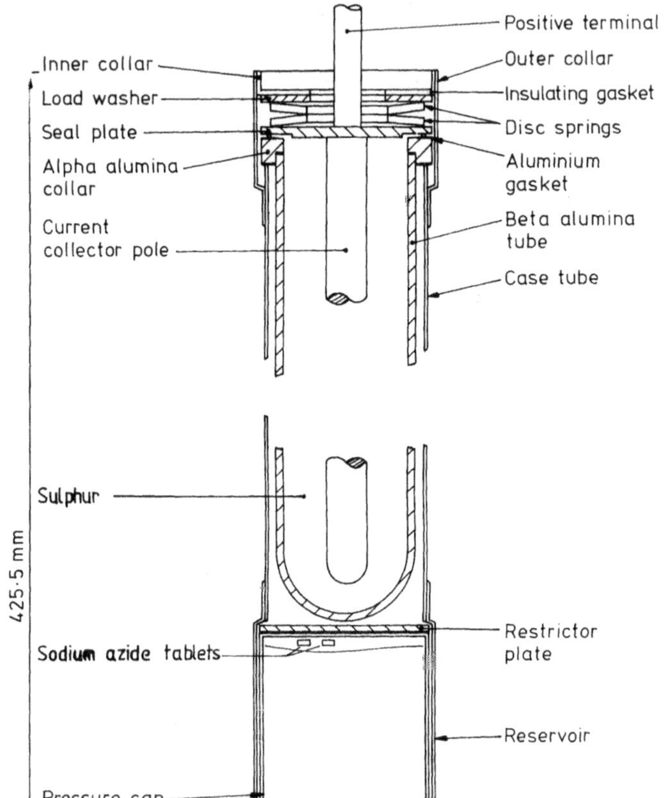

Fig. 4.5 A 100 Ah central sulfur cell design (Tilley, 1980). Reproduced by permission of British Rail Research.

(2) The central sulfur cell design more easily lends itself to the incorporation of safety features. Comparison of Figs 4.4 and 4.5 shows that in the event of beta alumina tube breakage, the sodium and sulfur reactants are better separated and their flow to each other is more restricted in the central sulfur design.
(3) A central sulfur cell contains a greater electrolyte area than a central sodium cell of equal total volume. Peak power output is proportional to the active surface area of the electrolyte, so for high rate applications such as vehicle traction use, where the Wh:W ratio is two hours or less, the central sulfur design gives a better match of energy and power.

The UK developers (CSPL, Harwell and BR) changed to the development of the central sulfur cell design in 1974, after failing to find a satisfactory solution to the case corrosion problem in central sodium cells. This change of cell design was made possible by the development of large diameter beta alumina tubes. Although the predicted specific energy for the central sulfur design was considerably less than that of the central sodium design, an overall battery specific energy of 100 Wh kg^{-1} and 100 kWh m^{-3} was considered possible and still commercially attractive.

Although the corrosion and safety problems were eased with this design, the limited space available for the seals assembly precluded the use of large diameter axial compression or thermocompression seals such as could be used for the central sodium design. Much development work by CSPL has resulted in a very compact seal arrangement that is contained within the cylindrical geometry of the cell. This development is described more fully in Chapter 8.

(c) *Multitubular cells*

An extensive study of multitubular cells has been undertaken by the General Electric Co. (Chatterji, 1976). GE constructed and tested 45-tube assemblies using 15 × 150 mm long beta alumina tubes. Problems with cracking of the alpha alumina header plate, reliability of the multiple glass sealing operation and alignment of the beta tubes were all overcome. The work at GE on central sodium multitubular designs was eventually stopped, as the multitubular design was intrinsically less reliable than single cell units and GE costing studies showed that equivalent single cell 'bundles' could be up to half the cost of multitubular cells. Other disadvantages were more difficult adaptation to assembly line production and a worse safety performance than single tube cells.

More recent work on central sulfur multitubular cells has been undertaken by Knoedler and Krapf (1983). In this work four-tube cells were constructed and operated; the cell design is shown in Fig. 4.6. The beta alumina tubes were 26 mm OD and 210 mm long and they were glass sealed to an alpha alumina header. The central current collectors were aluminium rods, 6 mm diameter, plasma sprayed with molybdenum, and structured sulfur electrodes were used. The capacity of the cells was 155 Ah and actual specific energy obtained at the

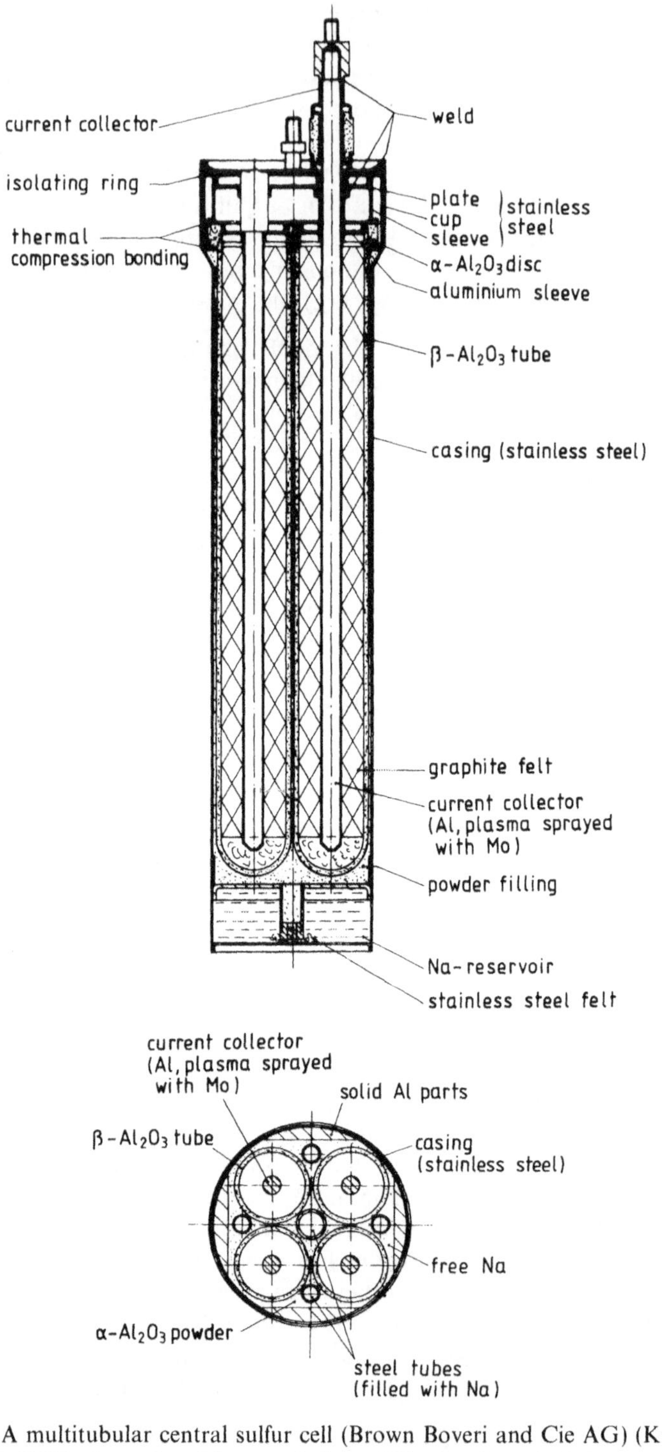

Fig. 4.6 A multitubular central sulfur cell (Brown Boveri and Cie AG) (Knoedler and Krapf, 1983). Reproduced by permission of Elsevier Sequoia S.A.

Table 4.1 Comparison of cell configurations

	Glass electrolyte	*Beta alumina electrolyte cells*			
	Dow	*Plate*	*Central sodium tubular*	*Central sulfur tubular*	*Multitubular*
Performance	Acceptable	Low but acceptable for very high rate applications	Very good	Acceptable	Good
Construction and sealing	Difficult	Difficult	Easy	Relatively easy	Difficult but acceptable for small assemblies
Corrosion	Still some problems	No solution demonstrated for flexing bipole construction	Cost effective solution not yet demonstrated	Corrosion resistant pole developed	As comments on central sodium and central sulfur
Safety	Questionable	Good	Questionable with large cells	Good	Not yet demonstrated
Cost prediction	Low	Acceptable	Depends on case corrosion solution	Acceptable	Unacceptable for large multitubular assemblies

two hour rate was 100 Wh kg^{-1} and 160 Wh l^{-1}. The main advantage of this type of design was that much higher specific maximum power values were projected than could be achieved for single cells of the central sodium design, without the problems of current collector corrosion. The potential problems of safety and sodium electrode performance still need to be investigated, and as with the GE design, although to a much lesser extent, the disadvantages of the high cost alpha alumina header and straightness requirement of the beta alumina tubes have to be overcome.

4.4 SUMMARY

The advantages and disadvantages of the main cell designs are summarized in Table 4.1. The following chapters give a more detailed account of the various electrode and cell constructions, cell performance, cell optimization, safety and battery performance of the various designs. Although there is unlikely to be one final preferred design it seems probable that this would be a single tube cell design. The central sodium cell can give batteries of high specific energy which undoubtedly will be needed for applications such as electric cars. The problems of case corrosion and the safety of large capacity cells, however, have yet to be completely solved. The central sulfur cell designs have been developed to a higher capacity than central sodium cells, and solutions to the safety and current collector corrosion problems have been demonstrated. Their specific energy, however, is inherently lower than central sodium cells, and seal construction is not as easy. For applications where a high specific energy is not so critical, such as commercial vehicles or load levelling, central sulfur cells can be considered.

4.5 REFERENCES

Bast, J. (1982) EPRI Report EM-2579, Project 128-6.

Chatterji, D. (1976) EPRI Report EM-266, Project 128-3.

Hattori, S., Yamaura, M., Kimura, S. and Iwabuchi, S. (1977) Proceedings SAE Congress, Detroit, paper 770281.

Knoedler, R. and Krapf, R. (1983) *J. Power Sources*, **10**, 263.

Levine, C. A. (1980) 4th EPRI/DoE Sodium Sulfur Battery Workshop, Heidelberg.

Levine, C. A. (1983) in *The Sulphur Electrode* (ed. R. Tischer), Chapter 8, Academic Press, New York and London.

Levine, C. A., Brown, W. E. and Heitz, R. G. (1968) US Patents 3 476 602 and 3 679 480.

Partridge, B., Jenkins, T. and McGuire, M. (1976) US Patent 3 982 959.

Piercy, R. and Sudworth, J. L. (1977) Symposium on Electrode Materials and Processes for Energy Conversion and Storage, *Proc. Electrochem. Soc.*, **77–6**, 950.

Silverman, H. P. (1974) EPRI Report 127-1.

Silverman, H. P. (1975) EPRI Report 127-2.

Silverman, H. P. (1976) EPRI EM-226, Final Report, Project 127.

Sudworth, J. L., Hames, M. D., Storey, M. A., Azim, M. F. and Tilley, A. R. (1973) in *Power Sources*, vol. 4 (ed. D. M. Collins), Oriel Press.

Tilley, A. R. (1980) 4th DoE/EPRI Sodium Sulfur Battery Workshop, Derby.

5 The sulfur electrode

J. L. SUDWORTH

5.1 INTRODUCTION

In this electrode, elemental sulfur, which at the temperature of operation is a viscous liquid with a measurable vapour pressure, must be reduced to polysulfide ions on discharge and reformed on charge. Sulfur, of course, is an electronic insulator and the electron transfer necessary for electrochemical reaction requires the presence of an electronic conductor. This material must not react with sulfur and, in cells in which the solid electrolyte is beta alumina, carbon is invariably used. To maximize the area available for the electrochemical reaction, the carbon is in fibrous form as either a felt or a mat and this fibrous matrix fills the whole volume of the sulfur electrode. In the glass electrolyte cell aluminium is used but, as this forms an electronically insulating film of aluminium sulfide, it must first be coated with either carbon or molybdenum. The high surface area is achieved by using aluminium foil, and as this is effectively a two dimensional electrode the glass electrolyte tubes are arranged in a regular, close packed array sandwiched between sheets of foil. In this way transport of sodium polysulfides and sulfur within the electrode is kept to a minimum.

Although initially the only electrolyte in this cell is the solid electrolyte, immediately discharge begins a liquid electrolyte, sodium pentasulfide, is formed, thus extending the electrode area. As discharge proceeds the lower polysulfides are formed and the composition of the sulfur electrode passes through the two phase region into the single phase region until the composition Na_2S_3 is reached. Thus the properties of the sodium polysulfides are vitally important to an understanding of the sulfur electrode.

5.2 PROPERTIES OF MOLTEN SODIUM POLYSULFIDES

Pearson and Robinson determined the phase diagram for the sulfur/sodium polysulfide system in 1930, but forty years elapsed before attempts were made to determine the properties of these highly reactive and corrosive molten salts. It was the publication by Kummer and Weber in 1967 of their work on the sodium sulfur cell which provided the stimulus for this work.

5.2.1 Preparation

Sodium polysulfides can be prepared by a number of different routes and these have been summarized by Cleaver (1983). Direct reaction of the elements under boiling toluene, xylene or mesitylene (Pearson and Robinson, 1930; Cleaver *et al.*, 1973; Fehér and Berthold, 1953) can be used but care must be taken to ensure that the vigorous exothermic reaction proceeds in a controlled manner. The composition of the polysulfides formed is given in Table 5.1. An alternative solvent is liquid ammonia (Bergstrom, 1926; Fehér, 1963). Vaporization of the ammonia (BP − 33 °C) removes the heat of reaction. Both these methods suffer from the disadvantage that it is difficult to remove all the solvent from the product.

Direct reaction of the elements can be avoided by dissolving sodium in

Table 5.1 Composition of sodium polysulfides prepared by reaction of sulfur with sodium in hydrocarbon solvents. Reproduced with permission from Cleaver (1983).

		Composition of product		
Solvent	*Boiling point (°C)*	*Cleaver et al., 1973*	*Fehér and Berthold, 1953*	*Pearson and Robinson, 1930*
Toluene	110	$Na_2S_{3.3-3.5}$	$Na_2S_{3.3-3.5}$	Na_2S_3 approx.
Xylene	140	$Na_2S_{3.5-3.8}$	—	—
Mesitylene	163	$Na_2S_{3.7-3.9}$		

ethanol to form sodium ethoxide and passing hydrogen sulfide into the solution to form sodium hydrosulfide. Addition of sulfur followed by boiling the solution whilst purging with an inert gas results in the formation of sodium sulfide and evolution of hydrogen sulfide. The sodium sulfide can be recovered by filtration.

Most of these methods have been used by the author but the simplest method, and the one which gives the purest product, has been found to be controlled discharge of a sodium sulfur cell. When prepared in this way the sodium polysulfides are free flowing liquids, varying in colour from deep red (Na_2S_3) to black (Na_2S_5), which can easily be pipetted from the cell. A metal-free cell for the preparation of sodium polysulfides has been described by Cleaver *et al.* (1973). On cooling, the polysulfides form clear amber or yellow glasses with a purity of better than 99.5%.

5.2.2 Structure and constitution

The structures of the solid polysulfides Na_2S_x have been determined by Föppl *et al.* (1962) and Tegman (1973). In Na_2S_4 the polysulfide anion consists of a chain of sulfur atoms with SSS bond angles of 109°. Na_2S_3 can be prepared in the solid state, but decomposes to Na_2S_2 and Na_2S_4 when heated above 100 °C. The structures of the molten polysulfides have not been determined, but it seems probable that the anions are similar to those in the solid. In the opinion of Cleaver (1983) all the sodium polysulfide melts contain an equilibrium mixture of the anions S_2^{2-}, S_3^{2-}, S_4^{2-}, S_5^{2-} and S_6^{2-} in proportions which depend on the stoichiometry and the temperature. The presence of S_3^{2-} and S_6^{2-} may appear inconsistent with the instability of Na_2S_3 and the non-existence of solid Na_2S_6; however, as Cleaver says, there are many examples of melts which contain ionic species differing from those of the solid phases with which they are in equilibrium.

Cleaver and Davies (1973a) have considered the question: do the sodium polysulfide melts contain free sulfur? By determining the EMF of the concentration cell with transference

$$C|Na_2S_x(l)|Na_2S_y(l)+S(l)|C$$

it is possible to show whether or not the sulfur in the melt is present as free sulfur. The EMF of this cell was found to be 270–275 mV which, as Cleaver and Davies show, indicates that all the sulfur in the melt is present as polysulfide ion and that no free sulfur is present.

5.2.3 Physicochemical properties

Cleaver *et al.* (1973) have determined the conductivity of sodium polysulfide melts. Their data are given in Fig. 5.1. There are several interesting points to be noted. First, the absolute value of the conductivity of all the sodium polysulfides is about twice that of beta alumina at 350°C and one thousand times that of the conducting glasses used in the Dow design; thus the sulfur electrode resistance need not be a major part of the cell resistance. Second, the

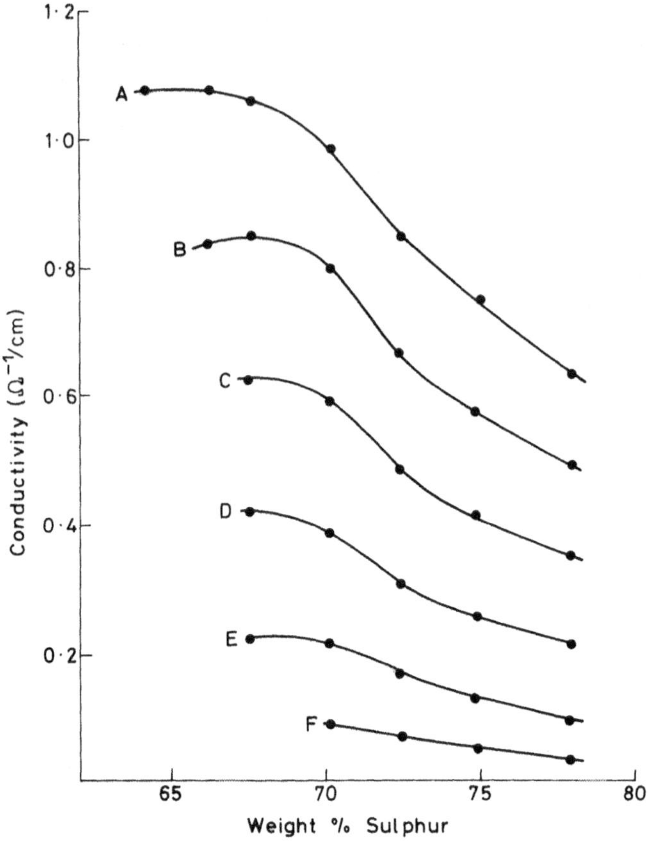

Fig. 5.1 The electrical conductivity of sodium polysulfide melts as a function of composition. (A) 420 °C; (B) 380 °C; (C) 340 °C; (D) 300 °C; (E) 260 °C; (F) 220 °C. (Reproduced from Cleaver *et al.* (1973) with permission of the authors and publishers, Pergamon Press Ltd).

conductivity of the sodium polysulfides increases as the sodium content increases, i.e. as the cell is discharged. Perhaps the most interesting feature of Fig. 5.1 is the curve at 220 °C. This is well below the liquidus and confirms the observation made by many workers, e.g. Gupta and Tischer (1972), that these melts are easily supercooled. This is of importance when considering the useful operating temperature range of the sodium sulfur cell.

Table 5.2 gives values of parameters in the empirical equation described by Cleaver *et al.* (1973):

$$\kappa = A_\kappa \exp[-E_\kappa/R(T-T_0)]$$

This equation can be used to calculate the conductivity of the polysulfides in the stated temperature ranges and at pressures up to 1 kbar.

Table 5.2 Values of parameters in the equation $\kappa = A_\kappa \exp[-E_\kappa/R(T-T_0)]$ describing the conductivity of fused sodium polysulfides at atmospheric pressure, in the stated temperature ranges

From Cleaver *et al.* (1973). *Electrochimica Acta*, **18,** 719. Reproduced by kind permission of the authors and publishers, Pergamon Press Ltd

Melt	*wt% S*	*Temperature range (K)*	A_κ *(Ω^{-1} cm^{-1})*	E_κ *(J mol^{-1})*	T_0 *(K)*	*Relative standard deviation* in* κ
$Na_2S_{2.1}$	60.0	728–840	5.478	3079	499	0.010
$Na_2S_{2.9}$	66.1	642–698	3.836	2478	458	0.010
$Na_2S_{3.0}$	67.5	582–693	7.033	5693	329	0.005
$Na_2S_{3.2}$	70.1	458–694	7.048	5854	330	0.013
$Na_2S_{3.8}$	72.3	428–694	7.056	6436	325	0.014
$Na_2S_{4.2}$	74.8	456–671	6.279	6163	341	0.022
$Na_2S_{5.1}$	77.8	477–681	5.815	6329	344	0.022

* The quantity given here is the standard deviation for a linear regression of ln κ on $1/(T-T_0)$.

The conductivity of sulfur saturated with Na_2S_5 has also been determined by Cleaver *et al.* (1973). It has been found to be about ten times higher than the conductivity of pure sulfur (conductivity 10^{-8} Ω^{-1} cm^{-1}) but lower than that of Na_2S_5 by a factor of 10^6, indicating that Na_2S_5 is only very slightly soluble in sulfur at 350 °C. Using an optical microscope, Suski and Kolacz (1980) studied the formation and dissolution of sodium pentasulfide in a cell consisting of a carbon fibre electrode in a thin film of sulfur in contact with a sodium ion conducting glass. They observed deposition of pentasulfide remote from the carbon electrode and concluded that the conductivity of sulfur saturated with sodium pentasulfide is electronic in nature. Interesting as these results are, the magnitude of this electronic conductivity is so much less than that of the ionic conductivity of the solid electrolyte and the sodium polysulfides that its effect will be insignificant in a practical cell.

Cleaver and Davies (1973b) have determined the viscosity of the sodium polysulfides over the temperature range 304–367 °C. At 350 °C the viscosity of melts ranges from 18.0 cpoise (Na_2S_4) to 18.6 cpoise (Na_2S_5). Thus compared with sulfur (viscosity 500 cpoise at 350 °C) the sodium polysulfides are relatively mobile liquids. The viscosity of these melts can be calculated accurately from the equation:

$$\eta = A_\eta \exp[E_\eta/R(T-T_0)]$$

the parameters are given in Table 5.3 (Cleaver and Davies, 1973b).

Table 5.3 Values of parameters in the equation $\eta = A_\eta \exp[E_\eta/R(T-T_0)]$, describing the viscosity of fused sodium polysulfides in the stated temperature ranges. Reproduced with permission from Cleaver and Davies (1973)

Melt	*wt% S*	*Temperature range (K)*	A_η *(cpoise)*	E_η *(J mol^{-1})*	T_0 *(K)*	*Relative standard deviation in η**
$Na_2S_{3.1}$	68.4	577–653	0.5647	8 428	332	0.020
$Na_2S_{3.3}$	70.0	589–647	0.3344	11 240	288	0.005
$Na_2S_{3.6}$	71.7	558–646	0.6351	7 915	342	0.012
$Na_2S_{3.9}$	73.1	533–652	0.4055	9 283	326	0.019
$Na_2S_{4.1}$	74.0	587–641	0.4079	9 333	328	0.008
$Na_2S_{4.3}$	75.1	572–675	0.8021	6 020	390	0.010
$Na_2S_{4.7}$	76.8	557–654	0.4733	8 181	356	0.015
$Na_2S_{5.2}$	78.5	620–648	1.717	3 125	465	0.012

* The quantity given here is the standard deviation for a linear regression of ln η on $1/(T-T_0)$.

The same workers have also determined the surface tension of the polysulfides, and at 350 °C this varies from 114 MN m^{-1} for Na_2S_5 to 172 MN m^{-1} for Na_2S_3. The corresponding value for sulfur is 46 MN m^{-1}. The equation relating surface tension and temperature is given by Cleaver and Davies as:

$$\gamma = B + C(T-600)$$

The parameters in this equation are given in Table 5.4.

The final property determined by Cleaver and Davies was the density. Plots of density against temperature were found to be linear, and Table 5.5 gives the values of the parameters in the equation

$$\rho = D + E(T-600)$$

The density of Na_2S_5 at 350 °C is 1.86 g cm^{-3}, which is higher than that of sulfur at the same temperature (1.66 g cm^{-3}). Thus sulfur forms the upper phase in the two phase system. An accurate value for the density of Na_2S_3 is needed to calculate the maximum weight of sulfur permissible in a cell of given volume.

The densities and other physicochemical properties of polysulfide melts at 350 °C are summarized and compared with those of sulfur in Table 5.6.

5.2.4 Thermodynamic data

Gupta and Tischer (1972) and Cleaver and Davies (1973a) have determined the

Table 5.4 Values of parameters in the equation $\gamma = B + C(T-600)$, describing the surface tension of fused sodium polysulfides in the stated temperature ranges. Reproduced with permission from Cleaver and Davies (1973b)

Melt	*wt% S*	*Temperature range (K)*	*B (mN m^{-1})*	*C (mN m^{-1} K^{-1})*	*Standard deviation (mN m^{-1})*
$Na_2S_{3.0}$	67.7	583–691	173.8	−0.060	0.8
$Na_2S_{3.1}$	68.5	588–673	165.3	−0.119	0.7
$Na_2S_{3.3}$	69.7	586–671	152.6	−0.065	0.6
$Na_2S_{3.6}$	71.5	589–677	141.6	−0.034	0.3
$Na_2S_{3.9}$	72.4	539–676	132.0	−0.036	0.3
$Na_2S_{4.1}$	74.0	556–641	128.2	−0.058	0.6
$Na_2S_{4.3}$	75.1	602–676	127.8	−0.067	1.5
$Na_2S_{4.7}$	76.3	562–671	127.2	−0.071	0.7
$Na_2S_{4.8}$	76.8	625–688	128.0	−0.091	1.7
$Na_2S_{5.2}$	78.4	616–661	115.3	−0.039	0.6

Table 5.5 Values of parameters in the equation $\rho = D + E(T-600)$, describing the density of fused sodium polysulfides in the stated temperature ranges. Reproduced with permission from Cleaver and Davies (1973b)

Melt	*wt% S*	*Temperature range (K)*	*D (g cm^{-3})*	*10^4E (g cm^{-3} K^{-1})*	*Standard deviation*
Na_2S_3	67.6	590–683	1.887	−5.65	0.004
$Na_2S_{3.3}$	69.7	576–689	1.901	−7.96	0.008
$Na_2S_{3.7}$	72.0	563–669	1.926	−5.47	0.003
$Na_2S_{4.4}$	75.4	571–680	1.869	−6.66	0.006
$Na_2S_{4.8}$	77.0	573–683	1.876	−7.16	0.005

important thermodynamic parameters for the sodium sulfur cell. Both determined the EMF of the cell

$$Na(l)/\text{beta alumina}/Na_2S_x(l)/C$$

over a wide range of sodium polysulfide compositions and at temperatures between 280 and 440 °C. There is good agreement between the two sets of data, and Gupta and Tischer's results are shown graphically in Fig. 5.2.

From these data it is possible to calculate the values of a number of thermodynamic functions. The EMF of the cell given above is determined by the following reaction:

$$2Na + (x-1)Na_2S_x = xNa_2S_{x-1} \tag{5.1}$$

Table 5.6 Physicochemical properties of sodium polysulfides and sulfur at 350 °C

	MP (°C)		*Density* ($g\ cm^{-3}$)		*Viscosity* (*cpoise*)		*Conductivity* ($\Omega^{-1}\ cm^{-1}$)		*Surface tension* ($MN\ m^{-1}$)
S	115.3	S	1.66	S	500	S	10^{-7}	S	46
Na_2S_5	258	$Na_2S_{4.8}$	1.86	$Na_2S_{5.2}$	18.6	$Na_2S_{5.2}$	0.39	$Na_2S_{5.2}$	114
Na_2S_4	285	$Na_2S_{3.7}$	1.91	Na_2S_4	18.0	Na_2S_4	0.48	$Na_2S_{4.1}$	127
Na_2S_3	235	Na_2S_3	1.87	$Na_2S_{3.1}$	18.5	Na_2S_3	0.69	Na_2S_3	172

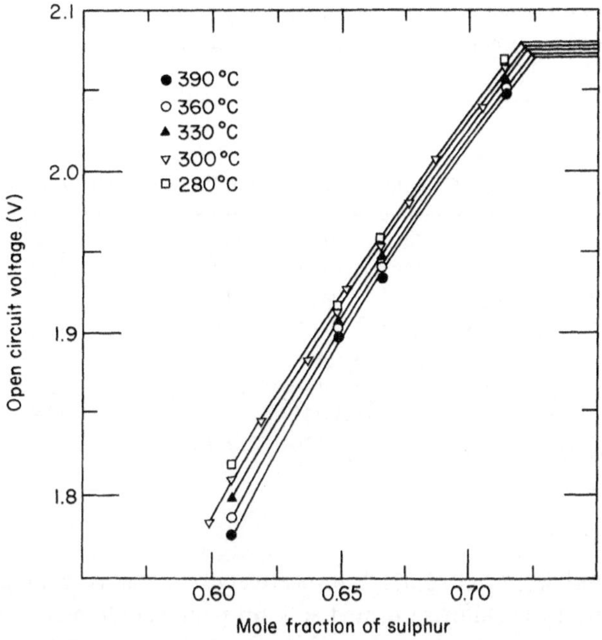

Fig. 5.2 The open circuit voltage of the sodium sulfur cell plotted against the melt composition at various temperatures. (Reproduced from Gupta and Tischer (1972) with permission of the authors and publishers, the Electrochemical Society Inc.).

As Gupta and Tischer point out, the thermodynamic functions – free energy, enthalpy and entropy – for this reaction are the differential energies and entropies of adding sodium to a large excess of polysulfide melt such that the polysulfide composition does not change. From these values and their temperature dependence the energies and entropies of the following reaction can be calculated:

$$2Na + xS = Na_2S_x \tag{5.2}$$

These are the energies and entropies of formation of the various sodium polysulfides.

Table 5.7 gives the data for the free energy of formation of the sodium polysulfides (reaction 5.2). Because the variation of EMF with temperature is small, the values of entropy and enthalpy of formation could not be determined accurately and these values are given only for reaction 5.1.

The data of Gupta and Tischer (1972) and Cleaver and Davies (1973a) result in a mean value for the weight fraction of sulfur (W) at the point of inflexion in the EMF/composition curve of 0.7863 at 360 °C and 0.7895 at 350 °C respectively, which corresponds to a depth of discharge of 56.85% and an EMF

Table 5.7 Thermodynamic data for sodium polysulfides calculated from open circuit voltage measurements. Reproduced with permission from Gupta and Tischer (1972)

Polysulfide composition (molten)	*−G(kcal)* 280°C	300°C	330°C	360°C	390°C	$-\bar{H}_1$ *(kcal)*	$-\bar{S}_1$ *(cal K⁻¹)*
Na_2S_3	95.7	96.0	95.8	95.7	95.7	47.8	4.1 ± 1
Na_2S_4	95.0	95.3	95.1	95.1	95.1	48.2	3.8 ± 2
$Na_2S_{3.7}$	94.6	94.9	94.6	94.6	94.6	46.3	4.9 ± 1.5
$Na_2S_{3.1}$	93.5	93.4	93.1	92.9	92.6	47.6	10.0 ± 2.5
Na_2S_3		93.1					
$Na_2S_{2.6}$		no data				47.8	24.6
$Na_2S_{2.4}$		no data				47.8	24.6
Na_2S_2 (solid)		89.6 (?)					

of 2.071 V and 55.7% and 2.076 V respectively. This can be compared with the composition of the sulfur saturated sodium pentasulfide melt determined by Gupta and Tischer (1972) which at 360 °C is 0.7848 weight fraction sulfur.

The depth of discharge (D) of a cell with respect to the composition corresponding to Na_2S_3 is given by:

$$D = 2.092 \frac{(1-W)}{W} \tag{5.3}$$

If a linear relationship between composition and EMF in the single phase region is assumed, the EMF of a cell in this region can be calculated using the following equation:

$$\text{EMF} = 2.071 - \left[\frac{0.305(D-0.5685)}{0.4315}\right] \qquad 1 > D > 0.5685 \tag{5.4}$$

This corresponds to an EMF of 1.766 V at composition Na_2S_3. Equation 5.4 simplifies to:

$$\text{EMF} = 2.4728 - 0.7068\,D \qquad 1 > D > 0.5685 \tag{5.5}$$

Equation (5.4) holds for temperatures in the range 280 to 400 °C if the first term (the cell EMF in the two phase region) is adjusted accordingly. A fuller treatment of the data of Gupta and Tischer (1972) and Cleaver and Davies (1973a) is given in the appendix.

The entropy data in Table 5.7 can be used to calculate the heat absorbed or

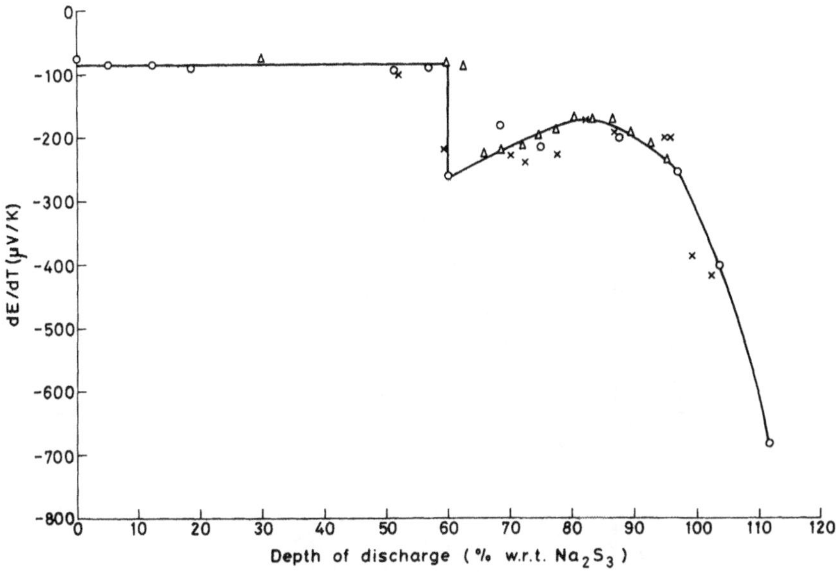

Fig. 5.3 Temperature coefficient of cell EMF plotted against melt composition. ○ Knoedler's data; △ Mitoff's data; × Davies's data.

evolved during the discharging and charging of the cell. More recent determinations of the temperature dependence of the cell EMF by Mitoff (1980), Davies (1972) and Knoedler (1984a), (1984b) have yielded more accurate data for the entropy of reaction 5.1, and these are given in Fig. 5.3.

The vapour pressure of sulfur over these melts is quite significant at sulfur rich compositions, and care must be taken to ensure that condensation of sulfur does not lead to changes in melt composition. Cleaver and Davies (1973a) have determined the variation of sulfur activity with melt composition and compared this with values calculated from the cell EMF. Their results are shown in Fig. 5.4.

5.2.5 Wetting phenomena

Although our knowledge of the physical properties of these melts has been increased greatly in recent years, the characteristics of the sulfur electrode in a sodium sulfur cell are determined not only (or even mainly) by the physicochemical properties of the melts but also by the interaction between the melts, sulfur and the electrode surface under the influence of an electric field.

An important effect which has been studied by Hall (1979) is the variation in wetting properties of liquid sulfur on a carbon surface immersed in an inert ionic melt. Optical techniques were used to determine contact angles of sulfur droplets at an electrically polarized carbon surface.

Using fused KSCN, which is a colourless melt at 200 °C, and which has

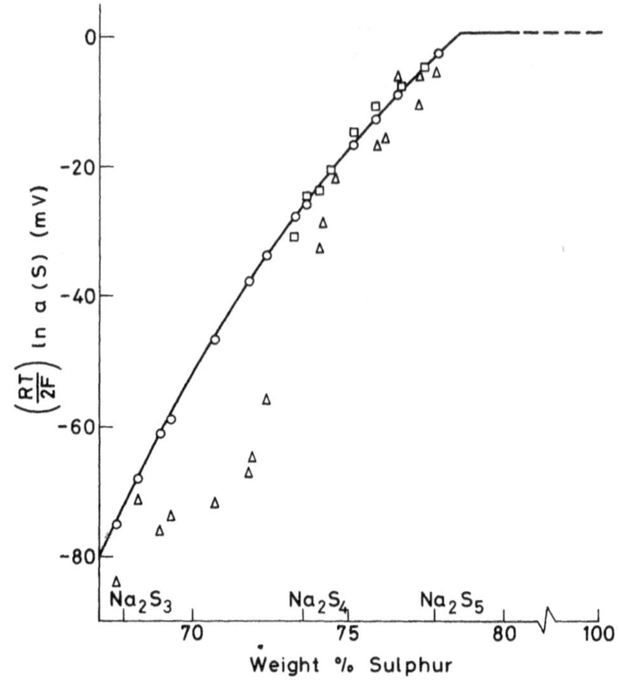

Fig. 5.4 Variation of sulfur activity with composition for sodium polysulfide melts at 360°C. □ results from transpiration experiments; △ results using electrochemical cells; ○ calculated values. After Cleaver and Davies (1973a).

physical characteristics similar to those of the sodium polysulfides, Hall was able to measure contact angles by microphotography and to observe variations of contact angle caused by pretreatment of the carbon surface and by polarization. The experiments were repeated, with similar results, in KCl/LiCl eutectic melt at 400 °C.

Hall showed that the contact angle of sulfur droplets electrodeposited on carbon fibres was dependent upon the previous history of the fibres: chemical pretreatment with solvents and dilute acid resulted in a contact angle of 0–30 ° for sulfur on carbon, whereas sulfur was deposited on untreated fibres in the form of hemispherical droplets with a contact angle of about 90°. He attributed the change in wetting characteristics to the formation or stripping of surface impurities. He also found a change in the wetting properties of a polished vitreous carbon electrode after an initial potential cycle, which he related to electrochemical stripping of such a layer. This led him to conclude that performance of the sodium sulfur cell after the first cycle would not be affected by chemical pretreatment of the carbon felt.

Having established the behaviour of sulfur in the absence of an electric field,

Hall went on to determine the effect of polarizing a polished vitreous carbon electrode in both melts. Microcinephotography of a sulfur drop, mechanically placed on the vitreous carbon electrode, revealed an extreme variation of sulfur wetting over a relatively narrow potential range. After initial polarization, the equilibrium value of the contact angle at open circuit reproducibly returned to $95° \pm 5°$. In the central part of the potential range, the contact angle always increased when the overpotential was made more negative and decreased when the overpotential was made more positive. This behaviour was observed in both KSCN and KCl/LiCl melts, and Cleaver (1983) believes it is likely that it would occur at the carbon/sulfur/polysulfide electrode. Hall observed that cathodic polarization greater than -0.5 V caused the contact angle to increase abruptly, resulting in the sulfur drop separating from the carbon surface. On anodic polarization the drop collapsed back on to the carbon surface and spread out.

Hall showed conclusively that changes in melt/carbon and melt/sulfur interfacial tension with electrode polarization were not responsible for the observed changes in wetting behaviour. He proposed instead that coulombic repulsion between the negatively charged electrode surface and the negatively charged long chain polymeric species adsorbed on the bulk liquid sulfur surface could account for the observed behaviour.

The significance of these results for the operation of the sodium sulfur cell will become clearer when the operation of the sulfur electrode in a sodium sulfur cell is discussed in Section 5.3.3.

5.2.6 Electrochemistry of polysulfide melts

Selis (1970) was the first to publish data on the electrochemistry of the sodium polysulfides. Using cyclic voltammetry with carbon and platinum electrodes in a sulfur saturated $Na_2S_{5.2}$ melt, he investigated the anodic and cathodic reactions. He obtained a linear dependence of i_{max} on the square root of sweep rate, suggesting diffusion controlled anodic and cathodic reactions.

South *et al.* (1972) extended the work of Selis and investigated three melt compositions – Na_2S_3, Na_2S_4 and $Na_2S_{5.2}$ – using voltammetry, chronopotentiometry and current interruption techniques at vitreous carbon electrodes. The cathodic and anodic voltammetric curves are shown in Figs 5.5 and 5.6 respectively.

In the cathodic curves, the first peak in all the melts is due to reduction of the polysulfide ions to lower polysulfides. South *et al.* showed this to be diffusion controlled as suggested by Selis. This has been confirmed by Armstrong *et al.* (1975) and Dzieciuch (1977) using rotating disc electrodes, by Tischer and Ludwig (1977) using voltammetry, and by Armstrong *et al.* (1976) using alternating current impedance measurements.

The limiting current region between 0.5 and 1.0 V is ascribed by South *et al.* to the formation of a film of Na_2S_2 on the electrode. This is supported by the high resistance values obtained from the current interruption measurements. Ludwig (1977) has accurately measured the potential at which the film redissolved and

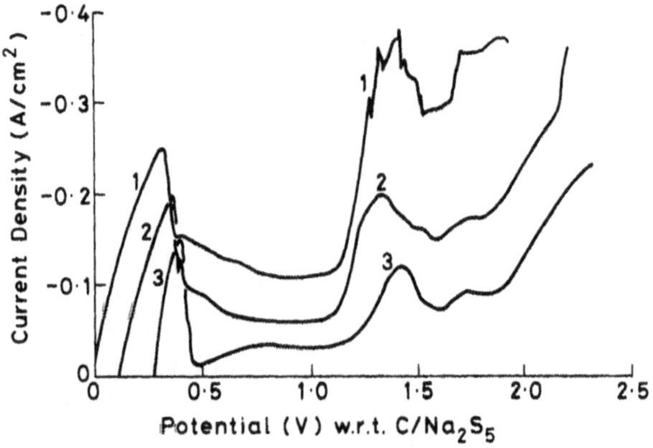

Fig. 5.5 Cathodic voltammetric curves for different melt compositions at 350 °C. Sweep rate 8.33 mV s^{-1}. (1) Na_2S_3 (2) Na_2S_4 (3) Na_2S_5. (Reproduced from South *et al.* (1972) with permission of the authors and publishers, the Electrochemical Society Inc.).

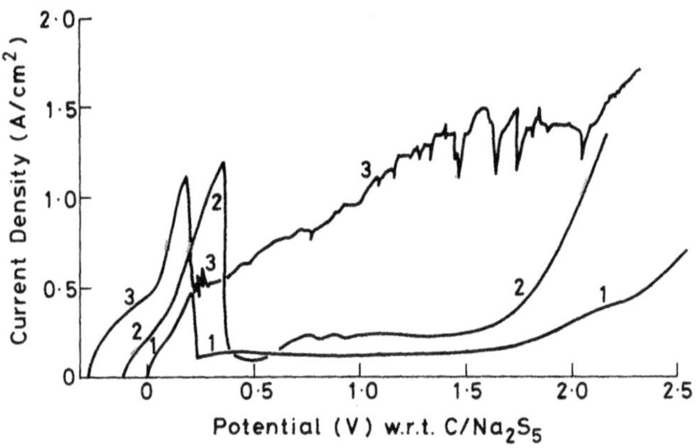

Fig. 5.6 Anodic voltammetric curves for different melt compositions at 350 °C. Sweep rate 50 mV s^{-1}. (1) Na_2S_5 (2) Na_2S_4 (3) Na_2S_3 from South *et al.* (1972). (Reproduced with permission of the authors and publishers, the Electrochemical Society Inc.).

shown it to be −0.3 V with respect to $Na_2S_{5.2}$. This corresponds to the potential difference between solid Na_2S_2/melt and liquid S/$Na_2S_{5.2}$ melt (Gupta and Tischer, 1972), confirming that the film is indeed Na_2S_2.

The second peak at about 1.5 V had also been observed by Tischer and Ludwig (1977) and Armstrong *et al.* (1975), and both agree with South *et al.* that it probably represents a change in composition or rearrangement of the film to a higher conductivity form (possibly Na_2S). The final rise in current at around 2 V is undoubtedly due to deposition of sodium metal.

(a) *Cathodic reactions*

The mechanisms of these electrode reactions have been considered by all the workers mentioned. In the linear region, before deposition of Na_2S_2 occurs on the electrode, the reaction is thought to involve the transfer of one electron (Armstrong *et al.*, 1976). Tischer and Ludwig (1977) propose that a sulfide ion is reduced to a lower sulfide, e.g.

$$S_4^{2-} + \varepsilon \rightarrow S_4^{3-}$$

$$S_4^{3-} + \varepsilon \rightarrow 2S_2^{2-}$$

In the Na_2S_5 melt there is definite evidence of a following chemical reaction (South *et al.*, 1972; Tischer and Ludwig, 1977). In Na_2S_4 melts the evidence is equivocal, and in Na_2S_3 melts there is no evidence of such a reaction. South *et al.* attributed this to reaction of the lower sulfides with free sulfur in the $Na_2S_{5.2}$ melt. As was pointed out earlier, however, Cleaver and Davies (1973a) have shown that free sulfur is not present in these melts, and the following chemical reaction must therefore be the reaction of lower sulfides (e.g. S_2^{2-} and S_3^{2-}) with higher sulfides (S_5^{2-}) to form other sulfides (in this case S_4^{2-} and S_3^{2-}) (Tischer and Ludwig, 1977).

Armstrong *et al.* (1975) consider it unlikely that the reaction occurring at the electrode covered with a film of Na_2S_2 is the same as that on the film-free electrode. Their rotating disc studies indicate that the film is continuous. This would not be expected to result in the observed decrease in the diffusion limited current, but they put forward as a possible explanation of this the idea that the electrode surface is completely covered by a film which is in equilibrium with a minority species in the melt. They show that this would explain the observed dependence of the current on rotation speed.

Armstrong *et al.* (1976) have derived approximate values for diffusion coefficients at 350 °C from impedance data. In Na_2S_3, $D_O = D_R = 9.8 \times 10^{-6}$ cm^2 s^{-1}; in Na_2S_5, $D_O = D_R = 14 \times 10^{-6}$ cm^2 s^{-1}. They also calculated the concentrations of the oxidized and reduced species and found them not inconsistent with the values expected if the dissociation of the two sulfides was a disproportionation into higher and lower sulfides. The values obtained were: Na_2S_3, $C_O = 1.54$–1.7 mol l^{-1}, $C_R = 1.0$–1.1 mol l^{-1}, Na_2S_5, $C_O = 1.28$–1.32 mol l^{-1}, $C_R = 0.8$–0.88 mol l^{-1}. Tischer and Ludwig (1977), using data from the rotating disc experiments of Armstrong *et al.* (1975) and chronopotentiometric data of South *et al.* (1972), calculated the product nC and D in both Na_2S_4 and Na_2S_5 melts. At 350 °C they obtained for Na_2S_5 melts the values $D = 6.3 \times 10^{-7}$ cm^2 s^{-1} and $nC = 18.2$ mol l^{-1}. Since C in Na_2S_5 cannot exceed 9.02 mol l^{-1}, n must equal 2. Obviously there is a considerable difference between the values obtained by Armstrong *et al.* and Ludwig (about an order of magnitude), and further experimental work is required to resolve these differences.

(b) *Anodic reactions*

Regarding the anodic processes, it might be expected that diffusion controlled oxidation of lower sulfides to higher sulfides would occur at potentials near to open circuit followed by oxidation of these higher sulfides to elemental sulfur at more positive potentials. This is in fact largely borne out by the experimental evidence.

South *et al.* (1972) obtained a linear dependence of I_{max} on the square root of sweep rate, indicating diffusion control. However, their chronopotentiometric data did not obey the Sand equation, which would indicate that the reaction is not diffusion controlled. Tischer and Ludwig suggest that the sweep rate range used in the voltammetric studies was too limited for good conclusions to be drawn regarding the constancy of $i_{max}/v^{\frac{1}{2}}$.

Armstrong *et al.* (1975) found that results obtained for the anodic process were less reproducible than for the cathodic process. Nevertheless a plot of reciprocal of current density against reciprocal of the square root of the rotation speed was linear at potentials close to the equilibrium potential, i.e. in the region where current increases linearly with potential. At more positive potentials this relationship was not obeyed, and at potentials as high as +1 V the current was virtually independent of rotation speed in Na_2S_5 melts.

The explanation of these results is as follows. In all melts the first part of the anodic peak, i.e. up to the inflexion, is due to oxidation of lower polysulfide ions to higher polysulfides, and is diffusion controlled. The remaining part of the peak is due to oxidation of sulfide ions to elemental sulfur. Brennan (1977a) has confirmed this by determining the true electrode potential at the point of inflexion. He found it to be in the range 2.045 to 2.074 V with respect to sodium/beta alumina. These potentials are very close to the known equilibrium potentials for sulfur saturated melts (2.076 V).

In melts not saturated with sulfur the elemental sulfur produced at the electrode can react with the melt and, as Fig. 5.6 shows, in Na_2S_3 melts no limiting current is observed. The current increases with increasing potential, but considerable current oscillations are present. In Na_2S_4 and Na_2S_5 a limiting current region is observed but at higher potentials there is a region where the current increases with increasing potential. The rotating disc studies of Armstrong *et al.* (1975) show this more clearly (Fig. 5.7).

The shape of the current potential curves in Fig. 5.7 is difficult to explain. The initial rise of current with increasing potential is definitely associated with production of sulfur but in a way which does not lead to passivation of the electrode. Brennan (1977a) suggests that this is because an intermediate sulfide radical ion S_5^- is formed at the electrode surface. This then diffuses away from the electrode before producing elemental sulfur by the reaction

$$2S_5^- \rightarrow 5/8S_8 + S_5^{2-}$$

Support for this idea is provided by Brennan's (1977b) observations of the

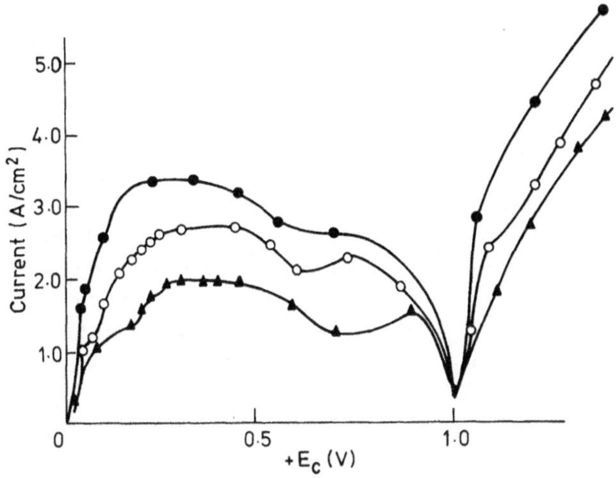

Fig. 5.7 Steady state current potential curves for Na_2S_5 at 350 °C in the anodic region. Rotation speeds ▲ 16.7 Hz, ○ 33.5 Hz, ● 70 Hz. From Armstrong *et al.* (1975). (Reproduced with permission of the authors and publishers, Pergamon Press Ltd).

anodic generation of sulfur from sodium polysulfide. In a thin film cell, using microscopy, he observed globules of sulfur growing in the melt remote from the electrode surface, suggesting the existence of intermediate species (such as S_5^-) which could diffuse from the electrode surface to the globule.

Some globules will undoubtedly be produced on or very near the electrode surface, but if these are non-wetting this need not lead to passivation of the electrode. However, as the electrode potential becomes more positive the observations of Hall (1979) on the effect of potential on contact angle of sulfur drops suggest that the contact angle would decrease leading eventually to coverage of the electrode by a film of sulfur. This would explain the minimum in the current observed at 1.0 V (Fig. 5.7).

The increase in current which occurs when the potential is increased beyond +1.0 V has not been satisfactorily explained. Ludwig (1977) attributes it to the generation of sulfur cations, but Brennan (1977b) noted that no transient arrests were observed in the voltage decay after interruption of anodic current, the open circuit potential immediately assuming the sulfur potential. This, he thought, indicated it was unlikely that cationic species were formed.

Armstrong *et al.* (1975) suggest that potential induced changes in contact angle could result in detachment of the sulfur droplets from the electrode surface. Hall's results on variation of contact angle with potential do not support this, and the cause of this phenomenon remains unexplained.

A possible explanation is coulombic repulsion of positively charged sulfur globules. If small quantities of cationic sulfur species were formed at the sulfur/electrode interface at very positive potentials, they might be absorbed on to the sulfur globules and confer upon them a positive charge. The consequent coulombic repulsion would lead to non-wetting contact with the electrode, freeing the surface for further reaction.

(c) *Sulfur electrodes in cells*

The various processes occurring in polysulfide melts are indicated qualitatively in Fig. 5.8 (Armstrong *et al.*, 1975). In a sodium sulfur cell the sulfur electrode potential is likely to remain in the linear region near the equilibrium potential for most of the charge and discharge. Here the electrode reaction is not complicated by the formation of passivating films and, although it is diffusion controlled, high current densities can be sustained.

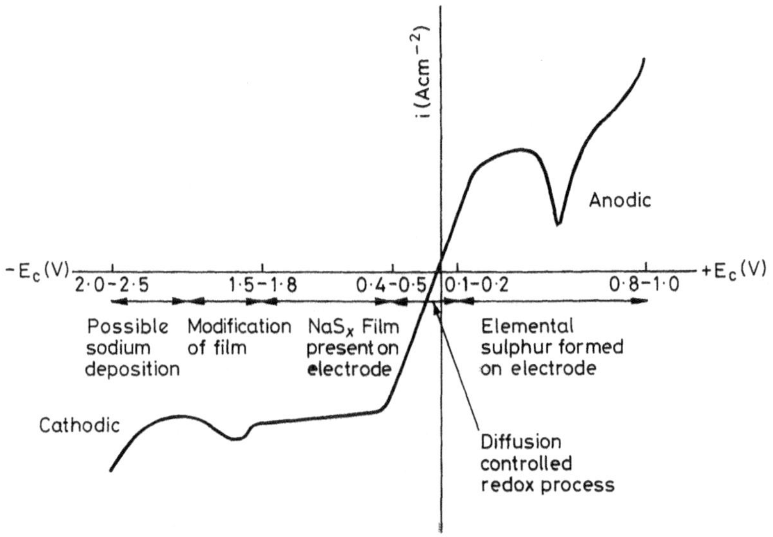

Fig. 5.8 Processes occurring at a smooth carbon electrode in sodium polysulfide melts. After Armstrong *et al.* (1975).

At the end of discharge the S_2^{2-} ion concentration increases and the deposition of a film of Na_2S_2 on the electrode surface eventually occurs. This results in a sudden increase in resistance but, since the limiting current is 25 mA cm^{-2} (South *et al.*, 1972) and the real electrode area is likely to be in excess of 10^3 cm^2, discharge can continue even at quite high currents. For a single cell discharging through a load, the electrode potential cannot exceed −1.78 V with respect to Na_2S_3/C reversible potential (at 350 °C). If, however, cell reversal occurs owing to current being forced through the cell (as might happen in a series chain), the electrode potential will exceed this value and sodium deposition will occur in the sulfur electrode.

At the top of charge the electrode potential will reach values where coverage of the carbon surface by sulfur will occur. Whether this is responsible for the extreme polarization observed at the end of charge in a sodium sulfur cell is questionable. The minimum value of the anodic current measured by Armstrong *et al.* (1976) was 0.2 A cm^{-2}. Similar values were measured by South *et al.* (1972). With a real electrode area in excess of 1000 cm^2, such current densities are unlikely to be achieved in cells, and some other explanation must be found for the sudden polarization which signals top of charge in a sodium sulfur cell.

5.3 PRACTICAL SULFUR ELECTRODES

5.3.1 Electrode materials

The main requirement of a practical sulfur electrode is an extended area of inert electronically conducting material at the surface of which the sulfur/polysulfide redox reaction can occur. Kummer and Weber (1967) used carbon felt and this has been almost universally adopted in cells having beta alumina solid electrolyte. Nevertheless, a number of other materials have been investigated and will be considered here together with carbon felt.

In the glass electrolyte cell, the large electrode area is provided by layers of aluminium foil between the hollow glass fibres. Sulfur fills the spaces between the fibres and the foil. Typical current densities on the foil and the glass electrolyte (Levine, 1983) are 0.7 mA cm^{-2} and 3 mA cm^{-2} respectively.

The important design parameter for the sulfur electrode in glass electrolyte cells is the spacing between the glass fibres and the metal foil. These spacings can be set during cell assembly, and typically the foil to foil spacing is 125 microns whilst the fibre to fibre spacing is 300–400 microns (Levine, 1983).

For a fibre diameter of 90 microns the average closest distance to the foil surface is less than 20 microns, although in fact each fibre is likely to be touching one of the foils. The close spacing of the foils allows the cell to be charged and discharged throughout the composition range. If the fibre to fibre spacing is increased substantially, some of the polysulfide melt can become isolated from the glass fibre, causing increased polarization (Levine, 1983).

Because aluminium forms an insulating sulfide film on its surface in a sulfur melt, it is necessary to coat the foil with carbon or molybdenum. Carbon can be applied from an aqueous suspension and molybdenum can be applied by vapour deposition, sputtering or metal evaporation.

In sodium sulfur cells with beta alumina electrolyte the space between the electrolyte and the current collector is filled with a porous matrix. South (1973) investigated the behaviour of a number of porous materials in a sodium tetrasulfide melt at 350 °C: carbon and graphite felts and cloths, carbon foam and stainless steel foam. Small pieces of porous material were clamped by a stainless steel ring behind a glass disc with an aperture of cross-sectional area

0.56 cm^2. This was then immersed in the melt together with a carbon reference electrode and a carbon counter electrode.

Some physical properties of a number of carbon felts are given in Table 5.8. These materials were found to have a cathode limiting current of up to 1.6 A cm^{-2} based on the cross-sectional area of the test electrode. The current density based on the total surface area would be much lower (typical values for specific area were 1 m^2 g^{-1}), indicating that only a small fraction of the surface is electrochemically active. Carbon foams with a mean pore size of 40 microns and 60 microns and a porosity of 41% were also tested, but these polarized quite quickly on application of the current, presumably owing to the low porosity (most felts are over 95% porous) and small pore size.

Table 5.8 Some physical properties of carbon and graphite felts

Type of felt	*Average thickness (uncompressed) (mm)*	*Bulk density (g cm^{-3})*	*Porosity (%)*	*Specific surface area* (m^2 g^{-1})*
Le Carbone RVG 1000 graphite	1.2	0.11	95.0	1.47
Le Carbone RVG 2000 graphite	3.5	0.06	97.3	1.62
Le Carbone RVG 4000 graphite	9.0	0.10	95.5	1.60
Le Carbone RVC 1000 carbon	1.3	0.10	95.5	0.627
Le Carbone RVC 2000 carbon	4.0	0.06	97.3	—
Le carbone RVC 4000 carbon	10.0	0.095	95.7	0.475
Carborundum GF graphite	6.3	0.067	97.0	0.628
Carborundum CF carbon	6.3	0.07	96.8	0.482

* As determined by Coulter Electronics using krypton absorption.

The metal foams were quickly corroded when polarized anodically and cathodically and cannot be considered as electrode matrix materials.

In the light of these results it is not surprising that the carbon and graphite felts used by Kummer and Weber have been widely adopted as electrode matrix materials in sodium sulfur cells with a beta alumina electrolyte. Because of the important role these materials play in the sodium sulfur cell a description of their main characteristics will be given here. The properties of the carbon felts are determined by the precursor materials used and by the extent of carbonization. The two precursors used are textile fibres and spun pitch fibres.

The most commonly used precursor for textile felts is viscose rayon fleece. When this is carbonized at 1200 °C a resilient carbon felt is produced. At higher temperatures the graphite content increases and above 2000 °C a graphite felt is obtained which, like the carbon felt, is resilient. The viscose rayon fleece consists mainly of fibres aligned in one direction, with a small number of fibres perpendicular to the plane of the fleece. This 'needling' of the fleece increases the resilience of the carbon or graphite felt, an important factor in ensuring good contact between the felt and the current collector or the beta alumina. These felts are anisotropic materials. They can be compressed to a quarter of their volume in the direction perpendicular to the plane of the majority of the fibres. Beyond this point, however, fibre damage occurs, the graphite felts being more susceptible to this than the carbon felts. Compression in the plane of the fibres is also possible but to a much lesser degree. The resistivity of the felts is also anisotropic, being much less in the plane of the fibres than in the direction perpendicular to the plane of the fibres. This has important implications for sulfur electrode design.

Pitch fibre based mats are prepared by carbonizing spun pitch fibres and are in the form of loose mats with little or no needling. They are much less resilient than the textile based felts and the fibres of some mats are composed of vitreous carbon rather than amorphous carbon or graphite. The resistivity of these felts measured perpendicular to the fibre orientation is about twice that of the rayon based felts, but the resistivity measured in the direction parallel to the fibres is much lower. The electrical resistivity and other characteristics of the pitch based felts are shown in Table 5.9. The pitch based felts have one other important characteristic – their cost is significantly lower than that of rayon based felts.

Carbon foams have been used as electrode matrices (Sudworth and Tilley, 1983), but suffer from the disadvantage that it is difficult to obtain good contact to the current collector. If this is carbon, the foam can be bonded to it (Weiner, 1976). For metal current collectors, interposing a layer of resilient carbon felt provides a means of maintaining electrical contact between the current collector and the foam (Weiner, 1977). Ranges of pore sizes are available, typically 10–100 pores per inch. A more consistent and reproducible electrode might be expected to result from the use of these materials, but difficulties of cell construction and the possibility of degradation with cycling (Silverman, 1976) have limited their use.

One other type of carbon electrode deserves mention – a dispersion of carbon black in sulfur. Weiwiorowski (1972) reported that molten sulfur containing 2% by weight carbon black has a conductivity of 0.1 Ω^{-1} cm^{-1} at 130 °C. South (1973) prepared sulfur electrodes from suspensions of carbon black in sulfur but found that a 2.4% by weight suspension gave high cell resistance (32 Ω cm^2 for a 6 mm thick electrode) and the conductivity of the suspension was 0.15 Ω^{-1} cm^{-1} at 130°C. By pressing a 1:1 mixture of sulfur and carbon black in an isostatic press South was able to obtain a cell resistance of 2.3 Ω cm^2 for a 2.8 mm thick electrode with a utilization of 95% at 118 mA cm^{-2}. This

Table 5.9 Properties of pitch based carbon mats

	Units	*Mat type* *A*	*B*		*C*	
Filament						
Diameter	microns	9	10.5		10.5	
Specific gravity	$g \cdot cm^{-3}$	2.0	1.6		1.6	
Tensile strength	psi (Gpa)	200 000 (1.4)	150 000 (1.05)		150 000 (1.05)	
Surface area	m^2/g	0.4	—		—	
Electrical resistivity	ohm cm	12×10^{-4}	10×10^{-3}		5×10^{-3}	
Thermal conductivity	$W \cdot m^{-1} \cdot K^{-1}$	—	0.017		0.035–0.045	
Mat		*A*	B_1	B_2	C_1	C_2
Area density	$g \cdot m^{-2}$	352	500	1000	500	1000
Bulk density	$g \cdot cm^{-3}$	0.036	0.07 min	0.07 min	0.07 min	0.07 min
Electrical resistivity (long.)	ohm cm	0.70	—	—	—	—
Electrical resistivity (trans.)	ohm cm	0.70	—	—	—	—
Thermal conductivity	$W \cdot m^{-1} \cdot K^{-1}$	0.035	—	—	—	—
Thickness	mm	13	5–6	10–12	5–6	10–12

A: Thornel VMA; B: Kureha KCF-100; C: Kureha KGF-200.
B_1: KCF-100 felt FT-50; B_2: KCF-100 felt FT-100.
C_1: KGF-200 felt FT-50; C_2: KGF-200 felt FT-100.

electrode suffered from two disadvantages: the low bulk density resulted in only 23% of sulfur by volume in the electrode compartment, and the cell resistance at the beginning of discharge was high – approximately 10 Ω cm^2. For these reasons, this type of electrode has not been adopted for sodium sulfur cells, although it is interesting to note that such electrodes were used in another high temperature cell, the sodium/antimony trichloride cell.

5.3.2 Electrode fabrication

When using carbon felts or mats for sulfur electrode fabrication, fibre orientation and the method of sulfur addition are the two main variables. For a low resistance electrode, the fibres should be oriented parallel to the current flow. This can easily be achieved in tubular cells by cutting felt rings with a central hole slightly smaller than the beta alumina tube (or the central pole in a central sulfur cell) and with an external diameter slightly larger than the cell container (or internal diameter of the beta alumina tube in a central sulfur cell). When the cell is assembled, the felt will then be compressed slightly, ensuring good contact with the current collector and beta alumina. Molten sulfur is then added to the electrode via a fill tube.

Although the ring electrodes give the most favourable orientation for low resistance, preforms in which fibres are perpendicular to the direction of current flow are much easier to make. South (1973) used two trough shaped preforms fabricated from rectangular pieces of felt to make a complete sulfur electrode. This type of preform has been widely adopted. Storey (1977) showed that it was possible to fabricate sulfur electrode preforms by injection moulding. Both disc electrodes for flat plate cells and trough shaped preforms were fabricated by this method. The method of electrode construction in which rings of felt are cut from sheet results in little felt damage and in electrodes in which the fibres are aligned in the most favourable direction for high conductivity. Using this technique, sulfur electrode resistances of 0.4 Ω cm^2 can be achieved (Sudworth, 1980).

The amount of sulfur in preforms is determined by the dimensions of the mould and the amount of felt, and can be controlled to within ±2.5%. For an unevacuated sulfur electrode, the amount of sulfur which can be accommodated is approximately 50% by volume at room temperature, since allowance must be made for expansion of the sulfur on heating to 350 °C and the increase in volume which occurs when the cell is discharged to Na_2S_3. The weight of sulfur permissible can be calculated from the data in Table 5.5 and the density of sulfur at room temperature (2.08 g cm^{-3}). From this it is clear that a fully charged sulfur electrode will contain about 30% voids, the distribution of which could be an important factor in the operation of the electrode. Jacquelin and Pompon (1980) have shown that in their electrodes the voids are evenly distributed throughout most of the electrode, but with a high concentration at the top of the electrode.

Further discussion of the sulfur electrode structure will be deferred until the next section, but the materials and methods of electrode fabrication described can be utilized in most electrode designs. The cell geometry, of course, will be an important factor in deciding fabrication method and electrode structure. Thus the electrode designs based on rings of carbon felt are particularly suitable for the central sulfur cell geometry because the small diameter pole results in minimum felt wastage. For central sodium cells with large diameter electrolyte tubes this electrode structure would be very wasteful of felt, and trough electrodes are usually used in these cells. Another factor influencing the method used for fabricating the electrode is the use of the non-uniform electrode structures which will be described in the next section.

5.3.3 Operation of the sulfur electrode

As was seen earlier in this chapter, a considerable body of knowledge has been amassed on the physical properties and electrochemistry of sulfur and the sodium polysulfides. We can use this together with experimental observations on sulfur electrodes in cells to build up a picture of the processes occurring in the sulfur electrode in a cell.

A typical sulfur electrode in a freshly activated cell with a beta alumina electrolyte will consist of a resilient matrix of carbon fibres slightly compressed between the beta alumina surface and the current collector surface. The matrix will be filled with molten sulfur – a highly viscous and non-conducting liquid – and throughout the electrode will be distributed, fairly uniformly, gas bubbles or voids depending upon whether the electrode has been evacuated. The EMF of the cell in this state can be any value between 2.076 and 2.8 V. The higher value presumably reflects the extremely low concentration of sodium sulfide in the virgin sulfur electrode. At 350°C a cell EMF of 2.8 V is equivalent to a sulfide ion activity of 10^{-6}. Discharge of the cell, even by a few coulombs, results in the EMF falling to 2.076 V, and this value is never again exceeded even after prolonged recharge. As will be discussed later, tetracyanoethylene has been added to the sulfur electrode to improve recharge, and one effect of this is to raise the cell EMF in the two phase region to ~2.10 V; no satisfactory explanation of this effect has been advanced.

(a) *Discharge*

In a virgin cell the electrochemical reaction must be limited to those sites where there is a three phase boundary – solid electrolyte, sulfur, carbon fibre. The total surface area of such sites must be less than the geometric area of the solid electrolyte, yet within a few milliseconds the cell potential reaches the value which it will retain throughout the first 60% of the discharge, i.e. throughout the two phase region. This is true for current densities in excess of 100 mA cm^{-2} based on the geometric area of the ceramic (South and Sudworth, 1973).

The electrochemical process occurring during the first few seconds of discharge is

$$5S + 2\varepsilon \rightarrow S_5^{2-} \quad (5.6)$$

The solubility of Na_2S_5 in sulfur is very low, and after a very short time sodium pentasulfide will appear as a second liquid phase. As noted earlier in this chapter, Suski and Kolacz (1980) have shown that sulfur saturated with sodium pentasulfide has some electronic conductivity: it is possible that, once the sulfur is saturated, formation of sodium pentasulfide will occur generally over the solid electrolyte surface.

At this stage in the discharge two competitive electrode reactions can occur, reaction 5.6 and the following reaction:

$$xS_5^{2-} + (10-2x)\varepsilon \rightarrow 5S_x^{2-} \qquad 5 < x > 1 \tag{5.7}$$

Reaction 5.6 can occur at two sites – initially at the three phase boundary described earlier, then at the three phase boundary sulfur/carbon/sodium pentasulfide. (This assumes that the electronic conductivity of sulfur saturated with Na_2S_5 is not important.) Because the reactant (sulfur) and the product (sodium sulfide) are immiscible and no other phases are present, no concentration gradients can exist and reaction 5.6 cannot be diffusion limited. This is borne out by current interruption experiments (South, 1973) which in the early stages of discharge result in potential/time curves displaying little evidence of concentration polarization (Fig. 5.9).

Reaction 5.7 has been shown to be diffusion limited and thus, in the early stages of discharge, reaction 5.6 will be favoured. As discharge proceeds, the solid electrolyte surface will become covered with sodium pentasulfide which

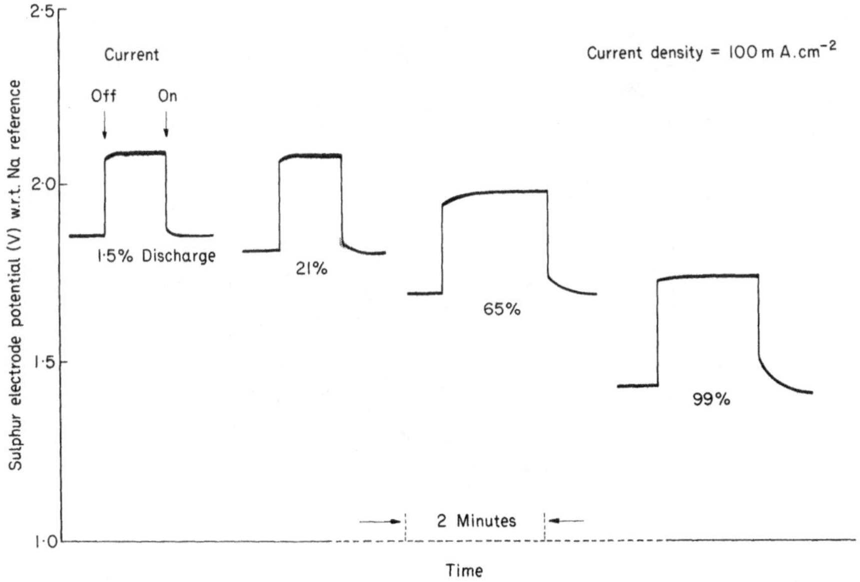

Fig. 5.9 Potential/time curves from current interruption experiments.

will extend progressively along the carbon fibres (or the carbon coated surface of the foil in glass electrolyte cells), fed by reaction 5.6 at the advancing three phase boundary. As this boundary moves progressively further into the electrode, the potential drop due to the ohmic resistance of the melt will increase: this will tend to favour reaction 5.7, which can occur close to the solid electrolyte surface. In glass electrolyte cells where electrode spacing is measured in microns, reaction 5.7 is not likely to be important in the two phase region. The lower sulfides formed in reaction 5.7 will diffuse to the phase boundary and react chemically with elemental sulfur to reform sodium pentasulfide. At moderate current densities, and temperatures around 350 °C, the rate of this diffusion and reaction is sufficient to maintain the polysulfide composition close to Na_2S_5. This is evident from Fig. 5.10, which was obtained by chemical analysis of the sulfur electrodes from cells in various states of discharge (South and Sudworth, 1975). Fig. 5.10 shows that up to 10% depth of discharge the sodium pentasulfide formed remains in a zone close to the solid electrolyte and also that much unreacted sulfur is present in this zone. For the remainder of the discharge until the melt becomes wholly single phase, the distribution of reaction products is random. In the single phase region the reaction is evenly distributed throughout the electrode and the composition of the polysulfides never becomes more sodium rich than Na_2S_3, and thus no solid sulfides are formed.

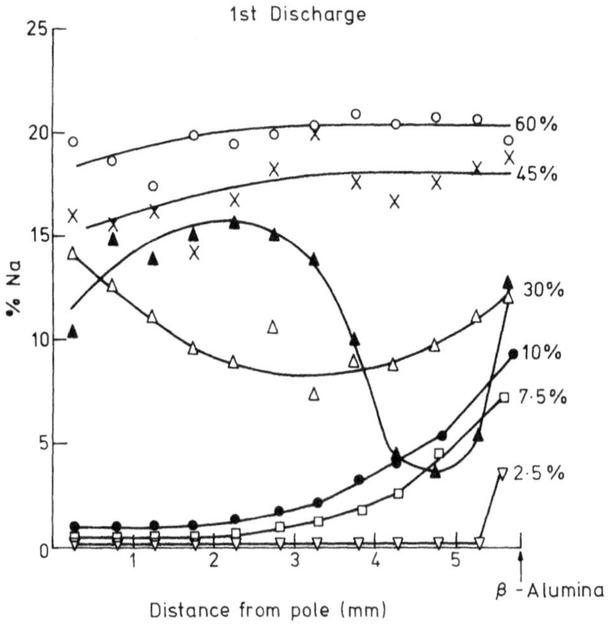

Fig. 5.10 Variation of polysulfide composition in sulfur electrodes from flat plate cells (first discharge).

Experiments by Minck (1975) gave somewhat different results. Using an alumina tube packed with porous graphite felt impregnated with sulfur and having a disc of beta alumina sealed to one end, he obtained the results shown in Fig. 5.11. The main difference between these and South's results (South and Sudworth, 1975) is the high ratio of sodium to sulfur in the region adjacent to the ceramic electrolyte. At 100 mA cm^{-2}, the same current density used by South, the composition is below the liquidus, i.e. solid Na_2S_2 is present. The temperature of the cells used by South and Minck, 350 and 300 °C respectively, could well be responsible for the different results; certainly the viscosity of sulfur is considerably greater at 300 than at 350 °C, which could lead to a reduced mobility of the sodium sulfides formed.

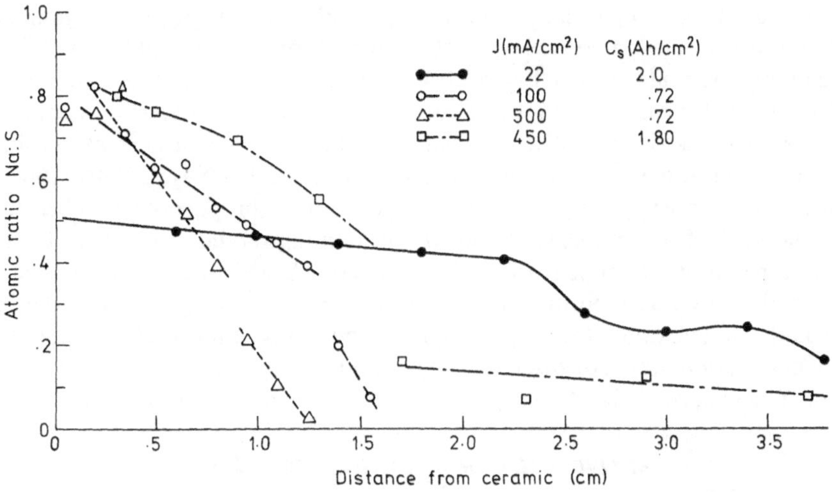

Fig. 5.11 Variation of polysulfide composition in a porous electrode. From Minck (1975).

Even if solid sodium disulfide is deposited on the carbon fibres, as in the case of Minck's experiment, there is no great increase in electrode polarization. This is in accord with the results obtained on smooth electrodes where a limiting current density of about 25 mA cm^{-2} is observed. The carbon felt used in the sulfur electrode provides a large surface area and thus will support high currents even when solid sulfides are present. A consequence of this is that a plain carbon felt sulfur electrode can be discharged almost to the composition Na_2S_2. The open circuit voltage remains constant at 1.74 V in the region where solid sodium sulfide is in equilibrium with the melt, resulting in a second two phase plateau region in the discharge curve.

In the glass electrolyte cell, discharge will occur first where the aluminium foil is touching the glass capillary, resulting in the formation of sodium pentasulfide. Eventually the space between the aluminium foil and the glass electrolyte will become completely filled with sodium pentasulfide melt. Because the interelectrode gap is very narrow and the current density is very low (2 mA cm^{-2}), there will be very little polarization associated with the sulfur electrode in this type of cell.

(b) *Charge*

Under normal circumstances the sulfur electrode at the beginning of charge will consist of a sodium trisulfide melt in a carbon felt matrix. The volume of the melt is at a maximum and the felt will be fully flooded with melt. From the work on smooth carbon electrodes, we know that in this melt very high current densities are possible before there is any indication of the current becoming limiting. This is reflected in the behaviour of sulfur electrodes in cells. South and Storey (1973) showed that very high charge rates were possible in the single phase region, but that the electrode passivated when the composition approached the two phase (sulfur/polysulfide) region. Fig. 5.12 shows the charge curves for a plain carbon felt sulfur electrode at a range of current densities. The interesting feature of these curves is the complete passivation of the electrode at current densities above 100 mA cm^{-2}, when the melt composition reaches Na_2S_5. Passivation of the carbon fibres by a layer of sulfur cannot explain this phenomenon because, as the work of South *et al.* (1972) shows, the limiting anodic current density on a smooth carbon electrode is approximately 100 mA cm^{-2}, i.e. four times greater than the cathodic limiting current density. The high surface area of the felt would result in quite a low current density, and this is confirmed by the fact that no passivation occurs on discharge when solid sodium disulfide is present. Thus some mechanism other than deposition of a sulfur film on the carbon must be

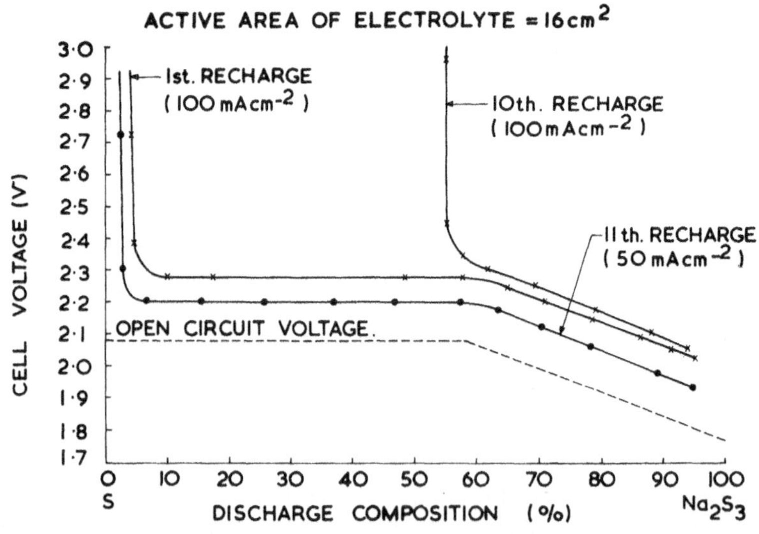

Fig. 5.12 Charge curves for plain carbon felt sulfur electrodes.

responsible for the passivation observed when sodium sulfur cells are charged into the two phase region.

The most likely explanation is that sulfur generated in the electrode adjacent to the solid electrolyte forms a coherent film on the surface of that electrolyte, thus blocking the transfer of sodium ions. This is supported by the results of the chemical analysis of an electrode removed from a cell which had polarized on charge. These are shown in Fig. 5.13, and it can be seen that the ratio of sodium to sulfur is very low adjacent to the solid electrolyte.

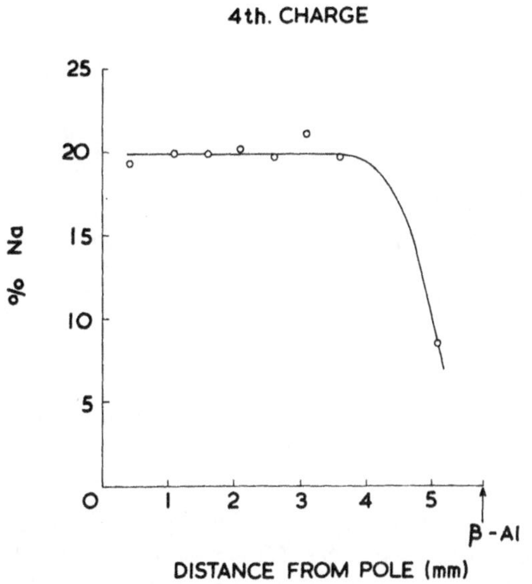

Fig. 5.13 Variation of polysulfide composition in a sulfur electrode from a flat plate cell after polarization on charge.

It would appear then that in an electrode which can be successfully charged to a composition approaching pure sulfur, generation of sulfur must occur remote from the solid electrolyte. This led to South and Sudworth (1975) suggesting that during charging in the two phase region a zone of sulfur advanced across the electrode, so that passivation of the solid electrolyte by sulfur coincided with complete conversion of sodium sulfide to sulfur. It has now been shown that this is not the case, at least in plain carbon felt electrodes. Thus South and Sudworth (1975) showed by chemical analysis of sulfur electrodes from plate cells that the distribution of reaction within the electrode is completely random (Fig. 5.14). This has been confirmed by Kleinschmager *et al.* (1977) using reference electrodes in the sulfur electrode.

It can be seen that the main problem with the sulfur electrode is the achievement of a high charge acceptance at a high charge rate. Ideally, this

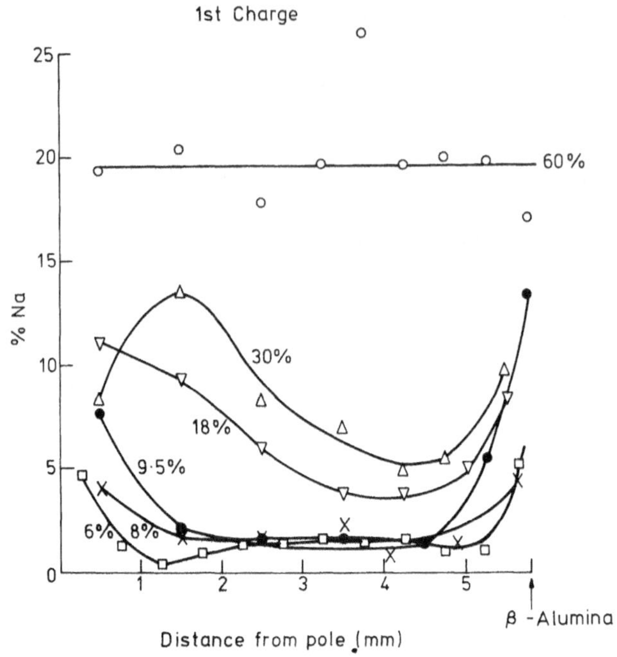

Fig. 5.14 Variation of polysulfide composition in sulfur electrodes from flat plate cells (first charge).

should be achieved without adversely affecting the discharge characteristics or increasing the electrode resistance. Cells with glass electrolyte charge back to almost pure sulfur, so it would appear that the glass fibres do not become completely coated with sulfur until almost all the sodium polysulfide has been oxidized.

5.4 THEORIES OF SULFUR ELECTRODE OPERATION IN CELLS WITH BETA ALUMINA ELECTROLYTE

Several quite different theories of sulfur electrode operation have been developed by different groups in attempts to explain the behaviour of the sulfur electrode, and these have led to electrode designs which allow high charge acceptance to be achieved at high charge rates. Two theories – potential distribution and capillary action – have received considerable attention and will be discussed in detail. Other models have been proposed, however, and these will be discussed briefly first.

5.4.1 Modification of the properties of sulfur

It is proposed that premature polarization of the electrode on charge is due to formation of a passivating layer of sulfur on the solid electrolyte. Thus removal of liquid sulfur from that region, by whatever process, will tend to prevent premature polarization, and this removal will be assisted if the viscosity of sulfur can be reduced. Simply increasing the temperature achieves this effect, and it is well known that the charge acceptance of a given sulfur electrode increases with temperature (see Chapter 10). The same effect can be achieved by adding small quantities of substances which break the sulfur chains (Doi, 1965). The materials used must be stable at 350 °C and have a low vapour pressure. One material which satisfies these criteria is selenium, and Bauer *et al.* (1976/1977) obtained some improvement in recharge characteristics by the addition of 1 mol%. The effect, however, is not very pronounced.

The low conductivity of sulfur is the cause of the problem experienced on recharge, and attempts have been made to increase this by the addition of substances which can form charge transfer complexes with sulfur. Weddigen (1977, 1978) showed that the addition of tetracyanoethylene to the sulfur electrode led to a significant increase in charge acceptance at high charge rates. He explained these results by postulating the formation of a charge transfer complex which allowed the sulfur species to be transported by electromigration. Using a half-cell with a reference electrode, he showed that the anodic limiting current normally observed in Na_2S_5 melts was not observed in the presence of tetracyanoethylene. The optimum amount of additive is 0.8 mol%, and Demott (1980) has confirmed Weddigen's results and obtained high charge acceptance at current densities up to 200 mA cm^{-2} as shown in Fig. 5.15. He also reported

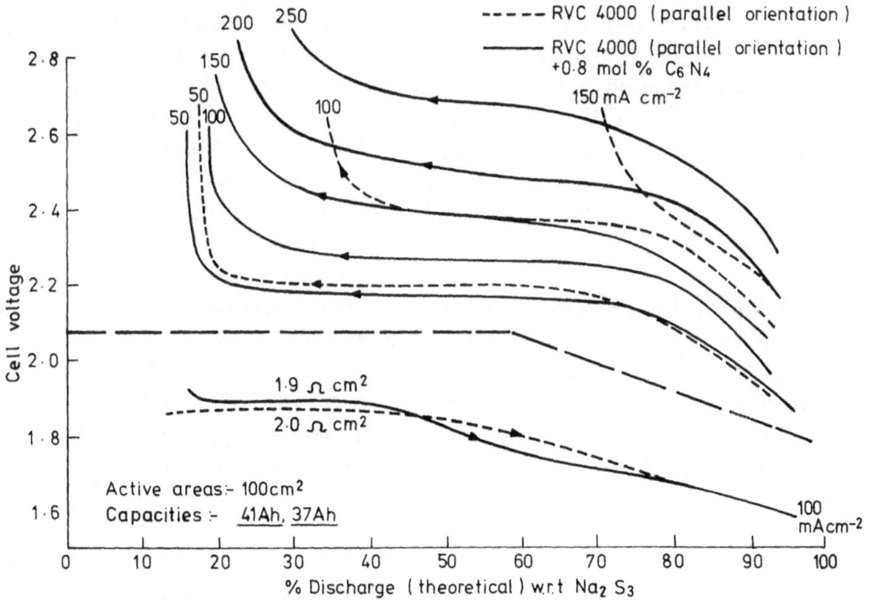

Fig. 5.15 Charge/discharge curves for a cell having 0.8 mol% tetracyanoethylene in the sulfur electrode.

that the open circuit voltage of the cell in the two phase region is 2.10 V, not the normally observed value of 2.076 V.

There are two aspects of the behaviour of electrodes containing tetracyanoethylene which are not fully understood – why elimination of the anodic limiting current on carbon should prevent premature polarization if this is caused by passivation of the solid electrolyte, and the reason for the high open circuit voltage. Two other features of these electrodes should also be mentioned. Demott (1980) has found that if aluminium is used as the substrate for the current collector, severe corrosion is found in the region above the melt. This is in contrast to the passivity normally exhibited by aluminium in polysulfide melts. Second, he has found that when tetracyanoethylene is added to an electrode which incorporates polysulfide wicks (see Section 5.4.4) the recharge performance is degraded, indicating that the material is affecting the polysulfide wetting behaviour of the electrode materials.

Addition of tetracyanoethylene can improve the recharge performance of the sulfur electrode without affecting the discharge characteristics, and it can be used with low resistance carbon felts. Although it appears that it cannot be used in cells with aluminium current collectors, it is certainly worthy of further investigation.

5.4.2 Mass transport by diffusion and convection

Minck (1976) attributed low charge acceptance to a low rate of removal of sulfur from the vicinity of the solid electrolyte, and he proposed a number of electrode designs which utilized felt-free channels to improve diffusion and convection. Some improvement was observed, particularly with thick electrodes. In some of these designs the solid electrolyte surface was only partially covered with felt, and Minck attributed the improved charge acceptance to maintenance of an ionic path to the uncovered solid electrolyte. He also constructed a cell in which the whole of the solid electrolyte surface was left uncovered by felt, a 1 mm gap being maintained between the felt and the solid electrolyte.

Although some improvement was observed with these electrode designs, there was a strong tendency for sulfur and sodium pentasulfide to separate into distinct layers, thus reducing the operating area of the solid electrolyte. Similar results were obtained by Le Méhauté and Bordet (1977).

It will be shown later that electrodes very similar in design to these described, but having a polysulfide wick in place of the felt-free channels, have excellent recharge characteristics. Thus, although electrodes having felt-free channels are of little practical interest, they do indicate the necessary conditions for obtaining good recharge performance.

5.4.3 The potential distribution model

Gibson (1974) considered the case of a flat cylindrical sulfur electrode consisting of carbon felt saturated with sodium polysulfide melt completely filling the space between an inert current collector, or pole, and the solid electrolyte. In any electrode, the electronic current injected at the pole becomes, via electrochemical reaction at the carbon/melt interface, wholly ionic when it enters the solid electrolyte. If the melt composition is assumed to be homogeneous throughout the electrode, the distribution of electrochemical reaction and the potential drop across the electrode for a given total current flow I_0 can be calculated from the values of felt resistance r, melt resistance R and the electrochemical reaction resistance ρ, where ρ (Ω cm) is the slope of the current/voltage characteristics of the redox process at the carbon felt/melt interface.

Qualitatively, a high $R:r$ ratio will result in the reaction being biased towards the solid electrolyte, a low ratio in it being biased toward the pole. This is shown quantitatively in Figs 5.16 and 5.17, which were computed using the relationship derived by Gibson. Fig. 5.16, in which the melt resistance is ten times the felt resistance, shows that the value $(\phi_c - \phi_m)$ (where ϕ_c and ϕ_m are the electrostatic inner potentials in the carbon and melt respectively) is at a maximum adjacent to the solid electrolyte, resulting in maximum reaction rate at this point. Fig. 5.17 shows that the opposite is the case when the melt resistance is one-tenth of the felt resistance.

According to this model, passivation of the solid electrolyte by elemental sulfur during charge will be less likely if a highly resistive carbon felt is used as the electrode matrix. A secondary effect of this will be to increase the total electrode resistance. This is shown in Fig. 5.18 (Gibson, 1974), where the electrode resistance is increased by more than a factor of two when the felt resistance is increased to 10 Ω cm from 1 Ω cm. The model requires that for good charge acceptance a high sulfur electrode resistance must be accepted. Gibson also predicted that the grossly heterogeneous nature of such electrodes would make them unsuitable for high charge and discharge rates and that this might also result in progressive attenuation of capacity.

In practice the performance of sulfur electrodes is better than that predicted by Gibson, and Kao and Wayner (1976) have shown how this can be explained by considering the effect of non-uniform concentration in the polysulfide melt. Different reaction rates will lead to different polysulfide compositions and thus to variations in reversible electrode potential throughout the electrode. These potential gradients will have the effect of reducing the reaction rate in the region where the greatest change in composition has occurred, and thus of shifting the reaction to the region where the reactants are most abundant. This will result in higher cell voltages at the same depth of discharge.

Although the non-uniform concentration model predicts that the effects of gross heterogeneity described by Gibson will not be so evident on discharge, it cannot affect the situation on charge when the composition exceeds Na_2S_5. Variation of melt composition in the two phase region is not possible, and both models should give the same result.

Both Gibson (1974) and Kao and Wayner (1976) assumed that the electrode is adequately described as the superposition of two continua – the electrolyte and the graphite matrix. In practice, discontinuities will occur owing to the

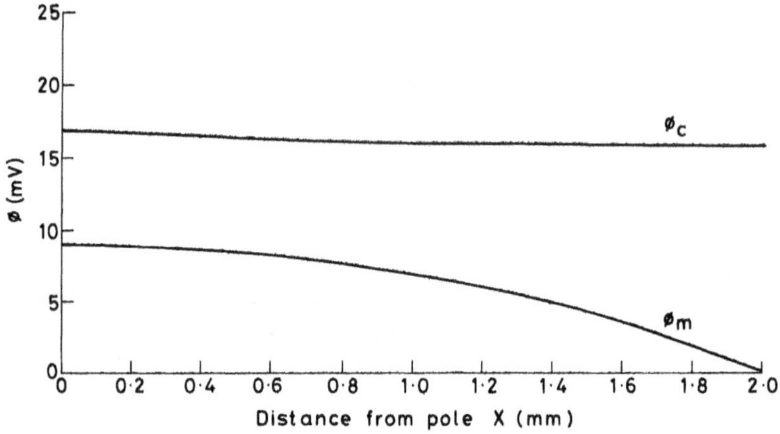

Fig. 5.16 Computed distribution of potential within the sulfur electrode. Felt resistance $r = 0.1\ \Omega\ \mathrm{cm}^{-1}$, melt resistance $R = 1.0\ \Omega\ \mathrm{cm}^{-1}$, $\rho = 0.02\ \Omega$ cm, $I_0 = 0.1$ A, sulfur electrode thickness 2.0 mm. After Gibson (1974).

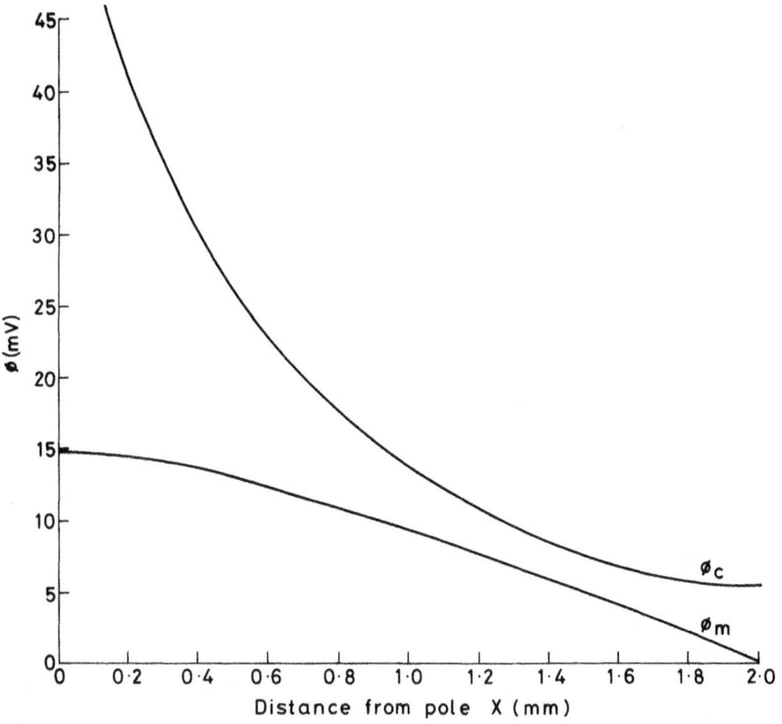

Fig. 5.17 Computed distribution of potential within the sulfur electrode. Felt resistance $r = 10\ \Omega\ \mathrm{cm}^{-1}$, melt resistance $R = 1.0\ \Omega\ \mathrm{cm}^{-1}$, electrode thickness 2.0 mm. After Gibson (1974).

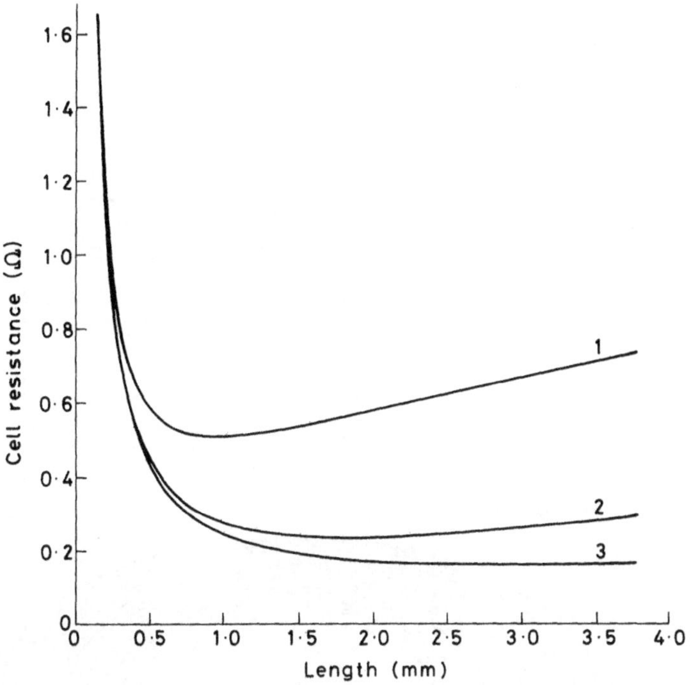

Fig. 5.18 Computed electrode resistance as a function of distance from pole. Melt resistance $R = 1.0\ \Omega\ \text{cm}^{-1}$. (1) Felt resistance $r = 10\ \Omega\ \text{cm}^{-1}$ (2) $r = 1.0\ \Omega\ \text{cm}^{-1}$ (3) $r = 0.1\ \Omega\ \text{cm}^{-1}$. After Gibson (1974).

presence of voids and insulating sulfur agglomerates. Jacquelin and Pompon (1978) have developed a model which takes these effects into account. The variation in the free volume in the form of small voids and in the amount of elemental sulfur, as the electrode is charged and discharged, is assumed to result in a random distribution of voids and sulfur throughout the electrode. The electrode is represented by a succession of thin annular neighbouring slices. The electrical network is shown in Fig. 5.19. The four components of the network are: r_n, the electronic resistance of the carbon felt; R_n, the ionic resistance of the melt (after correcting for the tortuosity due to the felt matrix, and for the presence of voids and sulfur); P, the non-linear polarization parameter, which was determined experimentally at a number of melt compositions using a carbon microelectrode; and E, the equilibrium potential of each slice.

The effect of felt resistivity on the distribution of reaction is shown in Fig. 5.20. In Fig. 5.20(a) (low resistivity felt) it can be seen that the electrode zone adjacent to the solid electrolyte gets progressively richer in sulfur relative to the zone adjacent to the container, as charging proceeds. This should be compared with Fig. 5.16, which was computed from Gibson's model. Similarly, Fig. 5.20(c) should be compared with Fig. 5.17.

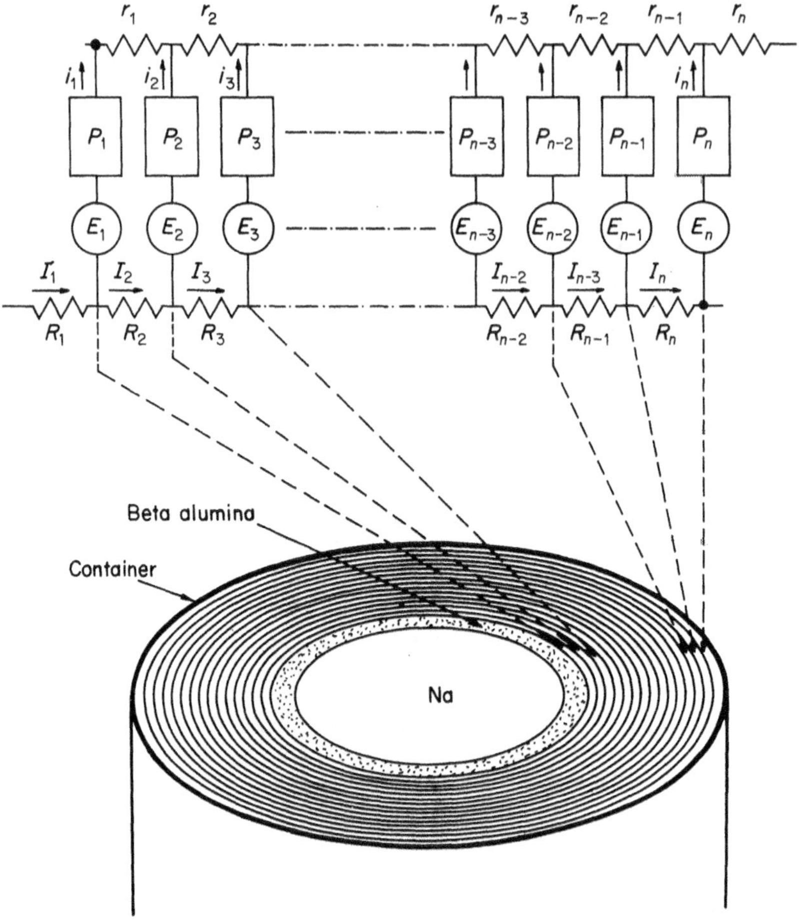

Fig. 5.19 Electrical network representing the sulfur electrode. Reproduced from Jacquelin and Pompon (1978) with permission of the authors and publishers.

Not surprisingly, both models predict similar results, but the model of Jacquelin and Pompon (1978) provides more information on changes in composition which occur during charge and discharge. This model predicts that, under certain conditions, discharge will proceed on a reaction front starting from the solid electrolyte, and this front will progress through the electrode until at composition Na_2S_5 it reaches the pole. Thereafter, discharge occurs uniformly throughout the electrode until the two phase (liquid/solid) region is reached. At this point the electrode potential becomes independent of the overall electrode composition and this mechanism for distributing reaction rate no longer operates. On charge, a zone of reaction again forms and moves through the electrode. Whether this reaction front starts at the solid electrolyte (resulting in premature polarization of the electrode) or at the pole (allowing charging to continue until the reaction front reaches the solid electrolyte) will depend to a large extent on the resistivity of the felt. This is shown in Fig. 5.21, where (a), (b), (c) and (d) correspond with the parts of Fig. 5.20. It can be seen that highly resistive felt is undesirable and that an optimum value of felt resistivity exists. Jacquelin and Pompon also point out that changing the electrochemical activity of the carbon felt has an effect: a low activity tends to dampen the front phenomena resulting in higher charge acceptance, but again this is at the expense of a higher resistance electrode.

Although the models of Gibson and Jacquelin and Pompon show that charge acceptance can be improved, this results in a high electrode resistance which may be unacceptable for many applications of the battery. Breiter and Dunn (1979a) have proposed an electrode design, based on an electrode model similar to that of Gibson, which overcomes this disadvantage. They refer to it as a gradated resistance electrode, and the essential elements are summarized in the following.

The model is based on the same assumptions as used by Gibson, i.e. linear electrode geometry, steady state conditions, and proportionality of reaction rate to the difference between the inner potentials in the carbon felt and melt. Its validity, therefore, is restricted to the one phase region, although it may be expected that it would provide some information on the likelihood of sulfur deposition adjacent to the solid electrolyte.

The essential feature of the gradated resistance electrode is the division of the electrode into two regions – a region adjacent to the solid electrolyte containing a high resistivity carbon mat, and the remainder of the electrode containing a low resistivity carbon mat. Breiter and Dunn used a slightly different mathematical procedure than Gibson, but when they compared their solution with the numerical results obtained with Gibson's model (but with the appropriate boundary conditions for the single mat configuration), they obtained perfect numerical agreement.

The effect of the dual mat configuration on reaction rate distribution is shown in Fig. 5.22, where R_m is the melt resistance, r_1 and r_2 the carbon mat resistance in regions 1 and 2 respectively, and r_c the carbon mat resistance for the single mat case. The reaction rate does not, as appears from Fig. 5.22, become zero; this is merely an artefact of the calculation. The relative slopes of the curves are the important factors, not the absolute values. As might be expected, the presence of the very high resistivity mat in region 1 results in a very much lower electrochemical reaction rate adjacent to the solid electrolyte. This in turn would be expected to reduce the likelihood of sulfur passivating the solid electrolyte surface.

Results from cells containing gradated resistance electrodes confirm that recharge to almost pure sulfur is possible under conditions which, with a single mat electrode, would lead to polarization on entry into the two phase region (Breiter and Dunn, 1979a).

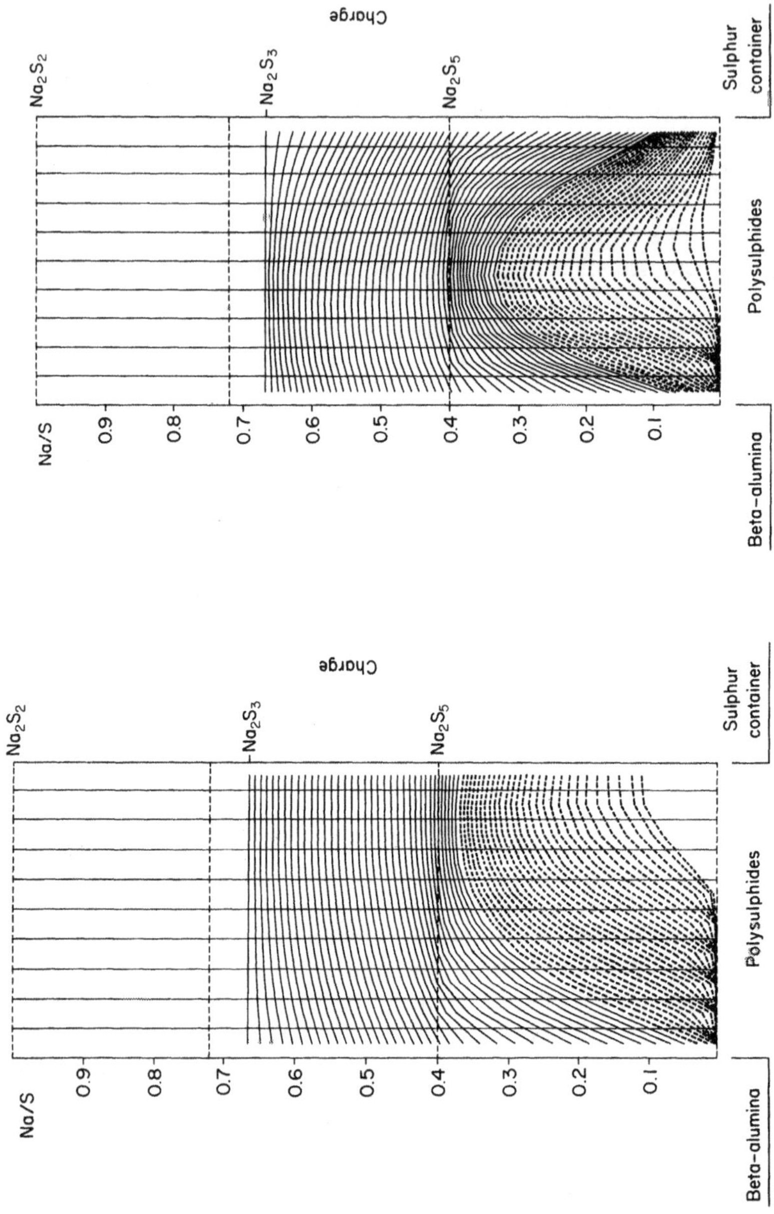
Na/S
0.9
0.8
0.7
0.6
0.5
0.4
0.3
0.2
0.1
Na_2S_2
Na_2S_3
Na_2S_5
Charge
Beta-alumina
Polysulphides
Sulphur container
(a)
(b)

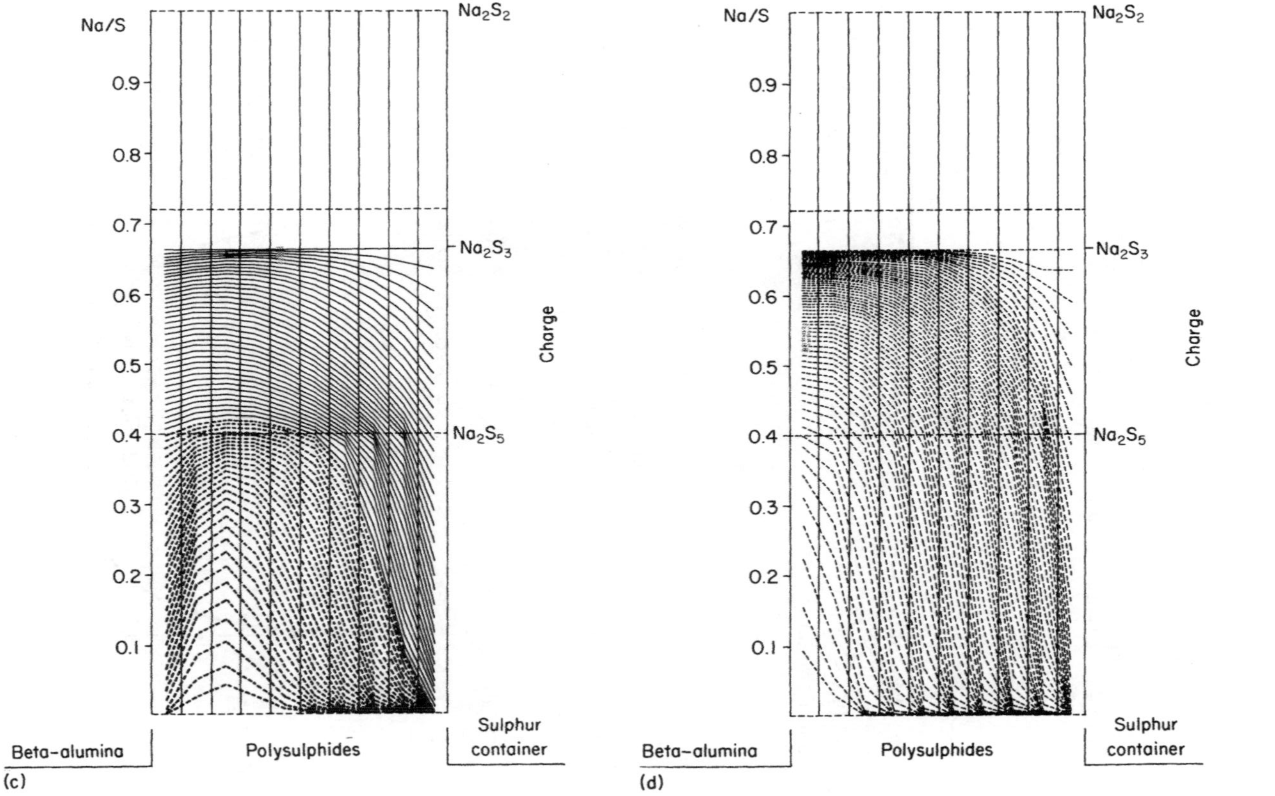

Fig. 5.20 Influence of felt resistivity on reaction rate distribution during charge. (a) Low resistivity (b) medium resistivity (c) high resistivity (d) very high resistivity. After Jacquelin and Pompon (1978).

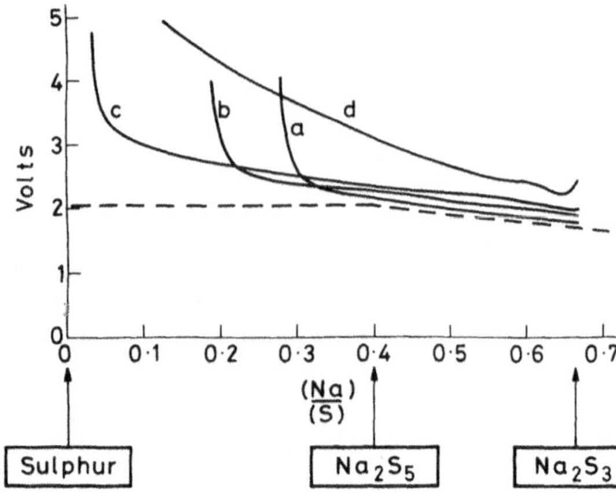

Fig. 5.21 Influence of felt resistivity on charge voltage. (a)–(d) as in Fig. 5.20. After Jacquelin and Pompon (1978).

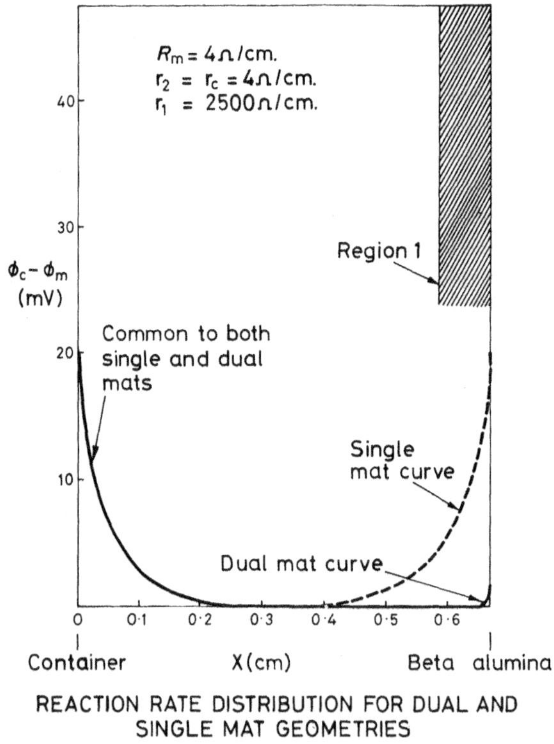

Fig. 5.22 Reaction rate distribution for dual and single mat geometries.

As noted earlier, the model applies only to the single phase region. A simple model has been developed by Breiter and Dunn (1979b) which describes operation of the electrode in the two phase region. They estimate the polarization resistance at the carbon/melt interface to be at least an order of magnitude less than the resistance of the felt and melt, and conclude that it can be neglected. Thus in this model the cell resistance is determined by the ohmic resistance of the felt and melt.

A number of assumptions are made concerning the distribution of reactants and products. The major assumption is that in the two phase region a reaction front progresses through the electrode from the solid electrolyte to the pole on discharge, the reverse occurring on charge. This is in accord with the calculations of Jacquelin and Pompon but not with the experimental results of South and Kleinschmager. The other assumptions are: that the distribution of reactants and products is homogeneous throughout the single phase region (this is supported by the experimental evidence reported by Piercy and Sudworth, 1977); that the highly resistive carbon fibre layer remains filled with Na_2S_5 during charge; and that the density of the sodium polysulfide remains constant throughout the single phase region but that the height of the melt is a function of the composition. The last two assumptions are also supported by experimental evidence, and the first model developed by Breiter and Dunn (1979a) for the gradated resistance electrode supports the assumption concerning the melt composition in the high resistance felt region.

Using this simpler model Breiter and Dunn were able to calculate cell resistances, and these are shown together with experimental results in Table 5.10 for a 6 mm thick electrode made up of two layers of carbon mat – a high resistance (2500 Ω cm) mat 0.8 mm thick adjacent to the solid electrolyte, and a low resistance (4 Ω cm) mat filling the rest of the electrode. R_{cell} was determined experimentally, R_{sul} by calculation. The value of $R_{cell} - R_{sul}$, which represents the solid electrolyte and contact resistances, is expected to be constant: this is found to be the case, indicating that the calculated values of sulfur electrode resistance are reasonable.

The models described allow the electrode resistance to be calculated and also aid the design of electrodes which are capable of charging into the two phase region. Piercy and Sudworth (1977) compared the resistance of a central sulfur cell with the resistance calculated using Gibson's model amended for the tubular cell geometry. The result is shown in Fig. 5.23, and it can be seen that there is good agreement. Gibson's model does not take into account variations in the melt composition, and thus predicts that the electrode resistance will be constant during discharge. Cell resistance as a function of depth of discharge is shown in Fig. 5.24 (curve 1). The resistance used was the difference between the load voltage and the theoretical open circuit voltage divided by the current, and thus includes concentration polarization. This probably explains the increase in resistance in the single phase region where concentration polarization is greatest. The results of Breiter and Dunn (1979b) show the opposite effect: the minimum resistance was found in the single phase region.

Table 5.10 Sulfur electrode calculations; cell 330, cycle 64. Reproduced with permission from Breiter and Dunn (1979b)

Q (Ah)	R_{sul} (Ω)	R_{cell} (Ω)	$R_{cell}-R_{sul}$ (Ω)
Discharge, two phase region			
1.8	0.077	0.180	0.103
3.6	0.069	0.167	0.098
5.4	0.064	0.157	0.093
7.2	0.060	0.150	0.090
Discharge, one phase region			
8.8	0.057	0.151	0.094
11.0	0.046	0.150	0.104
Charge, one phase region			
2.8	0.046	0.150	0.104
5.0	0.057	0.151	0.094
Charge, two phase region			
6.0	0.059	0.154	0.095
7.8	0.063	0.165	0.102
9.6	0.068	0.177	0.109
11.4	0.074	0.187	0.113

This is probably explained by the method used to determine cell resistance, current interruption, which tends to measure only the ohmic component of the cell resistance. This is shown in Fig. 5.24 (lower curve) (Piercy *et al.*, 1977), from which it can be seen that using an open circuit value measured after three seconds to calculate the resistance gives results more in line with those of Breiter and Dunn (1979b). This is not surprising as the model considers only the ohmic resistance of the felt and melt, and it shows the limitations of such models.

The effect of felt resistivity on the charge acceptance of homogeneous felt electrodes has been investigated by Hargreaves and Piercy (1978) and Demott (1980). Hargreaves and Piercy determined the resistivity of carbon felt (Le Carbone RVC 4000 uncompressed thickness 1.0 cm) and carbon mat (Union Carbide Thornel VMA uncompressed thickness 2.0 cm) as a function of thickness in a simple compression jig. Measurements were made with the felts in both parallel and perpendicular orientations. The results are shown in Figs 5.25 and 5.26. The bulk density of felt in cells is usually 0.15 g cm^{-3}. For one layer of RVC 4000 this corresponds to a thickness of 0.65 cm; to achieve the same density with Thornel VMA several layers of mat have to be used. At this density the resistivity of Thornel VMA in the perpendicular orientation is four times

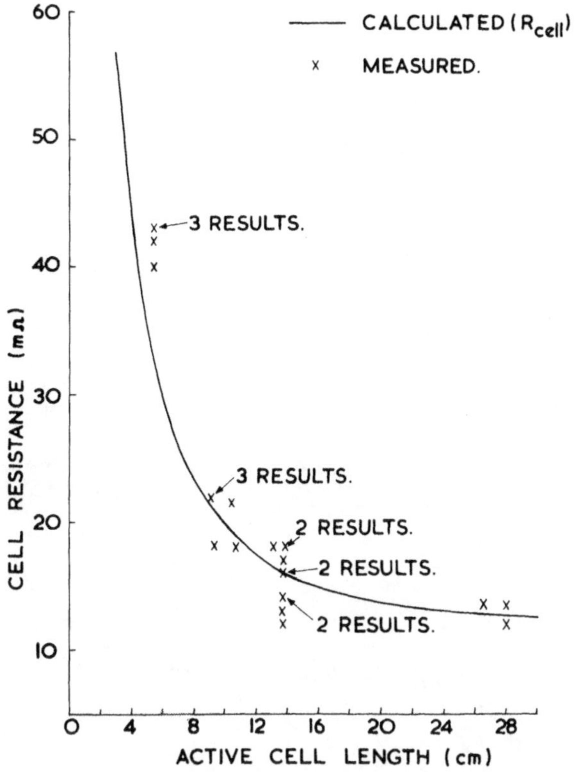

Fig. 5.23 Variation of cell resistance with electrode length – comparison of computed and experimental data.

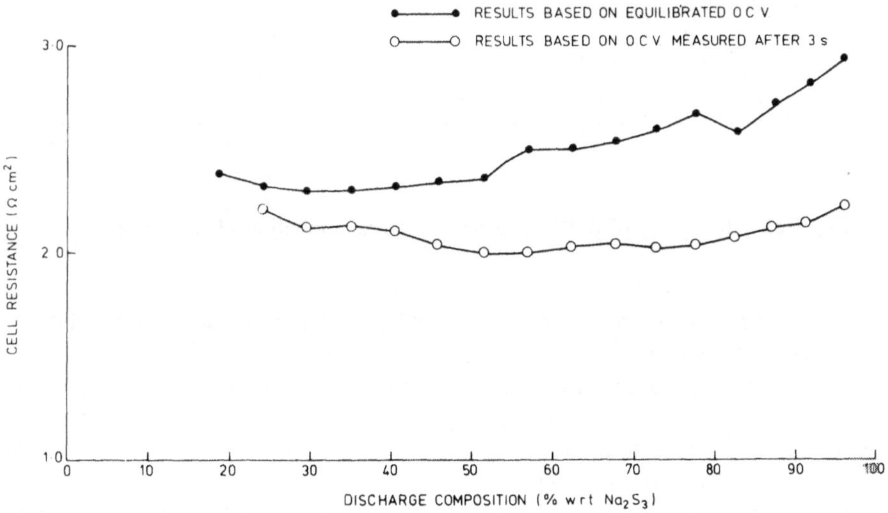

Fig. 5.24 Cell resistance as a function of state of discharge in 40 Ah central sulfur cell.

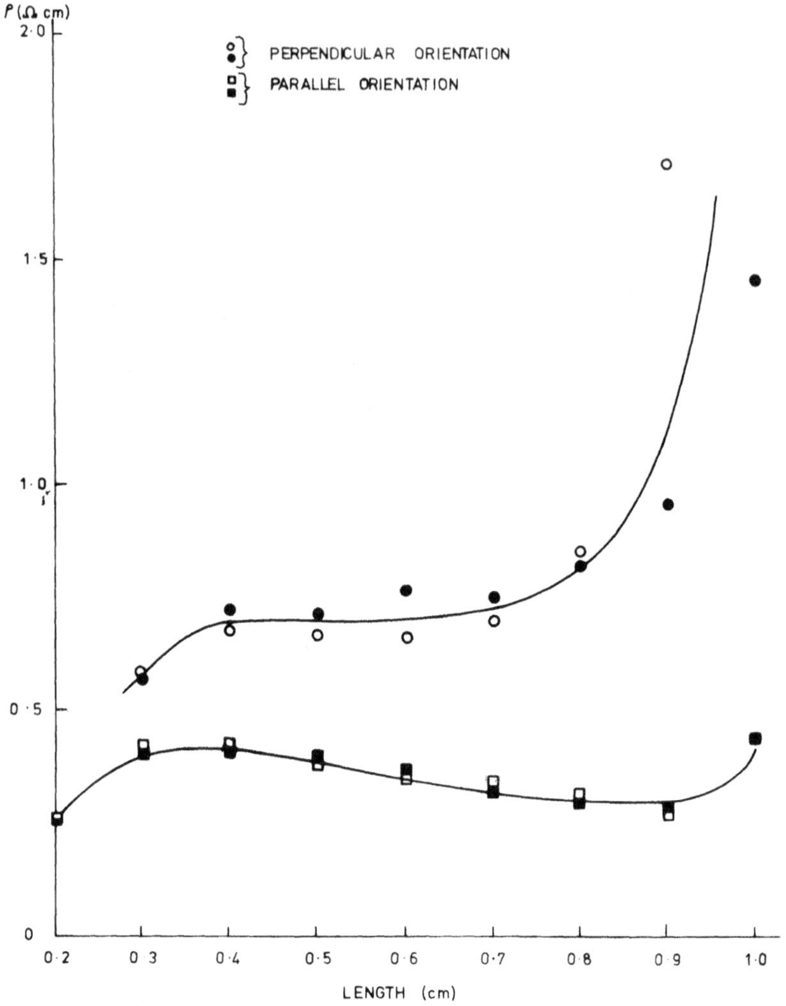

Fig. 5.25 Resistivity of RVC 4000 carbon felt as a function of thickness.

that in the parallel orientation, and for RVC 4000 the two orientations differ by a factor of over two.

The potential distribution models would predict a higher charge acceptance for cells in which the felt is in the perpendicular orientation than for cells in which the felt is in the parallel orientation. Figs 5.27 and 5.28 (Demott, 1980) show charge and discharge curves for such cells. In the case of the RVC 4000 the parallel orientation results in a slightly higher charge acceptance. The difference in resistivity does result in the expected difference in cell resistance, showing that the out-of-cell resistivity measurements are qualitatively correct.

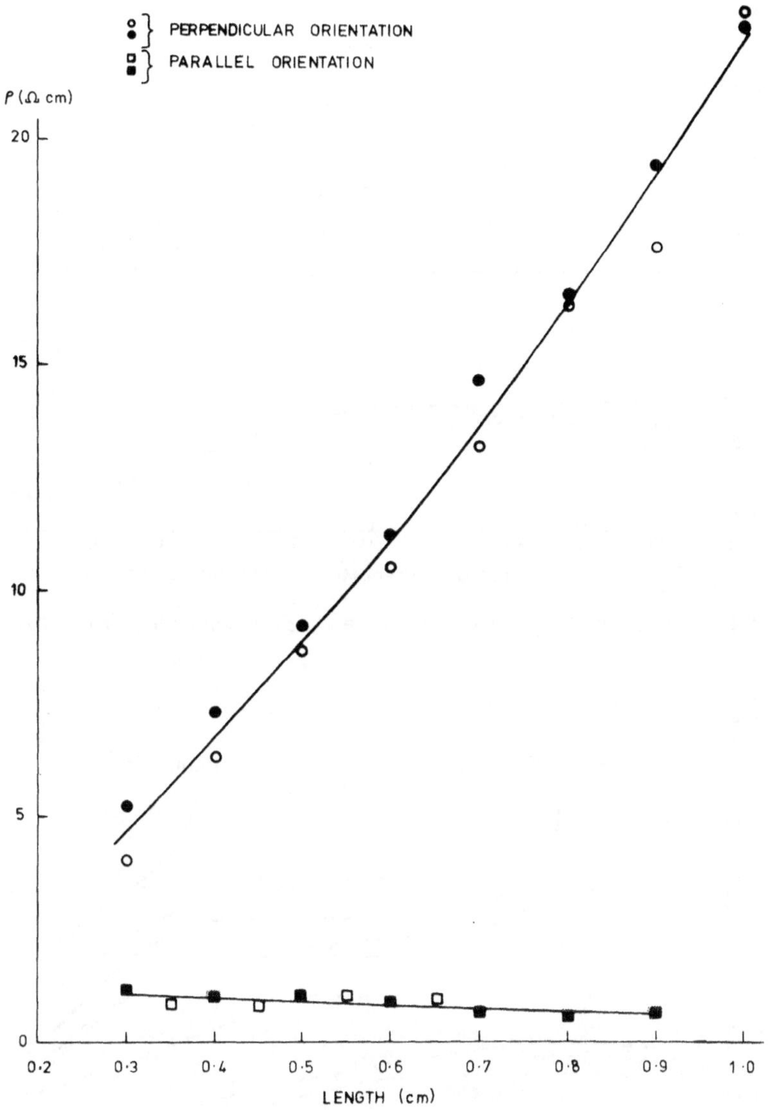

Fig. 5.26 Resistivity of Thornel VMA carbon mat as a function of thickness.

From the available experimental evidence, therefore, it would appear that although the models described are useful for predicting cell resistance, they are of little value in designing electrodes which will give a high charge acceptance. It is possible that the gradated resistance electrode model could be useful in this respect, but it will be shown later that the results of Breiter and Dunn (1979a)

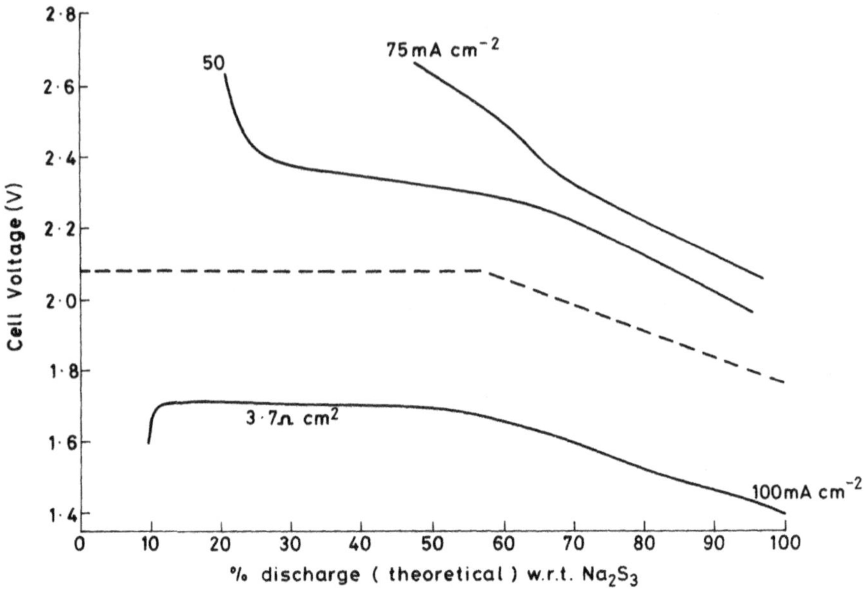

Fig. 5.27 Charge/discharge curves for cells with felt in perpendicular orientation.

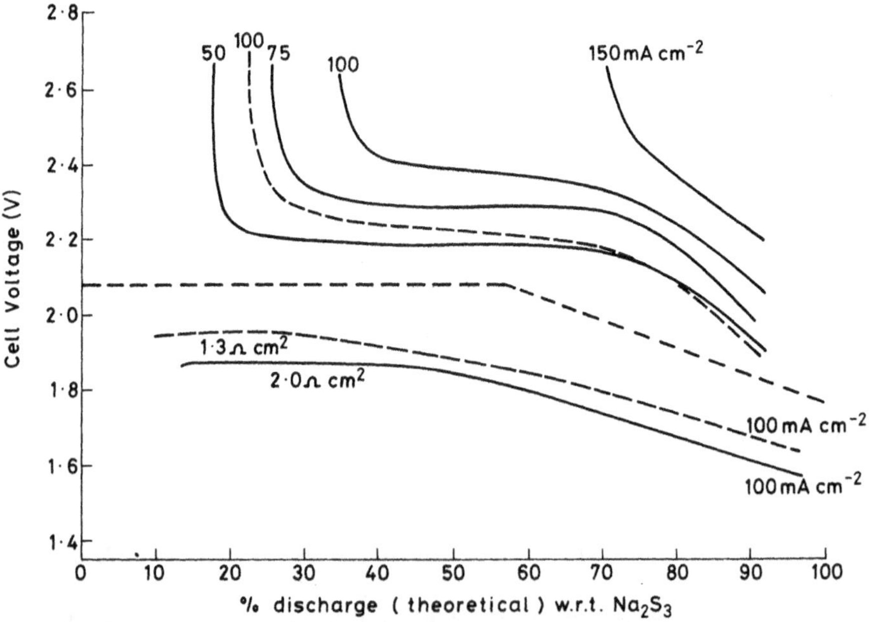

Fig. 5.28 Charge/discharge curves for cells with felt in parallel orientation. ——RVC 4000 carbon felt; --- Thornel VMA carbon felt.

can be explained in another way and that gradated resistance electrodes can show a lower charge acceptance than homogeneous electrodes.

5.4.4 Capillary action

The cause of premature polarization on charge is, as was noted earlier in this chapter, almost certainly passivation of the solid electrolyte surface by an insulating sulfur film. If all or part of this surface could be kept free of sulfur, passivation would be prevented. Since sulfur is immiscible with sodium pentasulfide, one way of achieving this would be to ensure that all or part of the surface was always protected by a sodium pentalsulfide barrier. Minck's (1976) experiments referred to earlier showed that some way of stabilizing this layer is required. A very effective way of immobilizing or retaining a liquid is to absorb it in a fibrous material which is wetted by that liquid, i.e. a wick.

The essential requirements of a polysulfide wick are as follows: it should be completely inert to sulfur and polysulfides; the material should be preferentially wetted by polysulfide in the presence of sulfur; and the effective 'capillary' radius should be small.

Ludwig (1977) has measured the contact angle of Na_2S_4 on three materials at 318 °C. The angles were: graphite 100°, vitreous carbon 100° and stainless steel 0–5°. In helium at 282 °C the contact angle of sulfur on all three materials was 20–30°. Although these experiments were not performed under completely rigorous conditions, they strongly indicate that graphite and vitreous carbon are preferentially wetted by sulfur, whereas stainless steel is preferentially wetted by polysulfide. This is further supported by the observation by Ludwig (1977) that a graphite anode in a polysulfide melt polarized after a period which depended upon the current density, whereas a stainless steel anode showed no tendency at all to polarize.

The results on carbon are in accord with those of Hall (1979) described earlier, and explain why a sulfur electrode consisting of only carbon felt immersed in sulfur/polysulfide melt is likely to polarize prematurely on charge: the carbon fibres, coated with sulfur, are in intimate contact with the solid electrolyte, and passivation of this surface is inevitable at high sulfur deposition rates. The use of a material which is preferentially wetted by polysulfides in the presence of liquid sulfur would be expected to improve charge acceptance. Metals such as stainless steel which have this property cannot be used because they are corroded during cycling. Treatment of the carbon felt with oxidizing and intercalating agents in an attempt to modify the surface properties has proved unsuccessful (Brennan, 1977c). Weiner (1976) described the addition of transition metal sulfides such as FeS to the sulfur electrode to improve charge acceptance. It was thought that these would modify the surface properties of the carbon in the required manner. Although some improvement was noted, the effect proved transitory and after a number of cycles the charge acceptance fell to a low value.

Ludwig (1976, 1978) and Weiner (1977) investigated the effect of coating the felt with alpha alumina. They found that a charge acceptance of 82% could be obtained at 100 mA cm^{-2}. The wetting properties of carbon felts coated with metal oxides have been investigated by Janz *et al.* (1980). The experimental method used was to immerse pieces of felt in sulfur, sodium polysulfide and mixtures of the two phases, determine the weight increase and, in the case of the two fluids, the relative amount of sulfur and sodium polysulfide. They used the adhesion tension ratio $(\gamma \cos \theta)$ sulfur/$(\gamma \cos \theta)$ polysulfide, calculated from the ratio of sulfur to polysulfide, to measure the relative affinities of fluids for the felt materials. For untreated felts they obtained values of 27 and 40 from measurements in single fluids and in mixtures of the two fluids respectively, thus confirming the affinity of the carbon surface for sulfur. These results also show that measurements in single fluids are a good guide to the competitive wetting situation in the two liquid phase mixture. The former measurements are experimentally much simpler.

Using these techniques, they showed that aluminium and chromium oxides deposited on the carbon fibre surface significantly reduced the adhesion tension ratio, so that with a felt containing 50 wt% alumina a value of 0.1 was obtained, i.e. the felt was preferentially wetted by sodium polysulfide. They also investigated the effect of sulfur purity and found that the wetting properties of polysulfides prepared from three grades of sulfur were closely comparable.

The results of Janz *et al.* (1980) explain the improved recharge performance observed by Ludwig (1976, 1978) and Weiner (1977) when they used carbon felt coated with alumina. The deterioration in charge acceptance with cycling could have resulted from the alpha alumina becoming separated from the carbon fibres.

Although Janz *et al.* did not report results on pure alumina fibres, these are available in the form of cloths, felts and papers and will be preferentially wetted by polysulfides. An intimate mixture of carbon and alumina fibres might be expected to operate in the same manner as a carbon fibre mixture coated with alumina. Jones and Robinson (1978) have reported that an electrode constructed of a 50:50 mixture of chopped carbon and alumina fibres gave good charge acceptance (90%) (using an unspecified regime) and retained this for an extended period of cycling. Diluting the carbon felt with a non-conducting material such as alumina will obviously result in a higher electrical resistance, and for this reason this type of electrode is of little practical interest.

If the cause of premature polarization on charge is passivation of the solid electrolyte by sulfur, then it should be possible to prevent this by interposing a thin layer of material, preferentially wetted by polysulfide, between the solid electrolyte and the bulk of the carbon felt electrode. Fally *et al.* (1973) used a 1 mm layer of insulating mineral wool (kaowool), and found that this resulted in improved charge acceptance.

Desplanches *et al.* (1977) found that both glass fibre and alpha alumina layers were effective, and that if all traces of water were removed by heating to 600 and 1000 °C respectively, attack by the polysulfide melt was prevented. Piercy investigated the effect of interposing a 1 mm layer of alpha alumina fibres (ICI Ltd Saffil low density mat) between the carbon felt and the solid electrolyte in a flat plate cell (Sudworth and Tilley, 1983). A few crystals of Na_2S_5 were added to initiate discharge. The charge curves are shown in Fig. 5.29, and it can be seen that good charge acceptance is obtained up to 200 mA cm^{-2}. Similar results were obtained with central sulfur tube cells (Demott, 1980). After a prolonged recharge, the fibre mat presumably fills with sulfur because the initial resistance on the following discharge is very high (Fig. 5.30). This effect can be prevented by not allowing the cell to charge back more than 85%, confirming that the mat becomes impregnated with sulfur.

The high initial discharge resistance is obviously a disadvantage of this type of electrode, and Piercey *et al.* (1977) investigated the effect of interposing between the solid electrolyte and the bulk felt matrix a thin layer consisting of the 50:50 mixture of chopped carbon and alumina fibres described above. Fig. 5.31 shows that the recharge characteristics, although not as good as the cell with a pure alpha alumina layer, were much better than would be expected for a plain carbon electrode. Furthermore, the initial high resistance at the

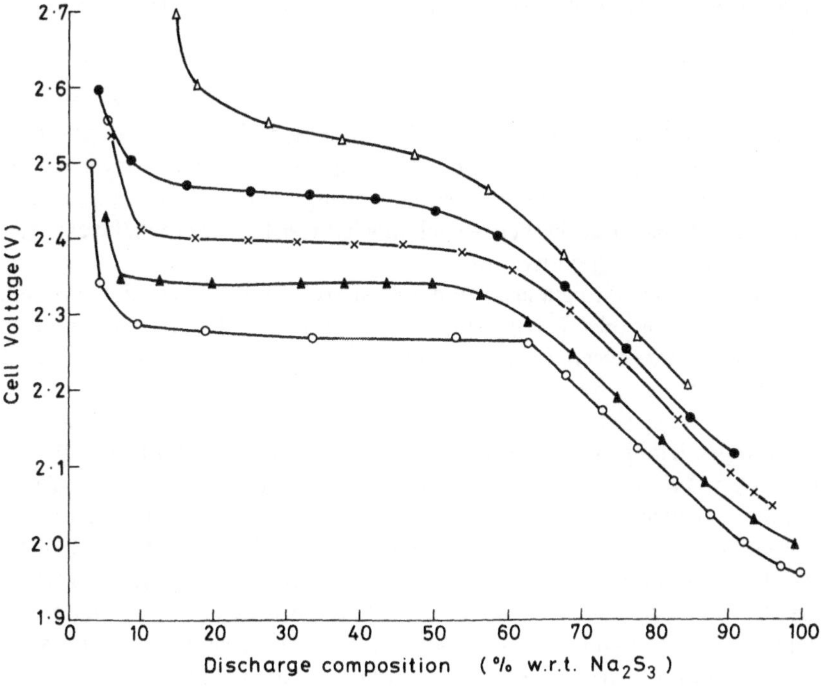

Fig. 5.29 Recharge curves for a flat plate cell containing a 2 mm thick porous insulating layer interposed between the solid electrolyte and the sulfur electrode matrix. ○ 100 mA cm^{-2}; ▲ 125 mA cm^{-2}; × 150 mA cm^{-2}; ● 175 mA cm^{-2}; △ 200 mA cm^{-2}.

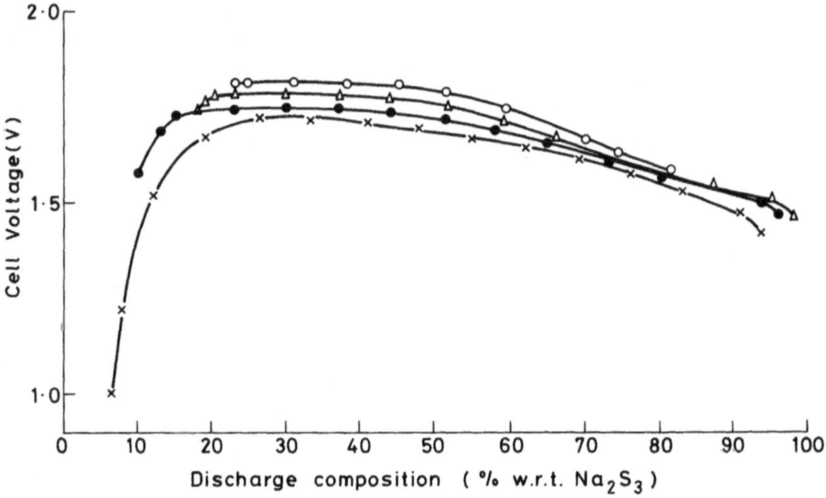

Fig. 5.30 Discharge curves for a central sulfur tube cell containing a 1 mm thick porous insulating layer interposed between the solid electrolyte and the sulfur electrode matrix. Current density 100 mA cm^{-2}. × cycle 1; • cycle 2; △ cycle 12; ○ cycle 20.

beginning of discharge was completely suppressed. The resistivity of the felt in the bulk of the electrode was 1 Ω cm and that of the mixed fibre layer ~2 Ω cm. This strongly supports the preferential wetting hypothesis, since according to the potential distribution theory such low felt resistivities would be expected to result in premature polarization.

Although interposing a layer of material preferentially wetted by polysulfide between the solid electrolyte and the bulk of the electrode results in a much improved charge acceptance, it cannot assist the transport of polysulfide from the bulk of the electrode to the surface. In this respect the electrode structure in which the polysulfide wettable material is dispersed through the electrode has advantages. The problems of high resistance (Jones and Robinson, 1978) and instability (Janz *et al.*, 1980) can be overcome by incorporating in the electrode radially oriented layers of materials such as alumina fibres which will act as polysulfide wicks. Such electrodes have been described by Minck (1978), and the charge/discharge curves for a central sulfur cell containing such an electrode are shown in Fig. 5.32. If such an electrode is fabricated from alternate rings of carbon felt or mat and alumina paper, the electrode resistance is very low owing to the favourable orientation: a value of 0.4 Ω cm^2 is obtained with a 10 mm thick electrode in a central sulfur cell, giving a total cell resistance of 1.5 Ω cm^2 (Demott, 1980). This electrode structure unfortunately results in accelerated corrosion of the current collector, which typically shows a ring pattern of discoloration corresponding to the position of the alternate rings of carbon and

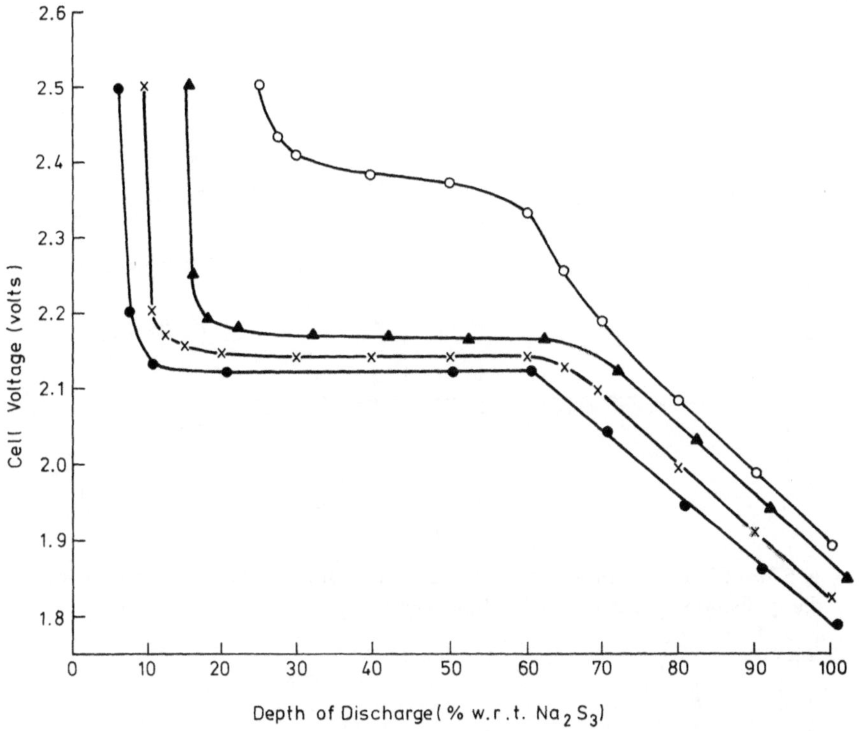

Fig. 5.31 Recharge curves for a central sulfur tube cell containing a 1 mm thick layer of 50:50 alpha alumina/carbon fibre interposed between the solid electrolyte and the sulfur electrode matrix. ● 30 mA cm^{-2}; × 50 mA cm^{-2}; ▲ 70 mA cm^{-2}, ○ 150 mA cm^{-2}.

alumina felt (Tilley and Wright, 1981). Similar discoloration occurs on the beta alumina tube.

An electrode structure developed by Wright (1982) combines the two concepts in one electrode. The electrode consists of alternate rings of carbon and alumina felt with a coating of mixed carbon and alumina fibres on the surface of the electrode adjacent to the beta alumina. This structure does not result in the pattern of discoloration of the beta alumina just described, presumably owing to the absence of current concentration effects. It also gives a better recharge performance (Fig. 5.33).

The problem of enhanced corrosion of the current collector led to the development of the carbon coated aluminium pole (Tilley and Wright, 1981) which is described in Chapter 6, and this has allowed the long term evaluation of this type of electrode. Fig. 5.34 shows how the capacity of a group of cells changed with cycling. The performance and durability of this electrode structure is compelling evidence that capillary action can be used to great effect

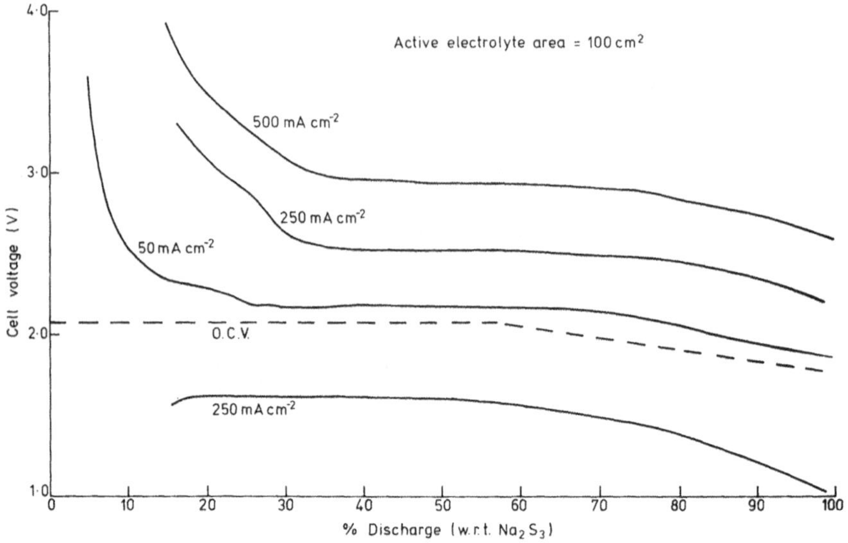

Fig. 5.32 Charge/discharge curves for a cell containing a sulfur electrode with radially disposed horizontal polysulfide wicks.

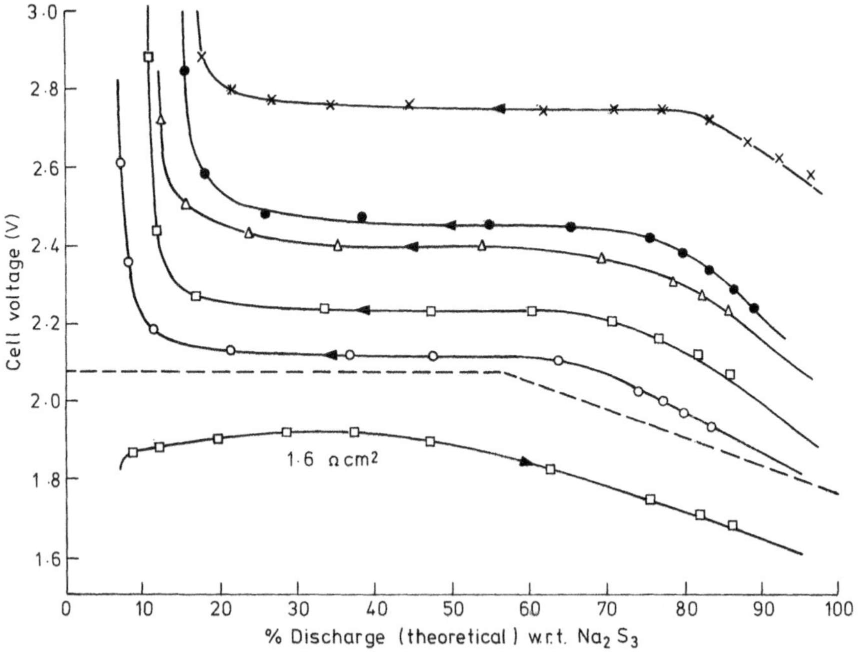

Fig. 5.33 Charge/discharge curves obtained from a cell containing a sulfur electrode with horizontal wicks and mixed fibre layer. ○ 30 mA cm^{-2}; □ 100 mA cm^{-2}; △ 200 mA cm^{-2}; ● 250 mA cm^{-2}; × 500 mA cm^{-2}.

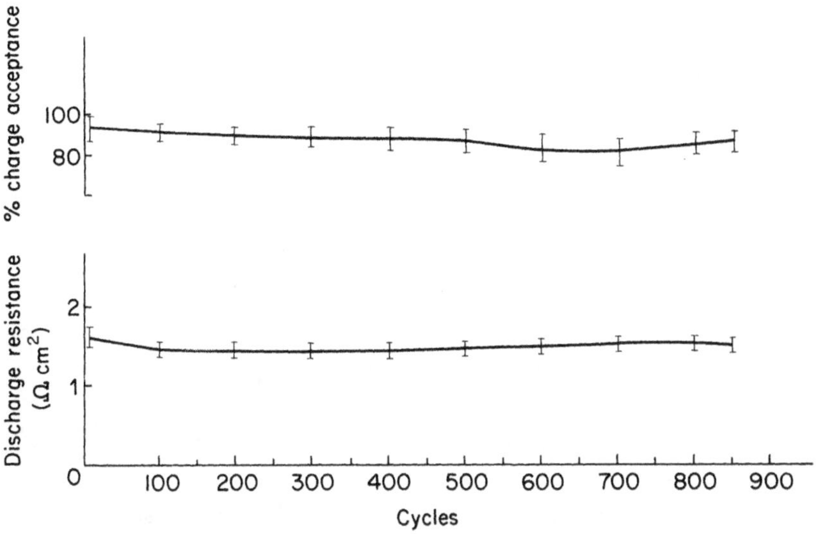

Fig. 5.34 Change in charge acceptance and resistance of a group of six cells containing sulfur electrodes of the type shown in Fig. 5.33.

in the design of the sulfur electrode. Furthermore, it is much more difficult to measure the capillary properties of carbon felts and mats with respect to sulfur and sodium polysulfide than the electrical properties. Thus the capillary properties have often been ignored, and it is possible that results which have been explained in terms of the potential distribution theory could well be due to capillary action. Thus the improvement in charge acceptance obtained by Breiter and Dunn (1979a) with a dual mat electrode was interpreted by them as supporting the potential distribution model. It could equally well be explained in terms of the wetting characteristics of the two mats: if the mat adjacent to the solid electrolyte is preferentially wetted by sodium sulfide, an improvement in charge acceptance would be expected. Demott (1980) has shown that an electrode fabricated from pitch based carbon mat (Union Carbide Thornel VMA) has recharge properties superior to those of an electrode fabricated from a viscose rayon based carbon felt (Le Carbone RVC 4000), even though the electrode resistance and thus the carbon felt resistance was higher for the RVC 4000 material. Further evidence against the potential distribution theory is provided by Fig. 5.35 (Demott, 1980), which shows that the inclusion of a high resistance layer adjacent to the solid electrolyte does not necessarily result in improved charge acceptance.

These points can only be resolved by a careful study of both the electrical properties and the wetting behaviour of the materials used in the electrode matrix. The methods developed by Janz *et al.* (1980) for determining the adhesion tension ratio should prove extremely valuable in this respect.

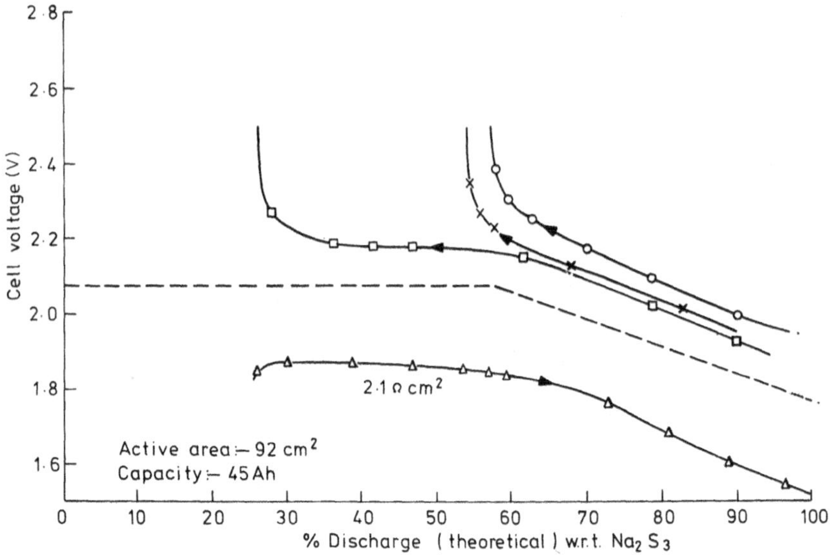

Fig. 5.35 Charge/discharge curves for a cell containing an electrode with a layer of high resistivity felt adjacent to the solid electrolyte. □ 30 mA cm^{-2}; × 50 mA cm^{-2}; ○ 70 mA cm^{-2}; △ 100 mA cm^{-2}.

5.5 SUMMARY

The operation of the sulfur electrode is dependent upon complex interactions between the solid electrolyte, sulfur, molten sodium polysulfides and the carbon fibre matrix or the coated aluminium foil. The physicochemical properties of the molten polysulfides are now well defined, thanks to the efforts of Cleaver and his co-workers, and the thermodynamics and electrochemistry of these melts have been thoroughly investigated by several workers, notably Gupta and Tischer, South, and Armstrong and Dickinson. The electrochemical reactions which occur at inert smooth electrodes in these melts are well characterized, although there are still aspects of the anodic reaction processes for which there is no satisfactory explanation, and the identity of the electrochemically active species has not been established beyond doubt. Also it is perhaps surprising that the expressions for the diffusion of dilute solutions of electroactive species in a supporting electrolyte should be obeyed by a binary molten salt.

In a practical electrode the situation is more complex. Thus for much of the discharge and charge two liquid phases, sulfur and sodium pentasulfide, are present, and it is the interaction of these two phases with the solid electrolyte and the solid electrode which determines the behaviour of the sulfur electrode. In the single phase region the possibility exists of calculating the distribution of reaction, and Gibson, Breiter and Dunn, Kao and Wayner, and Jacquelin and Pompon have modelled the electrode quite successfully. In the two phase region, however, these models are less applicable, and the critical factor here appears to be the wetting behaviour of sulfur and sodium polysulfides on the solid surfaces in the electrode compartment. Ludwig has shown that the wetting behaviour of sulfur and sodium polysulfides on the same surface can be quite different; Cleaver and Hall have shown that the contact angle of sulfur on carbon in polysulfide containing melts is potential dependent; and Janz has studied the competitive wetting behaviour of sulfur and sodium polysulfides on a variety of materials. These observations have led to the design of electrodes in which the relatively mobile polysulfides are transported throughout the electrode by capillary action. These electrodes contain alpha alumina fibres which act as wicks for the polysulfides, transporting them from the bulk of the electrode to the solid electrolyte surface and keeping the solid electrolyte from becoming covered by insulating sulfur until most of the polysulfide has been consumed. Such electrodes combine the desirable characteristics of low electrical resistance and high charge acceptance even at fast charge rates. Furthermore, they appear able to sustain these characteristics indefinitely, i.e. the capacity remains constant at a level close to the theoretical capacity irrespective of the number of charge/discharge cycles. The claim made by DeWitt (1967) that this liquid electrode should not suffer from the loss of capacity characteristic of many solid electrodes appears to have been vindicated. This can only be achieved in cells in which the current collector is not corroded by the sulfur/polysulfide melt, but in Chapter 6 we will see that such current collectors have been developed.

5.6 REFERENCES

Armstrong, R. D., Dickinson, T. and Reid, M. (1975) *Electrochim. Acta*, **20**, 109.

Armstrong, R. D., Dickinson, T. and Reid, M. (1976) *Electrochim. Acta*, **21**, 935.

Bauer, R., Haar, W., Kleinschmager, H., Weddigen, G and Fischer, W. F. (1976/1977) *J. Power Sources*, **1**, 109.

Bergstrom, F. W. (1926) *J. Am. Chem. Soc.*, **48**, 146.

Breiter, M. W. and Dunn, B. (1979a) *J. Appl. Electrochem.*, **9**, 291.

Breiter, M. W. and Dunn, B. (1979b) *J. Appl. Electrochem.*, **9**, 671.

Brennan, M. P. J. (1977a) *Electrochim. Acta*, **22**, 279.

Brennan, M. P. J. (1977b) *Electrochim. Acta*, **22**, 279 and 609.

Brennan, M. P. J. (1977c) Symposium on Electrode Materials and Processes for Energy Conversion and Storage, *Proc. Electrochem. Soc.*, **77-6**, 963.

Cleaver, B. (1983) in *The Sulphur Electrode* (ed. R. P. Tischer), Academic Press, New York and London.

Cleaver, B. and Davies, A. J. (1973a) *Electrochim. Acta*, **18**, 733.

Cleaver, B. and Davies, A. J. (1973b) *Electrochim. Acta*, **18**, 727.

Cleaver, B., Davies, A. J. and Hames, M. L. (1973) *Electrochim. Acta*, **18**, 719.

Davies, A. J. (1972) PhD thesis, Southampton University.

Demott, D. S. (1980) 4th DoE/EPRI Sodium Sulfur Battery Workshop, Derby.

Desplanches, G., Lazennec, Y. and Wicker, A. (1977) US Patent 4049884.

DeWitt, T. W. (1967) Symposium on Power Systems for Electric Vehicles, Columbia University, NY.

Doi, T. (1965) *Rev. Phys. Chem. Jpn*, **35**.

Dzieciuch, M. A. (1977) quoted by Tischer, R. P. and Ludwig, F. A. (1977) in *Adv. Electrochem. & Electrochem. Eng.*, **10**, 391.

Fally, J., Lasne, G., Lazennec, Y. and Margotin, P. (1973) *J. Electrochem. Soc.*, **120**, 1292.

Fehér, F. (1963) in *Handbook of Preparative Organic Chemistry* (ed. G. Brauer), 2nd edn (translated by Scripta Technica Inc.), Academic Press, New York, p. 341.

Fehér, F. and Berthold, H. J. (1953) *Z. Anorg. Chem.*, **273**, 144.

Föppl, H., Busmann, E. and Frorath, F. K. (1962) *Z. Anorg. Allgem. Chem.*, **314**, 12.

Gibson, J. G. (1974) *J. Applied Electrochem.*, **4**, 125.

Gupta, N. K. and Tischer, R. P. (1972) *J. Electrochem. Soc.*, **119**, 1033.

Hall, S. M. (1979) PhD thesis, University of Southampton.

Hargreaves, A. and Piercy, R. (1978) Unpublished results.

Jacquelin, J. and Pompon, J. P. (1978) in *Power Sources*, vol. 7 (ed. J. Thompson), Academic Press.

Jacquelin, J. and Pompon, J. P. (1980) 4th DoE/EPRI Sodium Sulfur Battery Workshop, Marcoussis.

Janz, G. J., Murphy, R. M. and Spink, W. A. (1980) *J. Appl. Electrochem.*, **10**, 789.

Jones, I. W. and Robinson, G. (1978) US Patent 4 118 545.

Kao, Y. K. and Wayner, P. C., Jun. (1976) *J. Electrochem. Soc.*, **123**, 632.

Kleinschmager, H., Haar, W., Weddigen, G. and Fischer, W. F. (1977) *Power Sources*, vol. 6 (ed. D. H. Collins), Academic Press.

Knoedler, R. (1984a) *J. Appl. Electrochem.*, **14,** 39.

Knoedler, R. (1984b) *J. Electrochem. Soc.*, **131,** no. 4.

Kummer, J. T. and Weber, N. (1967) Proc. SAE Congr. 1967 paper 670179, 1–6.

Le Méhauté, A. and Bordet, P. (1977) US Patent 4 037 028.

Levine, C. (1983) in *The Sulphur Electrode* (ed. R. P. Tischer), Academic Press, New York and London.

Ludwig, F. A. (1976) Proceedings of Symposium and Workshop on Advanced Battery Research and Design, Electrochemical Society and Argonne National Laboratory Meeting, Report ANL 76-8, B117.

Ludwig, F. A. (1977) *Adv. Electrochem. & Electrochem. Eng.*, **10**, 391.

Ludwig, F. A. (1978) US Patent 4 084 041.

Minck, R. W. (1975) quoted by Tischer, R. P. and Ludwig, F. A. (1977) in *Adv. Electrochem. & Electrochem. Eng.*, **10**, 391.

Minck, R. W. (1976) Proceedings of the Symposium of Advanced Battery Research and Design, Electrochemical Society and Argonne National Laboratory, ANL 76-8, B199.

Minck, R. W. (1978) Sodium/Sulfur Battery Review, Ford presentation to DoE, Newport Beach, Ca.

Mitoff, S. P. (1980) EPRI Report EM-1341.

Pearson, T. G. and Robinson, P. L. (1930) *J. Chem. Soc.*, 1473.

Piercy, R., Hargreaves, A. and Sudworth, J. L. (1977) Unpublished results.

Piercy, R. and Sudworth, J. L. (1977) Symposium on Electrode Materials and Processes for Energy Conservation and Storage, *Proc. Electrochem. Soc.*, **77-6**, 950.

Selis, S. M. (1970) *Electrochim. Acta*, **15**, 285.

Silverman, H. P. (1976) EPRI Report EM-266, Project 127.

South, K. D. (1973) quoted by Sudworth, J. L. and Tilley, A. R. (1983) in *The Sulphur Electrode* (ed. R. P. Tischer), Academic Press, New York and London.

South, K. D. and Storey, M. A. (1973) quoted by Piercy, R. and Sudworth, J. L. (1977) *Proc. Electrochem. Soc.*, **77-6**.

South, K. D. and Sudworth, J. L. (1973) 143rd meeting of the Electrochemical Society, Chicago, May.

South, K. D. and Sudworth, J. L. (1975) Society for Electrochemistry Meeting, Brighton.

South, K. D., Sudworth, J. L. and Gibson, J. G. (1972) *J. Electrochem. Soc.*, **119**, 554.

Storey, M. A. (1977) UK Patent 1 472 975.

Sudworth, J. L. (1980) 4th DoE/EPRI Sodium Sulfur Battery Workshop, Heidelberg.

Sudworth, J. L. and Tilley, A. R. (1983) in *The Sulphur Electrode* (ed. R. P. Tischer), Academic Press, New York and London.

Suski, L. and Kolacz, J. (1980) *J. Appl. Electrochem.*, **10**, 683.

Tegman, R. (1973) *Acta Cryst.*, **B29**, 1463.

Tilley, A. R. and Wright, M. L. (1981) 16th IECEC, Atlanta, 1, p. 841.

Tischer, R. P. and Ludwig, F. A. (1977) *Adv. Electrochem. & Electrochem. Eng.*, **10**, 391.

Weddigen, G. (1977) 28th ISE Meeting, Varna, Bulgaria, Abstract 114.

Weddigen, G. (1978) German Patent 2 633 455.

Weiner, S. A. (1976, 1977) Reports, Ford Sodium/Sulfur Battery Program, National Science Foundation, Washington DC, Contract NSF-C-805.

Wiewiorowski, T. K. (1972) *J. Electrochem. Soc.*, **118**, 1711.

Wright, M. L. (1982) UK Patent 2 097 992.

6 The sulfur electrode current collector

A. R. TILLEY

6.1 INTRODUCTION

During the development of the sodium sulfur cell, one of the major materials problems has been to develop a corrosion resistant positive current collector. Many low cost materials such as low carbon mild steel and aluminium are suitable for sodium containment, but because of the extremely corrosive nature of the sulfur and sodium polysulfides these are unsuitable for use as a sulfur electrode current collector. The main requirements for this component are good corrosion resistance and high electrical conductivity. Other important requirements are ease of fabrication, low weight and low cost.

In the central sodium cell design, the positive current collector is usually the outer cell case and so corrosion resistance is necessary to maintain the mechanical integrity of the cell. The difficulty in obtaining this formidable combination of requirements led the United Kingdom developers (Atomic Energy Research Establishment, Harwell, British Rail and Chloride Silent Power Ltd) to adopt the central sulfur tube cell design (Partridge *et al.*, 1976; Dell *et al.*, 1976). In this cell design the positive current collector is simply a rod or tube located in the centre of the sulfur electrode. The advantages of using the current collector in this form are:

(1) More expensive materials can be considered, since a smaller quantity of material is required;
(2) because the current collector does not also have to serve as the cell case,

mechanical failure no longer presents a safety problem and mechanically weaker materials can be considered;

(3) a wider range of coating methods can be used on the external surface of a rod or tube;

(4) disruption of corrosion resistant coatings, which can occur in central sodium cells owing to sealing operations such as welding, is less likely to be a problem when a rod or tube has to be joined to the cell;

(5) the outer cell case of a central sulfur cell can be manufactured from inexpensive mild steel tubing.

The main disadvantage of the central sulfur cell is that the current density at the surface of the current collector is much higher than on the cell case in an equivalent central sodium cell. This increase in current density can lead to enhanced corrosion.

Although corrosion resistance may be necessary to maintain mechanical integrity of the cell case, the main reason for requiring good corrosion resistance is to prevent degradation of cell performance. The corrosion products can cause this by increasing the cell resistance, reducing the cell capacity and degrading the beta alumina electrolyte. Cell resistance may be increased either by the formation of layers of corrosion product at the current collector/sulfur electrode matrix interface (Bones and Markin, 1977) or by full or partial blocking of the beta alumina surface by deposits of corrosion products (Weiner, 1975). Diffusion or ion exchange of impurity ions into the beta alumina lattice can alter its conductivity, and partial blockage of the ceramic electrolyte results in an inhomogeneous current distribution which may lead to excessively high local current densities and consequent ceramic failure (Weiner, 1975). Capacity loss can occur when sufficient sulfur is chemically bound in the metal sulfide produced by corrosion (Hattori *et al.*, 1977). This effect is not normally predominant, however, as very small amounts of corrosion products can alter the wetting properties of the sulfur electrode carbon felt matrix and so affect the recharge performance (see Chapter 5); in addition, capacity loss has been shown to be due to impurity films on the beta alumina surface (Weiner, 1975). Typical degradation of cell performance due to current collector corrosion is shown in Fig. 6.1.

In-cell behaviour of a current collector is therefore the most reliable test, not only because it takes into account the effects of any corrosion products, but also because in-cell testing is invariably more severe than out-of-cell coupon tests. This is due to the greater changes in sodium sulfide melt compositions which occur during charging and discharging of a cell, compared with the out-of-cell tests which are restricted to one polysulfide composition. It has also been shown that poor electronic contact between the current collector and the carbon felt matrix can cause extremely rapid corrosion (Bones *et al.*, 1975). The sulfur electrode and current collector should therefore be considered as one unit, because a change of the sulfur electrode matrix material or its construction can

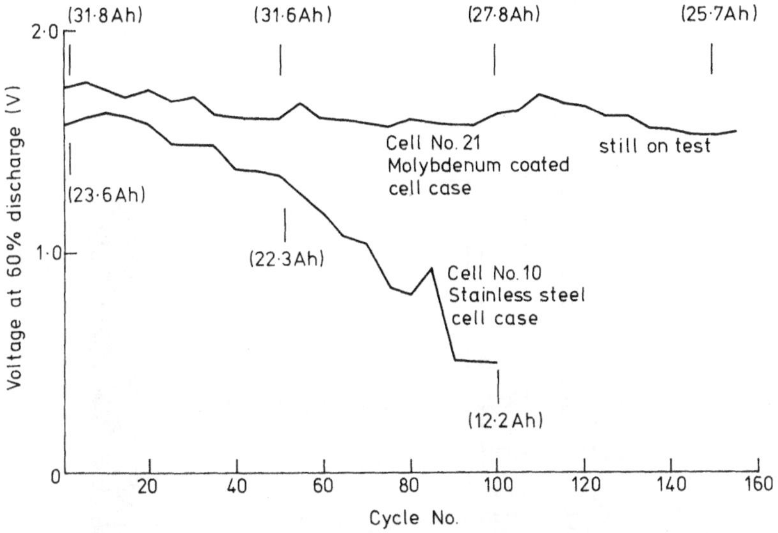

Fig. 6.1 A performance comparison of cells using stainless steel and molybdenum protected stainless steel current collectors. (Hattori *et al.*, 1977). Reproduced with permission from the Society of Automotive Engineers.

significantly increase the corrosion of a current collector that has previously been shown to be satisfactory. For instance, structured sulfur electrodes using alternate rings of carbon felt and alumina felt can cause rapid degradation of current collector coatings (Tilley and Wright, 1981), and the addition of tetracyanoethylene to the sulfur electrode to increase charge acceptance (Weddigen, 1977; Fischer *et al.*, 1978) considerably increases the corrosion of aluminium alloys that are unattacked in ordinary sulfur electrodes (Demott, 1980). The enhanced corrosion caused by a structured electrode is shown in Fig. 6.2.

Any out-of-cell testing therefore has to be taken only as a broad indication of a particular material's suitability. In-cell testing and the stability of cell performance with extended cycling is really the only satisfactory guide to current collector performance.

High electrical conductivity is essential, not only because a high resistance will increase the overall cell resistance, but also because any axial potential difference in the current collector will cause heterogeneity in the sulfur electrode. For long tubular cells these effects can be quite significant. For example, carbon ($\rho \sim 10^{-3}$ Ω cm at 25 °C) is only acceptable for short (<200 mm) central sulfur tube cells. High resistivity materials can be used in composite current collectors where the high resistivity corrosion resistant material is a thin coating or sheath on a low resistivity metal substrate.

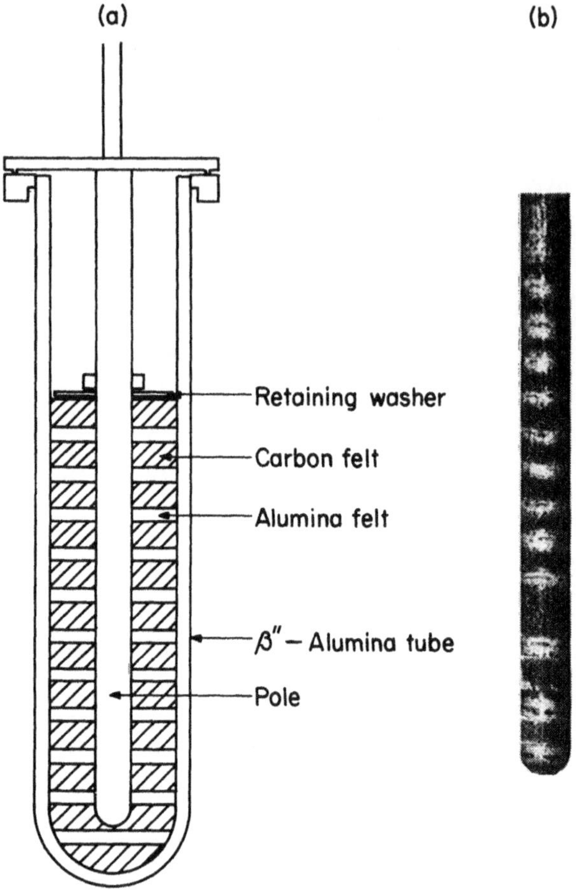

Fig. 6.2 Construction of a structured sulfur electrode and an example of enhanced current collector corrosion caused by this structure (Tilley and Wright, 1981). Reproduced with permission from the American Society of Mechanical Engineers.

The work on single materials is described briefly in Sections 6.2 and 6.3, and the development of composites is described in Section 6.4.

6.2 METALS AND ALLOYS

All metal sulfides have a negative free energy of formation at 350 °C (Mills, 1974) and only metals which form a stable coherent coating of sulfide can be considered for use in a cell. Chemical and electrochemical stability of the sulfide in the sulfur/polysulfide melt is essential, for if the sulfide dissolves in the melt or is electrochemically active a protective scale is not formed. To ensure slow growth of the corrosion layer, the diffusion rate of the metal ions in the sulfide layer should be low. A very large amount of work evaluating numerous metals and alloys has been undertaken by the developers. Static and dynamic tests have been undertaken at various melt compositions and purities, using weight loss or radioisotope methods, and workers at Ford (Weiner, 1973, 1974, 1975, 1976), General Electric (Bush, 1975; Chatterji, 1976; Mitoff, 1978, 1980) and Brown Boveri (Hartmann, 1978; Fischer *et al.*, 1978; Bauer *et al.*, 1976/1977) have evaluated virtually every metal and alloy that could be used. Only a few metals and alloys have been found to be stable or of an acceptable corrosion resistance. These are aluminium, chromium, molybdenum, some stainless steels and superalloys.

Aluminium forms a thin stable coating of Al_2S_3 which has a high electrical resistivity, approximately $10^9\ \Omega$ cm at 400 °C (Hartmann, 1978), and so aluminium alone cannot be used as a current collector. Aluminium does however provide an ideal passive substrate material for protective coatings (Hartmann, 1978; Tilley and Wright, 1981) and these are discussed under composites.

The rate of corrosion of chromium is quite acceptable and the solubility of Cr_2S_3 in Na_2S_4 is much lower than that for the sulfides of Fe, Cu, Ni (Weiner, 1975). Chromium metal, however, is expensive and difficult to fabricate and so can only be used as a coating on other metals; it is therefore discussed under composites.

Molybdenum forms a thin corrosion scale of MoS_2 which is electronically conducting, and it has been estimated that molybdenum can offer a good corrosion protection for many years without forming a thick sulfide layer (Mitoff, 1978). Molybdenum, however, is difficult to fabricate and weld and a solid container (or rod current collector) would be prohibitively expensive to be considered as a practical material; even thin foils would be too costly (Mitoff, 1978). In-cell tests have shown some peculiar polarization behaviour (Chatterji, 1976), and although in a more detailed study of the performance of molybdenum in cells a satisfactory performance was obtained by using an anodizing surface pretreatment, premature polarization on charge was still a problem in the long term (Markin *et al.*, 1978).

Many stainless steels and a range of superalloys have been tested (Weiner, 1977; Bauer *et al.*, 1976/1977; Kinsman and Oei, 1977), and those materials that contain an appreciable amount of chromium ($\sim 20\%$) all have a reasonable corrosion resistance and are attacked only slowly. Studies of the corrosion film composition on these alloys show that they develop a typical duplex corrosion layer (Weiner, 1977; Bauer *et al.*, 1976/1977). The inner layer is rich in chromium and it is this film of chromium sulfide, through which the more mobile ions have to diffuse, which controls the rate of corrosion. The corrosion rate of superalloys is about ten times lower than that of stainless steels, probably because of the lower diffusivity of nickel and cobalt compared with iron (Bauer *et al.*, 1976/1977). Although the corrosion rates in cells may be acceptable for the current collector for a life of many years, the corrosion products from some of these materials can still cause cell performance to deteriorate. The highly corrosion resistant superalloys and chromium and molybdenum are expensive and so more suited for use as protective layers or thin solid sheaths in composite construction.

6.3 NON-METALS

There are many non-metals which are electronically conducting and inert to sodium polysulfides, and extensive testing of various carbides, borides, nitrides, silicides and oxides has been undertaken (Chatterji, 1976; Weiner, 1975). Ideally, any material used should be incapable of being reduced by sodium, have a conductivity similar to metals and be mechanically strong and easily formed. The only material that comes close to these requirements and is intrinsically stable in polysulfide melts is carbon. Carbon rods have been used successfully as current collectors in central sulfur cells (Sudworth *et al.*, 1977; Tilley and Wright, 1981). The simple cell design utilized a screw thread attachment to connect the carbon current collector to the sulfur electrode closure plate, as shown in Fig. 6.3. This form of current collector was cheap and gave an extremely satisfactory long term cell performance. A plot of long term cell performance is shown in Fig. 6.4, and the degradation in cell performance is negligible compared with the rapid loss in performance shown in Fig. 6.1 for cells which used a stainless steel current collector. Vitreous carbon has also been successfully used as a current collector in flat plate cells (Piercy and Sudworth, 1977). The electrical resistance of carbon and graphite is however too high for practical, high capacity, high rate cells. For central sulfur cells, the length of a carbon or graphite current collector is limited to approximately 200 mm for a solid carbon rod diameter of 15 mm.

Two other non-metals, titanium carbide and chromium carbide, have been tested in central sulfur cells (Jones, 1976). In these tests, 12 mm diameter hot pressed rods of the carbides were used as central current collectors in cells using beta alumina electrolyte 22 mm ID × 200 mm long. An active current collector length of only 100 mm was used in these cells. Compared with cells using graphite current collectors, the carbide containing cells had a 25% lower cell resistance. No deterioration of capacity was found with extended cycling compared with a rapid fall off in capacity experienced with similar cells using stainless steel current collectors. Of the two materials, chromium carbide was more corrosion resistant but had a slightly higher electrical resistivity (2.5×10^{-4} Ω cm) and, as for carbon, this is too high for the material to be used as a solid rod for practical cells.

There are several other non-metals with higher electrical resistivities which are inert to sodium polysulfides, such as doped TiO_2 (rutile), SiC and $CaTiO_3$. These materials, like carbon, cannot be used as direct replacements for metal current collectors, but they are of considerable interest for composite current collectors.

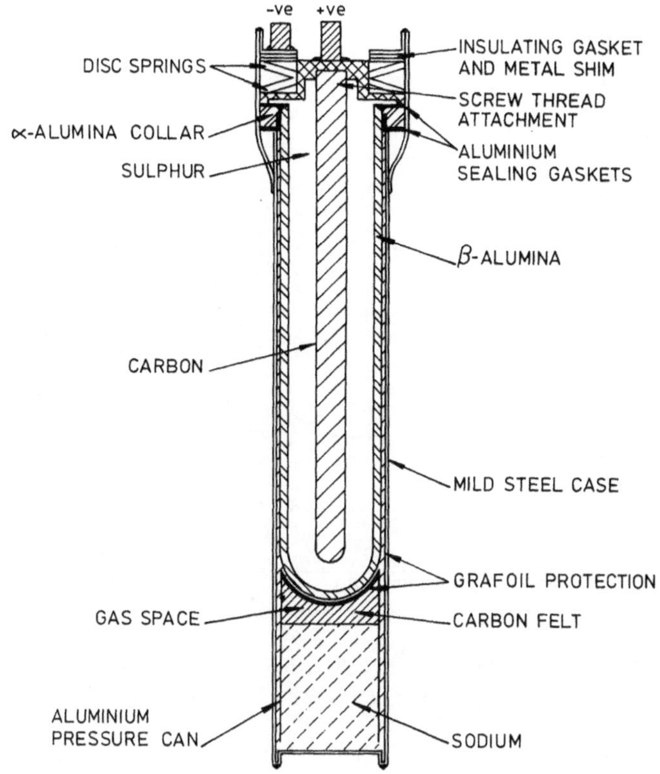

Fig. 6.3 A 40 Ah central sulfur cell design utilizing a carbon current collector (Tilley and Wright, 1981). Reproduced with permission of the American Society of Mechanical Engineers.

6.4 COMPOSITES

No single material, then, has been found that combines all the properties needed for a current collector in a practical cell design. Corrosion resistant materials such as chromium, molybdenum and some superalloys are expensive and difficult to fabricate, but they can be used as thin sheaths, liners or coatings on other metals. High resistivity, corrosion resistant materials can also be used in several ways as composites. If the materials have adequate mechanical strength and can be produced as thin walled tubes, then by coring these tubes with a low resistivity metal a composite is formed that can be used as a current collector in a central sulfur cell design. For central sodium cell designs, the thin walled tubes have to be coated on the outer surface with a low resistivity metal. Quite high resistivity materials can be considered for use in composites; for instance, at a wall thickness of 2 mm, materials with a resistivity of 0.5 Ω cm will only contribute 0.1 Ω cm^2 to the overall cell resistance. Even higher resistivity

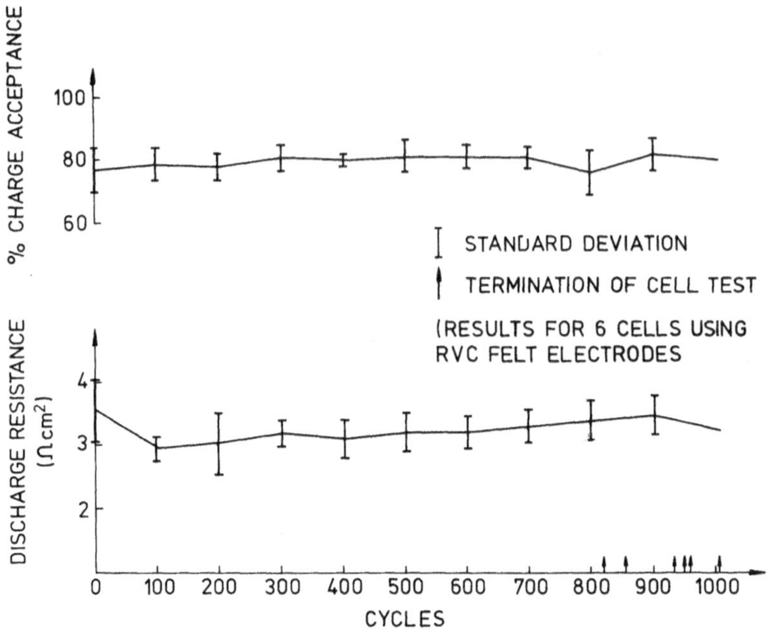

Fig. 6.4 Long term performance of cells using carbon current collectors (Tilley and Wright, 1981). Reproduced with permission of the American Society of Mechanical Engineers.

materials can be considered if they can be thinly coated directly on to a low resistivity metal. Consider a typical 100 Ah central sulfur cell using beta alumina electrolyte approximately 300 mm long and 33 mm diameter (Bird *et al.*, 1978): the central current collector is a 9 mm diameter aluminium rod coated with a 0.1 mm layer of corrosion resistant material, and even if this layer had a resistivity of 10 Ω cm it would only contribute 0.1 Ω cm^2 to the overall cell resistance (~5%). Voltage drop down such a current collector at the two hour rate (50 A) would only be approximately 30 mV, which would cause negligible heterogeneity in the sulfur electrode. Composite construction therefore appears an ideal way of utilizing the high cost materials which cannot be used in the solid form, and it increases the range of usable corrosion resistant materials to include high resistivity but inert materials such as carbon and doped TiO_2.

6.4.1 Constructional details

The constructional requirements of composite current collectors differ considerably for the two main types of cell design. The corrosion resistant material in a composite current collector can be used either as an applied coating or as a thin solid liner or sheath. For a central sulfur cell design, direct coating of a corrosion resistant material on to a low resistivity metal rod is relatively easy.

For a central sodium cell, however, coating the inside surface of cell case tubing can be difficult for coating methods such as flame and plasma spraying, ion spluttering, electroplating etc. Coating methods such as pack aluminiding or chromizing are, of course, relatively easy for both cell designs, although the cost differential can be considerable owing to the relative sizes of the articles.

High capacity, high specific energy central sulfur cell designs require current collectors greater than 200 mm long and less than 12 mm in diameter, so it is essential to use a low resistance core to avoid any appreciable voltage drop down the current collector at high rates. For materials that can be used in the solid sheath form, such as molybdenum, superalloys, treated steels, rutile and carbon, an aluminium or copper core is needed. The core–sheath contact resistance has to be low and remain stable with extended cell cycling. Oxidation and sulfur attack of the metal core has to be prevented if contact between the core and sheath is to remain stable. It is possible to use costly sealed and evacuated constructions when using metal sheaths but for a non-metal such as carbon this can be difficult owing to its inherent porosity and metal/non-metal sealing problems. Liquid metal interface systems have been proposed for use with carbon cored composites (Jones *et al.*, 1978a), and rutile tubes cored with copper using a lead tin alloy liquid interface have been tested in cells (May, 1979). Systems using a deformable interface such as carbon felt (Jones *et al.*, 1978b) and conducting powders (Johnson *et al.*, 1979) have also been patented. For small diameter metal sheaths, a simple coring system using liquid sodium has been tested (Tilley and Wright, 1981). The amount of sodium required is relatively small (approximately 8% of the Faradaic amount for a 100 Ah cell) so safety problems are negligibly increased. Sodium is ideally suited as a coring medium. It is low cost, and has a low density (0.87 g cm^{-3}) at 350 °C and an acceptable electrical resistivity ($\sim 18 \times 10^{-6}$ Ω cm at 350 °C; cf. aluminium 6.4×10^{-6} Ω cm at 350 °C) (Addison *et al.*, 1969). If high resistance materials are used to form the cell case in the central sodium cell designs it is much easier to overcome the problem of the high longitudinal resistance, as direct coating methods, such as flame spraying or electroplating of low resistance materials, can easily be used on the outer surface of the case (Weiner, 1976).

Composites that have been investigated can be divided into three main groups: coatings on corrodable substrates, coatings on aluminium, and metal coated or metal cored non-metals.

6.4.2 Coatings on corrodable substrates

(a) *Molybdenum*

Molybdenum has been used successfully as a thin liner in central sodium cells by Yuasa (Hattori *et al.*, 1977). Cells which used stainless steel cases showed a rapid decrease in cell capacity (50% after 100 cycles), but the molybdenum lined cells showed a much higher capacity retention, approximately 87% after 100 cycles, as shown by Fig. 6.1. The extent of the corrosion was measured and the MoS_2

layer was only 0.3–0.4 μm thick after 1000 cycles. The stainless steel substrate had been corroded in several places, however, and a metal-free case or a pore-free coating technique was thought to be needed. Molybdenum coatings on Rodar alloy (applied by sputtering to a thickness of 5 μm) have been found to be fairly corrosion resistant (Chatterji, 1976). The Rodar substrate was attacked, however, where defects occurred in the molybdenum coating. For long term protection, General Electric (GE) concluded, as Yuasa did, that pore-free coatings were essential (Mitoff, 1978). The latest cell tests at GE have used low carbon steel cell cases clad with molybdenum foil (Mitoff, 1980). The cells showed similar peculiar polarization behaviour to previous tests, and because the molybdenum foil was used as a simple liner the outer steel case showed substantial corrosion owing to inflow of polysulfide melt into the gap between the molybdenum foil and the steel case. Owing to its high cost and peculiar polarization behaviour in cells, molybdenum appears to be unsuitable even for use in composite current collectors.

(b) *Chromium*

Several types of chromium coating have been tested on various substrates. Chromium electroplated coatings have been investigated by Ford and GE (Chatterji, 1976; McClanahan, 1978; Ludwig, 1980). The difficulty in obtaining perfect defect-free coatings causes rapid subsurface corrosion of non-passivating substrates such as steel, and this results in cracking and spalling of large areas of the chromium coating. Improved electroplated chromium coatings on more corrosion resistant substrates, such as stainless steels, do show a slightly improved performance, and graphite/sodium silicate coatings (Tiodizing) have been used on electroplated chromium surfaces in an attempt to overcome the coating defect problem (McClanahan, 1978; Ludwig, 1980). Plasma sprayed chromium coatings on steel substrates have been tested by GE (Mitoff, 1978) and similar behaviour to molybdenum was observed.

Chromizing has been extensively studied in an attempt to obtain pore-free chromium coatings that are bonded to the substrate material. Chromizing of steel can be achieved by a chemical vapour deposition process, or by the more commonly used pack chromizing method which has been used in industry for many years. In the basic pack process, the steel components to be chromized are packed with chromium or ferrochromium alloy powder, an activator (usually an ammonium halide) and an inert filler such as alpha alumina powder. The pack is heated in an inert atmosphere at temperatures between 800 and 1100 °C. The ammonium halide reacts with the chromium powder and generates chromium halides. The chromium halide vapour diffuses to the surface of the steel components and deposits chromium which then diffuses into the steel. The thickness of coating is adjusted by the treatment duration and temperature; typically a pack chromizing treatment will take up to 24 hours. The structure of the chromized layer depends on the chromizing conditions and the carbon content of the mild steel substrate. A single phase layer of ferritic chromium enriched steel is produced if the carbon content of the mild steel is less than 0.1%. When the carbon content of the steel is between 0.1% and 0.4% it is possible to form chromium carbide in the grain boundaries of the chromized layer. This causes chromium depletion in the regions adjacent to the grain boundaries, and allows intergranular attack when exposed to corrosive media. Chromizing higher carbon content steels (>0.5%) produces a surface layer of chromium carbide, and this type of treatment is used frequently to impart wear resistance to components; it is not suitable for corrosion resistance as the carbide coating is easily crazed. A duplex layer structure consisting of carbides and nitrides covering a layer of ferritic chromized steel can also be produced if a high concentration of chromium species is produced in the chromizing. These four different structures are shown in Fig. 6.5. The use of chromized steel as a current collector has been patented and a large amount of work studying the corrosion resistance of the various chromized structures has been published (Park *et al.*, 1979a; Mitoff, 1978, 1980; Wicker, 1980). General Electric found that all cells with chromized steel cell cases showed good charge/discharge behaviour; in contrast to molybdenum, no peculiar polarization behaviour was observed. Cells with the single layer chromized steel did show a decrease in capacity with extended cycling, similar to cells with AISI 347 stainless steel cases; the cell cases had lost up to 0.1 mm thickness after three months testing. The chromized duplex layer cases, however, gave excellent cell performances for up to 17 months and the corrosion scale only grew to a thickness of 2 μm in one year. The scale composition was different to that found with the single chromized layer structure. The duplex layer scale was mostly Cr_2S_3, whereas the single layer scale was mainly FeS_2 (which probably accounted for the loss in cell capacities).

Park *et al.* (1979b) have described a method of bonding and chromizing a foil of a chromium alloy to mild steel for use as a container casing material. The foil is spot welded to the inside bore of the mild steel cell case, and during the chromizing bonding occurs between the case and foil.

Two major problem areas have been identified in utilizing chromized steel containers (Mitoff, 1980): the problem of defects present in the chromized layer, and the relatively high cost of the chromizing process. The coating defect problems – cracks, voids, non-uniformity of coating and embedded ceramic particles in the coating – are serious because the mild steel substrate can corrode rapidly once it is exposed. Localized attack resulting from cell geometry or poor felt contact could also lead to penetration of the coating at one point and expose the substrate. Eliminating these problems is difficult and will require an extremely well controlled chromizing technique, possibly even a change from the established pack process, to eliminate defects caused by the inert filler particles.

One further disadvantage of chromized coatings appears to be their rather poor high temperature performance. The workers at the Compagnie Générale d'Electricité in France have reported results from extensive static and dynamic

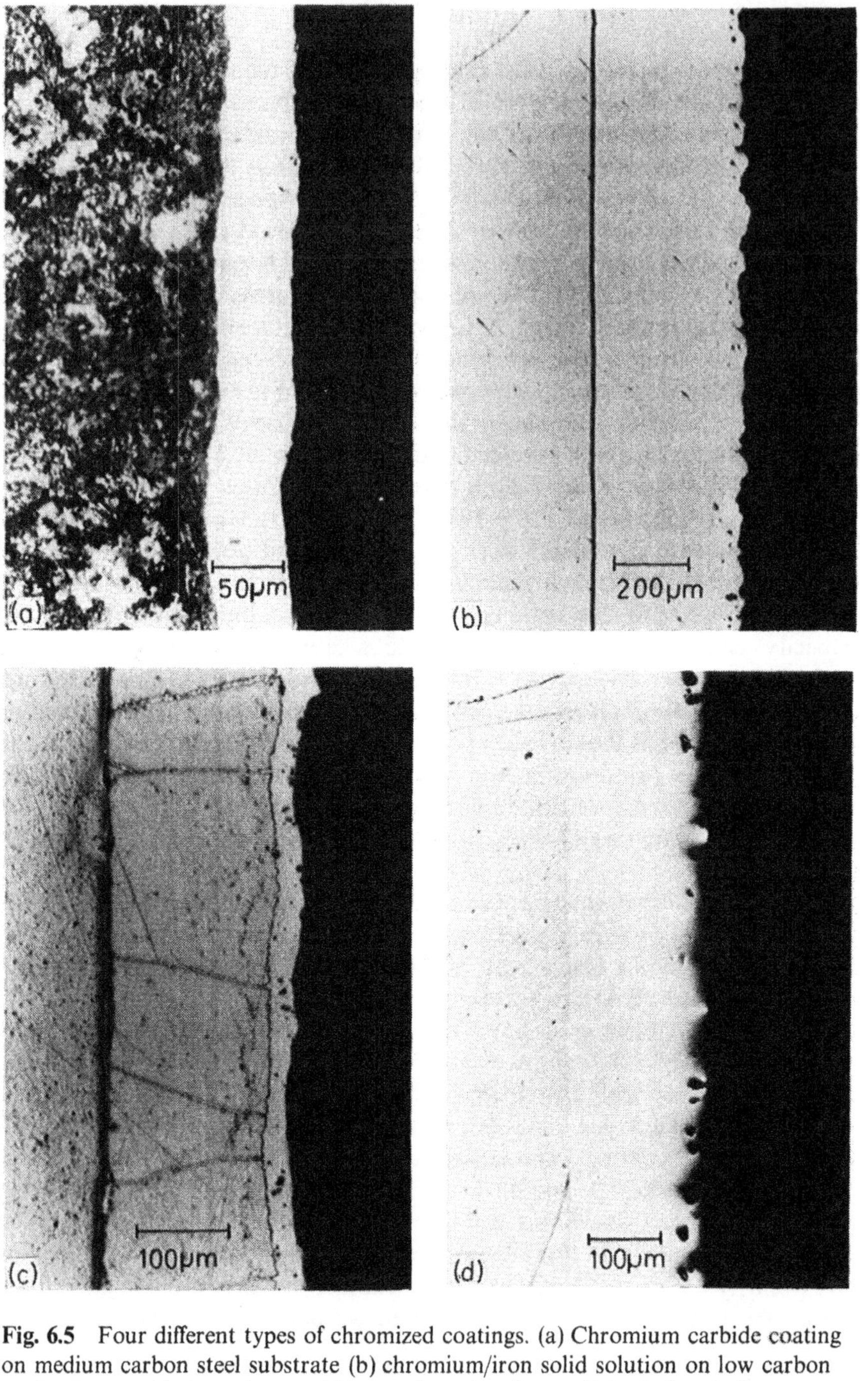

Fig. 6.5 Four different types of chromized coatings. (a) Chromium carbide coating on medium carbon steel substrate (b) chromium/iron solid solution on low carbon steel substrate (c) chromium carbide on chromium/iron solid solution on low carbon steel substrate (Duplex) (d) chromium/iron solid solution coating on Armco iron substrate. Photomicrographs kindly provided by Adrian Kempster, Diffusion Alloys Ltd, Hatfield, England.

tests on various chromized, electroplated chromium and electroplated and diffused coatings on 0.13%, 0.088% and 0.22% carbon steels (Wicker, 1983). In the one year static tests in various melts at 400 and 440 °C the chromized layer on all the samples was destroyed, and at 350 °C good results were only obtained in a Na_2S_5 plus sulfur melt. Rapid corrosion occurred on all the chromized samples at temperatures above 350 °C in the dynamic tests; the best behaviour was given by a duplex coating which was obtained when the carbon content of the steel was approximately 0.1%. The chromium plated samples also behaved poorly in the static tests at the higher temperatures, the external porous chromium layer being completely removed at 440 °C. Dynamic tests were not possible with the chromium plated and chromium plated and diffused samples, owing to cell assembly problems that required development such as thermo-compression seals using plated steel. The chromium plated and diffused samples performed well in the static tests in all melt compositions and at temperatures up to 440 °C. This good behaviour was attributed to the defect-free quality of the diffusion layer.

Significant cost reductions in the materials required for chromizing can be made by changing from expensive research grade chromium powder to commercial powders (Mitoff, 1980). The cost of coating cell tubing for the GE 168 Ah load levelling cell is quite high compared with other coating methods, especially as the pack chromizing process does not easily lend itself to automated production. For central sulfur cells, the cost of chromizing smaller diameter current collectors is much lower. Typically, a chromized mild steel sheath approximately 600 mm × 12 mm OD (for lots of 500 000) is approximately one third of the cost of a chromized cell case. The sheath, however, has then to be cored with a low resistance metal to reduce the overall longitudinal resistance. This can be fairly expensive if evacuated construction with an aluminium or copper rod core is used.

(c) *Stainless steel and superalloys*

Mild steel cases lined with AISI 347 stainless steel foil and 50:50 nickel chrome foil have been tested by GE (Mitoff, 1980). The test results were similar to the molybdenum foil lined cases, i.e. substantial subscale formation of the mild steel support can occur. General Electric have patented the construction of composite cell cases where the corrosion resistant foil liner materials are formed into sealed containers (Breiter *et al.*, 1980). Although this type of composite case would prevent the subscale corrosion, it would be difficult and expensive to construct.

(d) *Oxide films on metals*

Some metals and alloys will form oxide films which are stable in molten polysulfides and can protect the metal substrate from corrosion. Workers at Ford have evaluated oxide coatings on AISI 446 and E-Brite stainless steels and a large number of nickel based superalloys, such as Inconel 600, 601, 625, 671 and Incolloy 800, 825 and 840. Static corrosion tests showed that these oxidized materials showed a greater resistance to corrosion in polysulfide melts than the untreated materials (Weiner, 1976; Dubin, 1977). Evaluation of several chromium containing iron and nickel based alloys with oxide films showed that the corrosion resistance was roughly proportional to the chromium content. High chromium containing alloys such as Inconel 671 had a surface oxide layer that was essentially pure chromium oxide. It is this layer that makes treated Inconel 671 extremely corrosion resistant compared with the untreated material (Kinsman and Winterbottom, 1981). The disadvantage of this type of protective surface is its low electrical conductivity, and consequently the electrical resistance of the coatings thick enough to be effective is too high.

A protective nickel oxide film has been claimed to overcome these problems (Bindin, 1979). In this Chloride Silent Power Ltd (CSPL) work it was found that many nickel chromium alloys were suitable substrates on which the nickel oxide protective films could be formed; in particular, Inconel 600 was found to be convenient to use. Heat treatment of this material in an atmosphere containing an oxygen compound such as CO_2, CO or SO_2, but without any uncombined oxygen, can produce a protective oxide film comprising mainly nickel oxide with minor proportions of iron and chromium oxides. Many cells using this oxidized Inconel have been constructed and tested. In one set of tests aluminium cored oxidized Inconel 600 current collectors were used in central sulfur tube cells, and extremely good capacity utilization for over 100 electrical cycles was demonstrated (Bindin, 1979).

(e) *Aluminiding*

Aluminiding is an established industrial process used for improving the oxidation and sulphidation resistance of steels and superalloys. A typical pack aluminiding process involves heating a sample surrounded by aluminium, ammonium fluoride and aluminium oxide for a few hours at over 1000 °C. The depth of the aluminium coating is controlled by the temperature, time and aluminium activity in the pack. General Electric have two patents on the use of metal aluminide coatings for sodium sulfur cell reactant containers (Chatterji and Dubin, 1977 and 1978). In these patents, alloy samples 53.7% Fe, 29% Ni, 17% Co, 0.3% Mn were aluminided and then cycled in molten sodium polysulfide for 775 hours at 30 mA cm^{-2}. No evidence of chemical attack, flaking or scale formation was found, whereas identical alloys not aluminided showed extensive reaction and degradation. In-cell test results (Mitoff, 1978) have also shown that aluminided steel containers are more corrosion resistant than uncoated containers. The formation of a resistive surface layer of Al_2S_3 did, however, significantly increase cell resistance, and although it was thought possible by GE to avoid this problem by selection of an optimum composition of aluminium in the surface layer, no further work has been reported.

(f) *Summary*

A composite current collector using a corrodable substrate can only be effective if the coating or cladding is conducting and totally impervious to the polysulfide melt. With processes like pack chromizing or chrome plating, it is extremely difficult to eradicate defects formed during the coating process. Additional coatings of carbon have been tried to overcome these problems (Mitoff, 1980; McClanahan, 1978). However, these extra coatings of carbon, if effective, appear to make the first chromium coating redundant! An additional layer of carbon has been shown to be effective in overcoming the problems associated with structured sulfur electrodes, and this is discussed in Section 6.4.3(d). A successful composite current collector requires a passive substrate material such as aluminium. Aluminium is ideal as it is cheap, easy to fabricate, light (density 2.3 g cm^{-3}) and has an excellent electrical conductivity ($\rho = 6.4 \times 10^{-6}$ Ω cm at 350 °C). Its most important feature, however, is that it passivates in sulfur/polysulfide melts with the formation of a resistive layer of Al_2S_3 (10^9 Ω cm at 350 °C). Because of this feature, coating methods that result in porous or cracked coatings can be used without the danger of large subscale formation and spalling of the coating layer. It is also worth noting that in many instances an aluminium covering or coating of materials is essential to lower the longitudinal resistance of a current collector to an acceptable level (Bindin, 1979). For central sulfur cell designs it is far simpler and cheaper to coat a protective layer directly on to aluminium than to core a thin walled tube with aluminium and maintain a stable contact resistance between the core and tube.

6.4.3 Coatings on aluminium

(a) *Molybdenum*

One of the earliest coated aluminium current collectors was used in the Dow cell (Levine *et al.*, 1978). Vacuum vapour deposited or sputtered coatings of molybdenum have been used. Coating thicknesses of 200–300 Å appear adequate, and although this layer of molybdenum reacts to form molybdenum sulfide during cell operation, no changes in cell operation characteristics are seen (Levine, 1983). Molybdenum coated aluminium has been unsuccessful in conventional sodium sulfur cells. Plasma sprayed coatings were used in tests at GE and surprisingly failed, owing to heavy subscale formation which caused the coating to separate from the aluminium substrate material (Mitoff, 1978). Tests at British Rail with flame sprayed molybdenum coatings on an optimized aluminium alloy substrate material did not show any substrate attack, but cell resistances increased sharply with extended cell cycling similar to the effects seen with solid molybdenum current collectors reported earlier (Tilley and Wright, 1981). The molybdenum coated aluminium is probably satisfactory in the Dow cell because of the low operating current density on the current collector, which can be up to two orders of magnitude less than in conventional sodium sulfur cells.

(b) *Chromium carbide*

To establish the most suitable aluminium alloy for a substrate, workers at British Rail tested several alloys in rod form and as extended surface area current collectors which used fins or attached wires. Examination of these various current collectors after several weeks of in-cell testing showed that, in most cases, the aluminium alloy was hardly attacked; in several instances no trace level of aluminium could be found in the polysulfide melt. Although the cell resistances were significantly reduced with extended area current collectors, they were still far higher than for cells which used carbon current collectors.

The experiments showed that one easily available commercial aluminium alloy (HE30) was suitable. This had the approximate composition 0.5% Mg, 0.8% Si, 0.04% Cu, 0.5% Mn, 0.25% Fe with the balance aluminium. Several coatings were tested on this aluminium alloy in central sulfur cells. The current collectors were rods approximately 16 mm diameter × 160 mm long. A brief summary of some of these materials is given in Table 6.1 (Tilley and Wright, 1981). Plasma sprayed chromium carbide current collectors appeared to be

Table 6.1 The performance of various coating materials tested on aluminium substrates. Proceedings of the 16th I.E.C.E.C. conference, Atlanta, USA. Reproduced with permission of the American Society of Mechanical Engineers from Tilley and Wright (1981).

Material	*Coating method*	*Electrical cycling performance*
Molybdenum	Flame spray	50 cycles, high resistance
Carbon	PAVD	10 cycles, high resistance
Cr_2C_3	Ion plating	300 cycles, high resistance
Cr_2C_3	Plasma spray	1000 cycles, acceptable performance
$NiCr/Cr_2C_3$	Plasma spray	1000 cycles, acceptable performance
Nb doped TiO_2	Plasma spray	200 cycles, high resistance
Cr_2C_3	Detonation coating	200 cycles, good performance, very expensive
TiO_2/Al_2O_3	Flame spray	10 cycles, high resistance

satisfactory for several hundred cycles, and the British Rail workers changed from carbon rod current collectors to chromium carbide coated aluminium for all further cell development work. Cell construction was also improved, as the screw thread join used previously between the carbon and the sulfur electrode closure was eliminated by friction welding the aluminium rod to the stainless closure. Typical performance of cells using coated aluminium current collectors is shown in Fig. 6.6.

The main disadvantage of the plasma spray coating was the relatively high cost, which was mainly due to the chromium carbide spraying powder, and the low deposit efficiency (~5%).

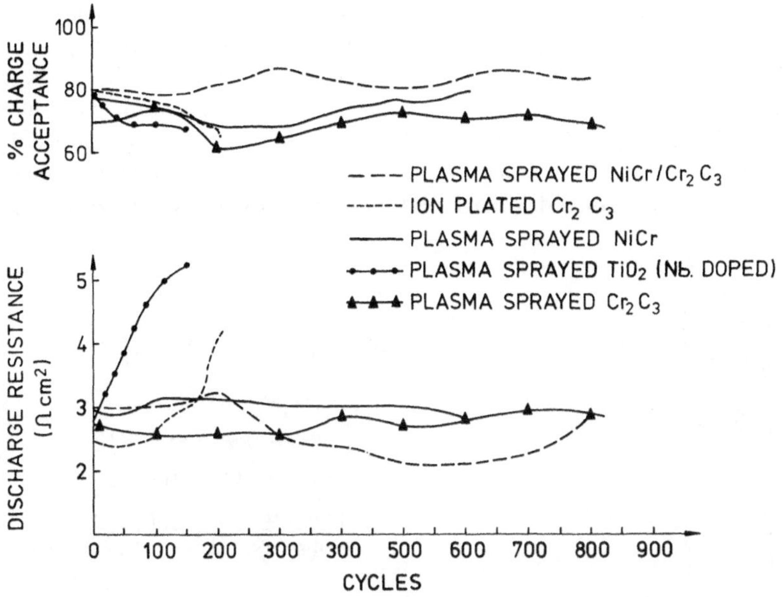

Fig. 6.6 Typical performance of cells using coated aluminium alloy current collectors (Tilley and Wright, 1981). Reproduced with permission of the American Society of Mechanical Engineers.

(c) *Superalloy coatings*

As discussed in previous sections, the corrosion rates of chromium rich alloys with iron, cobalt or nickel as basic metals are lower by over two orders of magnitude than those of pure iron, nickel, cobalt and even chromium.

Oxidized layers on some of these alloys can also enhance their corrosion resistance. Workers at Brown Boveri found that plasma sprayed coatings of superalloys on aluminium gave similar results in out-of-cell tests to the solid materials (Hartmann, 1978). Some test results using central sodium tube cells have been published by Brown Boveri using these coatings. In one such test, a prime coat of 96% nickel and 4% aluminium was used on the aluminium before applying a superalloy of composition 56.7% nickel, 18.4% cobalt, 12.3% chromium, 3.25% molybdenum, 5% aluminium and 4.33% titanium. During the test, cell capacity remained constant for over 100 electrical cycles (Hartmann *et al.*, 1980).

Workers at British Rail had found during the early studies on aluminium alloys that a mixed plasma sprayed coating of nichrome and chromium carbide had performed satisfactorily for several hundred cycles, as shown in Fig. 6.6. Nichrome powders could be flame sprayed, and a flame sprayed coating (20% Cr, 80% Ni) was developed which gave a coating porosity and adherence equivalent to plasma sprayed chromium carbide. The nichrome powder was considerably cheaper than the chromium carbide powder and, because spherical particle shape powder could be obtained, better flow properties and closer feed control were achieved. These features, combined with a flame pinch control, increased the efficiency to around 53%. The total coating materials cost (powder and gas) was reduced to a few pence for a 100 Ah central sulfur cell current collector. In-cell test results were excellent; stable capacity and resistance were obtained for over 1000 electrical cycles, as shown in Fig. 6.7.

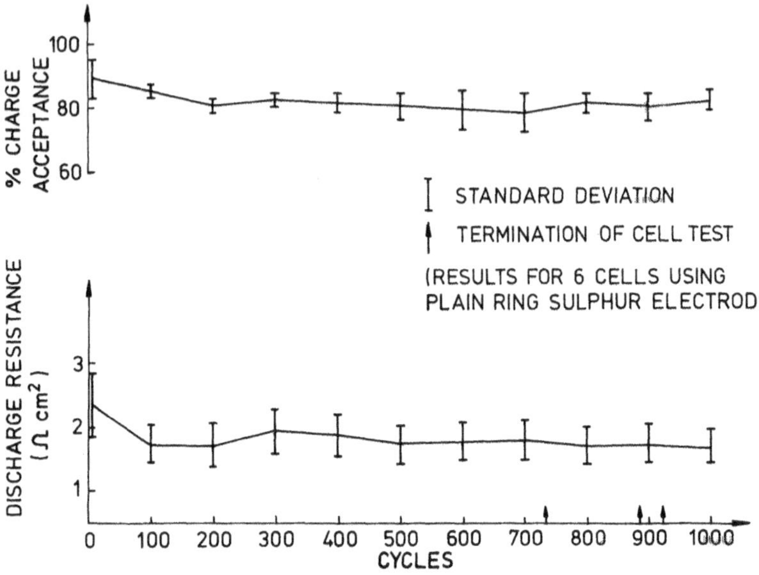

Fig. 6.7 Long term performance of cells using nichrome coated aluminium current collectors (Tilley and Wright, 1981). Reproduced with permission from the American Society of Mechanical Engineers.

(d) *Carbon and graphite coatings*

Several types of carbon and graphite coatings on aluminium have been tested. One method of coating aluminium with graphite is to use a graphite loaded resin. General Electric found that in out-of-cell tests, phenolic and poly (arylacetylene) resins were suitable. In 1976 Mitoff noted that, in contrast to steels, when areas of aluminium substrate were exposed by pinholes in the resin graphite coating there was no extensive subsurface corrosion and flaking of the coating. Graphite paper and cloth were also bonded to aluminium and these appeared to be satisfactory in out-of-cell tests (Mitoff, 1976; Dubin, 1980). In later work General Electric have concentrated on chromized mild steel as a current collector, and in-cell tests using graphite foil inserts on chromized containers have been reported (Bast, 1982). The work concluded that the use of graphite foil was not worthy of further consideration.

In contrast, Brown Boveri and British Rail workers have reported excellent results using carbon coatings. In the Brown Boveri work (Hartmann and Kleinschmager, 1979) a film of graphite 0.05 mm thick was fixed to the inner surface of a central sodium cell wall with an adhesive of phenol formaldehyde resin and carbon powder. In-cell tests at 360 °C showed no change in the coating or the electrical characteristics for over 1000 hours. A considerable safety advantage is also claimed for such coatings: whereas stainless steel or chromized steel are corroded by the high temperature polysulfides formed on fracture of the beta alumina electrolyte, the carbon based coating is unattacked (Hartmann, 1980).

At British Rail (Tilley and Wright, 1981) a graphite layer was used to overcome the effects of a structured sulfur electrode. Nichrome coated aluminium current collectors performed satisfactorily only with plain ring sulfur electrodes (Fig. 6.7). The change to structured, spun pitch sulfur electrodes, to improve the recharge characteristics and lower the cost, enhanced current collector corrosion. Typical corrosion caused by a structured sulfur electrode has been shown previously in Fig. 6.2. A simple wrap of flexible graphite paper prevented this effect, and in later work the application of the graphite paper was improved by isostatically pressing it on to the nichrome coated aluminium current collector (Wright and Tilley, 1981). By using isostatic pressing (at approximately 240 MPa (35 000 psi)) the graphite paper is increased in density and rendered impervious to the sulfur/polysulfide melt, and good adhesion to the nichrome coating is achieved. Typical performance of cells using this type of construction is shown in Fig. 6.8, and a comparison of

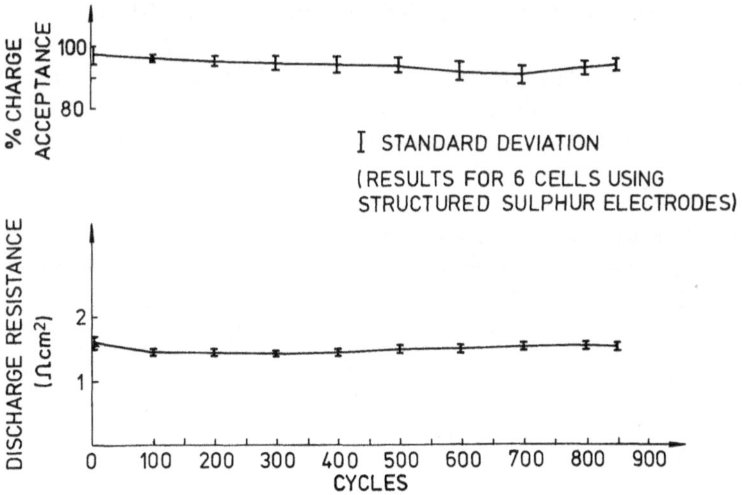

Fig. 6.8 Long term performance of cells using nichrome/isopressed graphite coated current collectors (Tilley and Wright, 1981). Reproduced with permission from the American Society of Mechanical Engineers.

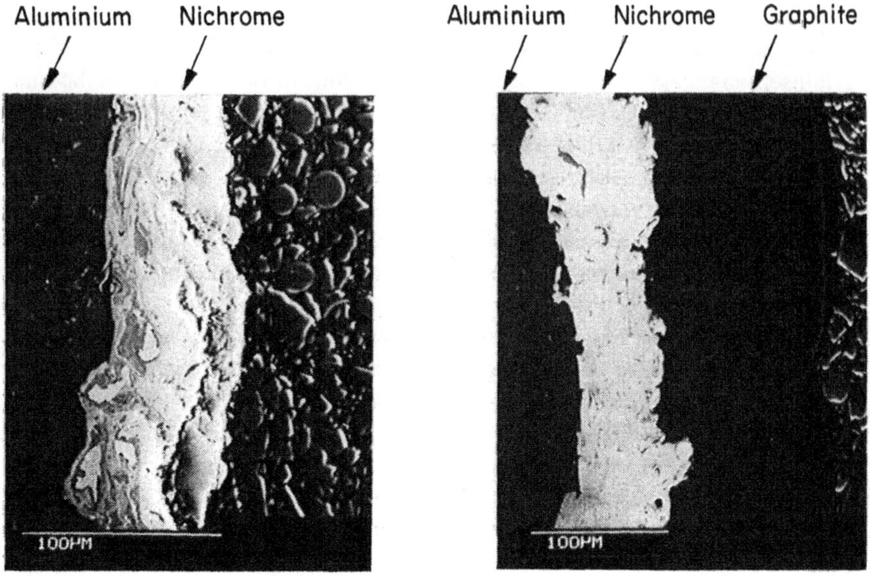

Fig. 6.9 A comparison of sectioned current collectors which have completed 700 electrical cycles (1750 Ah cm^{-2} current collector area) showing the effectiveness of the isopressed graphite coating (Tilley and Wright, 1981). Reproduced with permission from the American Society of Mechanical Engineers.

sectioned current collectors after 700 electrical cycles (1750 Ah cm^{-2} current collector area) is shown in Fig. 6.9. This work was extended to cell tests which contained current collectors using the isopressed flexible graphite layer passed directly on to an aluminium pole. Cells with this type of construction have been operated for over 1000 cycles with virtually no change in cell performance.

(e) *Rutile*

Rutile will be discussed in detail in Section 6.4.4(b), as the bulk of the work on conducting ceramics has been directed to producing thin walled tubes of the material. Thin coatings of niobium doped rutile on aluminium have been tested jointly by AERE Harwell and CSPL and at the British Rail laboratories. In-cell tests have been carried out on plasma sprayed coatings. The British Rail results are shown in Fig. 6.6 and the Harwell/CSPL results were very similar, the cell resistance increasing to 100 mΩ from an initial 20 mΩ after only 95 electrical cycles (Markin *et al.*, 1981). The Harwell and CSPL workers, however, did find that a plasma sprayed layer of doped rutile was beneficial if it was applied on nichrome coated aluminium. This type of construction with a nichrome interstrate gave very good in-cell test results. Virtually constant cell resistance was obtained for 560 electrical cycles (Markin *et al.*, 1981). The extra corrosion protection afforded by the conducting ceramic layer may be required for certain types of sulfur electrode structures, as with the British Rail isopressed graphite/nichrome interstrate aluminium current collector.

(f) *Summary*

As we have seen, there are several coated aluminium current collectors that give more than adequate in-cell performance. The easy fabrication of aluminium also lends itself to simple incorporation into cell designs. In central sulfur cells, aluminium intercell terminal connections can be simultaneously friction welded on to sulfur electrode closures along with the current collector (Bird, 1980). For the central sodium cell design, an aluminium cell case may be undesirable owing to the safety problems. Cell temperatures above the melting point of aluminium have often been reached in cell and battery safety tests. For the central sodium cell, aluminided and aluminized mild steel would be obvious alternatives to aluminium. The General Electric work using aluminided mild steel showed that a high aluminium content on the surface could be obtained by aluminiding, and that when tested in cells this was resistant to corrosion owing to the formation of an electrically insulating layer of Al_2S_3. A further coating over the aluminided layer is needed and, although the sprayed coatings would be difficult to apply to the inner wall of a cell case, the various graphite layers could easily be applied.

6.4.4 Metal coated and metal cored non-metals

(a) *Carbon and graphite*

As shown in Fig. 6.3, carbon was quite satisfactory for early development work using small central sulfur cells. For commercial use larger capacity cells are required; typically the beta alumina tube length has to be increased to 300 mm. For carbon to be used in larger cells it has to be cored to reduce the longitudinal resistance. Methods of coring carbon have been patented (Jones *et al.*, 1978a, 1978b) and these involve either complete liquid metal cores or solid metal cores such as aluminium with a liquid metal interface or a deformable conductive interface such as graphite felt. For coring to be effective, a stable low carbon-to-core interface resistance has to be maintained for long periods. This requires that the carbon or graphite porosity be eliminated and a leak tight method of joining carbon to the rest of the cell be found. Commercially available methods of reducing the porosity of carbon and graphite are available, such as CVD treatment and resin impregnation, although repeated treatments are needed before the porosity is completely eliminated. Molten salt impregnation methods have also been successful (Sudworth and Gibson, 1977). For central sulfur cells, however, the jointing problems and the high cost and low mechanical strength of long, thin wall, carbon or graphite sheaths have prevented any further development of this type of current collector.

Central sodium cells have been constructed and proposed which use carbon or graphite as the outer case material (Ludwig, 1979; Ritchie, 1978). As with the problems for the central sulfur cell, the porosity, the longitudinal resistance and sealing to the rest of the cell have to be overcome. Some methods of achieving these features have been proposed, such as sealing the surface porosity with pyrolytic graphite and metal plating or embedding wire mesh in the graphite body to provide axial conductivity (Ludwig, 1979), but no practical cell designs have been tested.

(b) *Conducting ceramics – G. R. Miller*

A large class of oxide ceramics can be readily made into semiconducting materials with room temperature resistivities as low as about $10^{-3}\ \Omega$ cm. Lanthanum salts such as lanthanum chromite and lanthanum manganites, as well as a wide variety of other perovskite mixtures, are normally p-type conductors, some with very low resistivities; they have a low temperature coefficient of electrical resistivity over a very wide temperature range. Most, however, are formed from quite costly raw materials and are difficult to process. As impervious coatings on metal shunts they may be more useful because cost limitations are less restrictive. These p-type conductors exhibit conductivities that are independent of oxygen partial pressure in the region around that of normal ambient atmospheres and down to oxygen pressures as low as 10^{-16} atmospheres, where the conductivity decreases quickly with oxygen pressure decrease. As a result, at high temperature where exposure to high oxygen partial pressure occurs, these materials can have advantages. Further, some of the p-type perovskites can have coefficients of thermal expansion which are close to ferritic steels, thus making them attractive as coatings on to those low cost steels.

Ceramics based on titanium oxide include common materials such as rutile and the titanates of calcium, strontium and magnesium. These can also be produced in a highly conductive form by additions of donor dopants. The resistivity of donor doped rutile at room temperature is as low as $8 \times 10^{-2}\ \Omega$ cm with little dependence on temperature between 25 and 1300 °C. Because of the higher electron mobilities in the calcium and strontium titanates, room temperature resistivities less than $1 \times 10^{-2}\ \Omega$ cm can be attained in donor doped materials (Yang *et al.*, 1981). Resistivities of these titanates increase by about a factor of three when the temperature is raised to 600 °C. A significant point to be made is that at high temperatures the conductivity of most donor doped titanates is unstable in highly oxidizing environments, decreasing with an increase in oxygen partial pressure (Johnson and Miller, 1977). However, this degradation in conduction characteristics is so slow as to be undetectable at temperatures below about 500 °C. Further, the use of these materials on well designed composite current collectors for the sodium sulfur cell would provide for reasonable protection from high oxygen partial pressures in any case. The obvious advantages of the n-type conductors are the inherent low cost of raw materials and their ease of fabrication. As such they can be fabricated into load bearing elements rather than simply being used as conductive, corrosion resistant coatings.

Potential current collectors constructed from conductive ceramic oxides have been described (Sudworth and Hames, 1974; Johnson *et al.*, 1977), and are illustrated in Figs 6.10 and 6.11. It was recognized at an early stage that the conductance of an oxide ceramic must be enhanced by the addition of a metal current collector (as for instance an aluminium core for a central sulfur cell or a nickel or aluminium coating for a central sodium cell container). In both of these designs the conducting ceramic is the load bearing element so that no special processing is required to avoid pinholes that coatings often exhibit. Processing techniques which allow for hermeticity over body thicknesses greater than about 1 mm are the only requirements, and these conditions are easily met. A concern involves the nature of the seal to the rest of the cell. With the ceramic acting as the load bearing element for the composite current collective device, it is difficult to avoid the construction of a metal to ceramic seal. Glassing of metal and ceramic parts and thermocompression bonding are the most likely prospects.

A most critical aspect of the composite current collector is the minimization of contact resistances not only between the corrosion resistant ceramic and the polysulfide melt, but also between the metal shunt and the ceramic. In experiments where liquid solder was used as a conductive interface with a solid

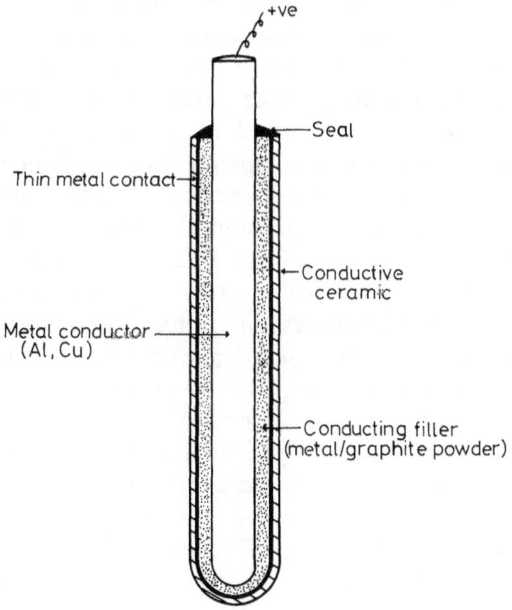

Fig. 6.10 Schematic of a conductive ceramic current collector for use in a central sulfur cell.

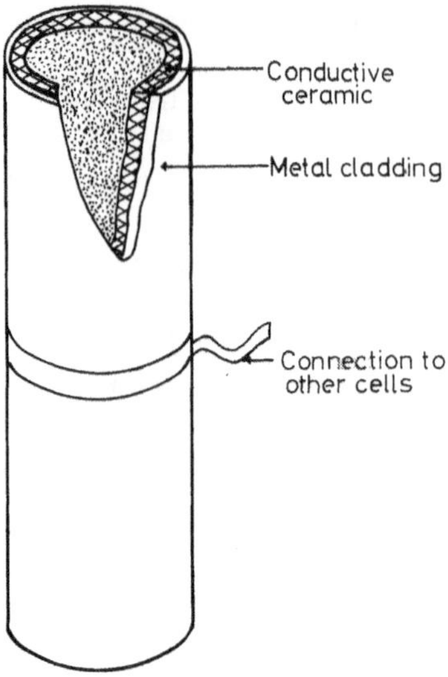

Fig. 6.11 Schematic of a conductive ceramic container for a central sodium cell.

metal shunt material, or between a solid metal shunt and an electroplated nickel coating on the surface of tantalum doped rutile (May, 1979), it was found that the interfacial resistance could easily be the greatest resistive component in a composite ceramic oxide current collector. That is, a liquid solder/metal contact or a liquid metal/conductive ceramic contact is highly resistive and is correlatable to a lack of wetting of the surface of the solid conductor by the liquid metal conductor. This effect of liquid metal/solid metal interfacial resistance has also been confirmed in experiments using liquid mercury contacts to a nickel coating on to tantalum doped rutile (Miller, 1980).

The solution to the contact resistance problem is apparently rather straightforward. Excellent contact can be maintained between electroplated nickel and conductive rutile as well as other titanates. The electroless nickel platings have processing advantages but produce a more brittle coating owing to the addition of large amounts of phosphorus to the plate. Normal electroplating from a Watts solution on to the surface of the ceramic material is easily carried out and the coating adherence is quite adequate up to at least 600 °C. At high temperatures in air the oxidation of the thin nickel coating can result in peeling of the coating from the conductive ceramic surface.

Contact with the metal shunt such as an aluminium rod (central sulfur) or aluminium sheaths (central sodium cell containers) can be quite adequate if graphite foil or fine metal powders are placed between the nickel coating and the thicker rod or sheath.

There has been only one adequate test of a ceramic oxide in a sodium sulfur cell. In a central sodium cell, the Ford Aerospace group used a tantalum doped rutile container with an electroplated nickel shunt (Paquette, 1981). The container was of 36 mm inside diameter and was about 130 mm in length, and used a 10 mm thick sulfur electrode. Sealing to the remainder of the cell was through a bolt type 430F flange which was glass sealed to the rutile tube. The cell was operated for 202 normal charge/discharge cycles with a mean utilization of about 80% of capacity and a cycle average resistance of 2.2 Ω cm^2 (including electrolyte and sulfur electrode resistance). This reasonably low resistance value seems to indicate that the preferential wetting of the oxide by sulfur (which would push the carbon fibres from the surface of the ceramic) was not a serious problem. After 180 cycles the capacity decreased with the decrease apparently due to a fracture of the flange to ceramic seal. After cooldown the rutile container was also fractured.

Post mortem examination showed a tiny amount of etching of the rutile in some areas, which appeared to be the actual removal of grains by spalling, indicating that a very small amount of grain boundary corrosion had occurred. This was likely due to an improper ceramic heat treatment, as had been predicted (Miller and Johnson, 1978). A proper postsintering heat treatment can probably avoid this problem entirely.

The only other known cell tests of conductive ceramic current conduction were carried out on rather high resistance conductive rutile (about 5–10 Ω cm at room temperature). Whenever the resistivity has been found to be this high, more corrosion occurs, especially grain boundary corrosion. Care must be taken to fabricate a homogeneously doped material with the resistivity that is appropriate for the quantity of dopant added to the material. It is quite clear that processes which provide a ceramic with improper conductivity (higher or lower than that expected, solely from dopant concentrations) will result in a less corrosion resistant substance.

In summary, it is becoming clear that conductive ceramic oxides show promise for applications as a current collector in a sodium sulfur cell. The amount of testing that has been carried out is, however, miniscule and only conductive rutile has received any attention at all. Since most conducting oxides exhibit maximum conductance and corrosion resistance under either cathodic or anodic conditions (but not both), the material selection has to be made at present without knowing many of the material parameters or the behaviour of the material in sulfur or polysulfide melts. It is also clear that the processing of the conductive ceramic must be carried out in a precisely controlled fashion for the combination of low resistivity and high corrosion resistance to be attained.

6.5 SUMMARY

The search for a suitable current collector has been one of the most difficult problems to solve in the development of the sodium sulfur cell. An enormous amount of early work on detailed corrosion studies in out-of-cell tests has been done but the results have been of limited use, and it is now accepted that the most conclusive results are obtained from testing materials in cells. During in-cell testing, the electrochemical potential as well as the stoichiometry of the melt varies periodically during operation, and the changes in melt composition cause changes in solubility of the corrosion products, in wetting properties, chemical aggressiveness and viscosity.

Similar materials have given different results when tested by different developers. This can often be attributed to different testing conditions, such as different sulfur electrode constructions. It has been shown how small changes in sulfur electrode structures can drastically affect current collector corrosion, and certain sulfur electrode structures that give good even electrical contact to the current collector, but not necessarily good cell performance, can enable cells to operate satisfactorily with quite poor current collectors. The development of the current collector and the sulfur electrode have to proceed together as one complete unit.

As the solution to the current collector problem is one of the keys to sodium sulfur battery commercialization, the patent literature is quite large compared with the normal published data. Several developers, however, appear to have arrived at practical solutions to the current collector problem that meet the requirements of good in-cell performance, long life, low cost and easy fabrication.

6.6 REFERENCES

Addison, L. C., Creffield, G. K., Hubberstey, P. and Pulham, R. J. (1969) *J. Chem. Soc.* (A), 1482.

Bast, J. (1982) EPRI Report EM-2579, Project 128–6.

Bauer, R., Haar, W., Kleinschmager, H., Weddigen, G. and Fischer, W. F. (1976/1977) *J. Power Sources*, **1**, 109.

Bindin, P. J. (1979) UK Patent Application GB 2013022.

Bird, J. M. (1980) UK Patent Application GB 2030758.

Bird, J. M., Tilley, A. R. and Sudworth, J. L. (1978) 13th IECEC, San Diego, Paper 789201.

Bones, R., Brook, A. and Markin, T. (1975) in *Power Sources*, vol. 5 (ed. D. H. Collins), Academic Press.

Bones, R. and Markin, T. (1977) 152nd Electrochemical Society Meeting, Atlanta.

Breiter, M. W., Chatterji, D., King, R. N. and Park, D. (1980) US Patent 4219128.

Bush, J. B. (1975) EPRI Report 128-2.

Chatterji, D. (1976) EPRI Report EM-266, Project 128-3.

Chatterji, D. and Dubin, R. (1977) US Patent 4048390.

Chatterji, D. and Dubin, R. (1978) US Patent 4123566.

Dell, R. M., Sudworth, J. L. and Jones, I. W. (1976) 11th IECEC Conference, Nevada, Paper 769085.

Demott, D. S. (1980) 4th DoE/EPRI Sodium Sulfur Battery Workshop, Derby.

Dubin, R. R. (1977), 152nd Electrochemical Society Meeting, Atlanta.

Dubin, R. R. (1980) US Patent 4189531.

Fischer, W. F., Haar, W., Hartmann, B., Meinhold, H. and Weddigen, G. (1978) *J. Power Sources*, **3**, 299.

Hartmann, B. (1978) *J. Power Sources*, **3**, 227.

Hartmann, B. (1980) 4th DoE/EPRI Sodium Sulfur Battery Workshop, Mannheim.

Hartmann, B., Haar, W., Fischer, W. and Bauer, R. (1980) US Patent 4 216 275.

Hartmann, B. and Kleinschmager, H. (1979) UK Patent Application GB 2 018 500.

Hattori, S., Yamaura, M., Kimura, S. and Iwabuchi, S. (1977) Proceedings SAE Congress, Detroit, paper 770281.

Johnson, O. W. and Miller, G. R. (1977) *J. Am. Ceram. Soc.*, **56**, 706.

Johnson, O. W., Miller, G. R. and Beutler, P. S. (1979) US Patent 4 160 069.

Jones, I. W. (1976) UK Patent 1 455 031.

Jones, I. W., Robinson, G. and Bird, T. L. (1978a) UK Patent 1 513 681.

Jones, I. W., Robinson, G. and Bird, T. L. (1978b) UK Patent 1 513 685.

Kinsman, K. R. and Oei, D. G. (1977) 152nd Electrochemical Society Meeting, Atlanta.

Kinsman, K. R. and Winterbottom, W. L. (1981) *Thin Solid Films*, **83**, 417.

Levine, C. (1983) in *The Sulphur Electrode* (ed. R. Tischer), Academic Press, New York.

Levine, C., Dean, D. and Wu, M. (1978) 2nd Annual Battery and Electrochemical Conference, Sodium Sulfur Battery, United States Department of Energy, Washington DC.

Ludwig, F. A. (1979) US Patent 4 166 156.

Ludwig, F. A. (1980) Sodium/Sulfur Battery Review, Ford presentation to DoE, Newport Beach, Ca.

McClanahan, M. L. (1978) Sodium/Sulfur Battery Review, Ford presentation to DoE, Newport Beach, Ca.

Markin, T. L., Bones, R. J., Scott, K. T., May, G. J. and Robinson, G. (1981) UK Patent 1 582 845.

Markin, T. L., Junkison, A. R., Bones, R. J. and Teagle, D. A. (1978) in *Power Sources*, vol. 7 (ed. J. Thompson), Academic Press.

May, G. J. (1979) *J. Mater. Sci.*, **14**, 633.

Miller, G. R. (1980) Sodium/Sulfur Battery Review, Ford presentation to DoE, Newport Beach, Ca.

Miller, G. R. and Johnson, O. W. (1978) in *Processing of Crystalline Ceramics* (eds H. Palmour, R. F. Davis and T. M. Hare), Plenum.

Mills, K. C. (1974) *Thermodynamic Data for Inorganic Sulphides, Selenides and Tellurides*, Butterworth, London.

Mitoff, S. P. (1976) EPRI Report EM-266, Project 128-3.

Mitoff, S. P. (1978) EPRI Report EM-683, Project 128-4.

Mitoff, S. P. (1980) EPRI Report EM-1341, Project 128-4, 5.

Paquette, D. G. (1981) Sodium/Sulfur Battery Review, Ford presentation to DoE, Newport Beach, Ca.

Park, D., Breiter, M. W. and King, R. N. (1979b) UK Patent Application GB 2 020 884.

Park, D., Chatterji, D. and Breiter, M. W. (1979a) UK Patent Application GB 2 017 389.

Partridge, B., Jenkins, T. and McGuire, M. (1976) US Patent 3 982 959.

Piercy, R. and Sudworth, J. L. (1977) Symposium on Electrode Materials and Processes for Energy Conversion and Storage, *Proc. Electrochem. Soc.*, **77-6**, 950.

Ritchie, E. E. (1978) Sodium/Sulfur Battery Review, Ford presentation to DoE, Newport Beach, Ca.

Sudworth, J. L. and Gibson, J. G. (1977) UK Patent 1 491 905.

Sudworth, J. L. and Hames, M. D. (1977) UK Patent 1 471 914.

Sudworth, J. L., Tilley, A. R. and Bird, J. M. (1977) 152nd Meeting Electrochemical Society, Atlanta.

Tilley, A. R. and Wright, M. L. (1981) Proceedings of the 16th IECEC, Atlanta, paper 819394.

Weddigen, G. (1977) 28th ISE Meeting, Bulgaria, Varna, Abstract 114.

Weiner, S. A. (1973, 1974, 1975, 1976, 1977) Reports, Ford Sodium/Sulfur Battery Program, National Science Foundation, Washington DC, Contract NSF C-805.

Wicker, A. (1980) Compagnie Générale d'Électricité, Research Report RP 109-7, prepared for EPRI.

Wicker, A. (1983) EPRI Report EM-2947, Project 109-7.

Wright, M. L. and Tilley, A. R. (1981) UK Patent Application GB 2 068 632.

Yang, Z. Z., Lee, C. C. and Miller, G. R. (1981) Electronics Division meeting of the American Ceramic Society, Newport Beach, Ca.

7 The sodium electrode

D. S. DEMOTT and A. R. TILLEY

7.1 INTRODUCTION

The development of a satisfactory sodium electrode for the sodium sulfur cell has concentrated on two areas of work: interfacial polarization and sodium supply. In order to ensure that the sodium electrode does not significantly contribute to the overall cell resistance, the sodium/beta alumina interfacial polarization has to be kept to a minimum throughout the cell operating life, and the sodium metal has to be maintained in contact with the full area of the beta alumina surface during cell operation.

7.2 POLARIZATION AT THE SODIUM/BETA ALUMINA INTERFACE

In the initial stages of sodium sulfur cell development it was not anticipated that polarization problems would arise from the sodium/beta alumina interface. Liquid sodium metal should make an ideal reversible electrode; provided it could be maintained in contact with the whole operating area of the beta alumina throughout discharge and charge of the cell, it was expected to be the least problematical component of the cell.

It was reported at a fairly early stage, however, that under certain conditions considerable polarization can occur at the sodium/beta alumina interface (Sudworth *et al.*, 1973). Using a beta″ alumina composition of 0.5% Li_2O, 2% MgO, 8.9% Na_2O, balance Al_2O_3, two types of experiment were performed.

It was suspected that the purity of the sodium, in particular its oxide content, could be important, and the first type of experiment was designed to compare the polarization behaviour of pure sodium with that of sodium containing dissolved sodium oxide. Two experiments were carried out on beta″ alumina tubes with sodium in contact with both the internal and external surfaces. The test cell for the first experiment contained two concentric beta″ alumina tubes as shown in Fig. 7.1. The larger tube was filled electrolytically from the external sodium and the smaller tube was then also filled by electrolysis from the intermediate sodium. Thus both surfaces of the inner tube, on which the measurements were made, were in contact with sodium which was free from dissolved oxide. AC resistances were measured between 250 and 300 °C over the frequency range 200 Hz to 20 kHz and DC cyclic galvodynamic polarization

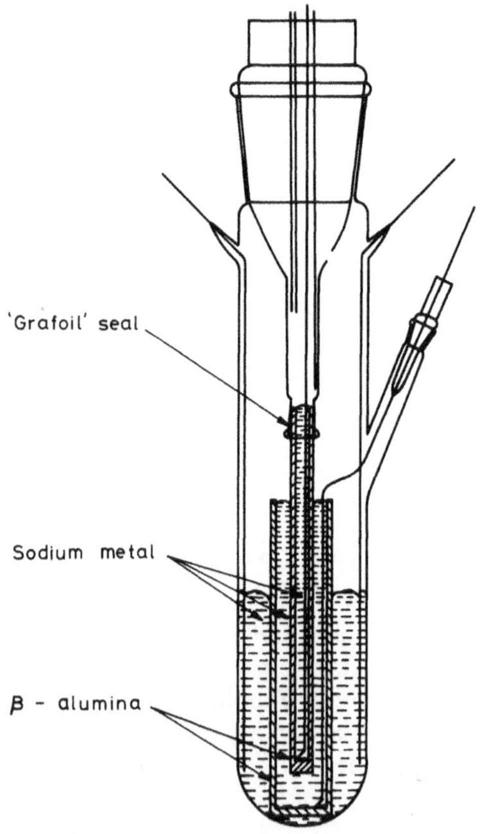

Fig. 7.1 Test cell for sodium/beta alumina/sodium tests using two concentric beta alumina tubes (Sudworth *et al.*, 1973). Reproduced with permission from North Holland Publishing Company.

curves were obtained at current densities up to 900 mA cm^{-2}. The initial DC polarization curves were non-ohmic, asymmetric and sweep rate dependent at all temperatures. The AC resistances decreased with increasing frequency and at 20 kHz were roughly one fifth of the DC values. No significant changes were observed after holding the temperature at 350 °C for several hours. However, after a period of DC cycling both the AC and DC resistances were lowered, although the values were still not in agreement and the DC values remained non-ohmic and asymmetric.

The second experiment of this type was similar except that the outer beta″ alumina tube was omitted; the tube under test was immersed in sodium saturated with sodium oxide at 125 °C and filled with sodium electrolytically. At 250 °C the DC polarization curves were similar to those for purified sodium but at higher temperatures they became more ohmic and more symmetrical. The AC resistances were also less frequency dependent and were closer to, but still less than, the DC values. These results were unaltered after maintaining the cell at 350 °C for several hours and after a period of DC cycling.

The second experiment involved incorporating a sodium reference electrode within the solid electrolyte. This was done by drilling a small radial hole in a thick beta″ alumina disc close to one face into which a platinum wire was sealed with glass. The disc was then assembled in a sodium sulfur cell. The hole was adjacent to the sodium electrode and a small amount of sodium was generated in it by electrolysis from the sodium electrode. The resistivity of the disc had previously been measured using $NaNO_2/NaNO_3$ electrodes so that the expected resistance between the reference and sodium electrodes could be calculated. Initial results indicated no detectable polarization at the sodium electrode/beta″ alumina interface, but after 16 hours on open circuit at 350 °C considerable polarization appeared. This remained on discharging the cell, fell to zero during cell recharge and reappeared when charging was discontinued.

Thus this early work showed that substantial, but very variable, polarization can appear under certain conditions at the sodium/beta″ alumina interface. Intuitively it appeared surprising that it was more severe when the oxide content of the sodium was lower; however, the results are consistent with the observations of Bradhurst and Buchanan (1961) who showed that wetting of ceramics is enhanced by dissolved oxide.

At the same time Armstrong *et al.* (1973) had been examining the sodium/beta alumina interface by complex plane analysis. They measured the impedance of the cell, sodium/beta alumina/sodium using beta alumina discs over the frequency range 0.01 Hz to 50 kHz at temperatures between 150 and 350 °C. It was found that with polished surfaces intimate contact with liquid sodium was established rapidly whereas with a rough surface the liquid sodium is essentially in contact only with the projections which constitute the major part of the surface. The wetting of rough surfaces was found to be temperature dependent and effective contact areas of 10^2–10^3 times the geometric area were estimated.

7.2.1 Wetting studies

Gibson (1977) performed some interesting experiments on the wetting of beta alumina by liquid sodium by means of X-radiography of sodium/beta alumina/sodium cells quenched from their operating temperature. He observed that at temperatures below 300 °C wetting did not occur; this was confirmed by electrical measurements which indicated that contact resistances far exceeded the resistive contribution of the electrolyte. At 350 and 400 °C contact angles were much lower as were the measured resistances. Both the contact angle and resistance measurements indicated that wetting, once initiated, was maintained even when cells were cooled to 200 °C. He also found that wetting was further improved by passage of current and that the cell resistivity then became equivalent to the electrolyte resistivity over the whole temperature range.

Wetting studies have also been performed more recently by Viswanathan and Virkar (1982) using the sessile drop technique. Although their results were in broad agreement with those of Gibson, the contact angles they measured covered a much wider range; they propose that this is due to distortion of the sodium surface during the quenching technique used by Gibson. However, their results show that the wide range of contact angles is due to the effect of moisture. Contact angles ranged from 160° at 260 °C, representing almost complete non-wetting, for an unbaked sample, to 0° at 360 °C for a sample baked at 425 °C in vacuum for 60 hours prior to measurement; a sample exposed to 100% relative humidity for 24 hours gave a contact angle of 103° at 360 °C. They propose that poor wetting due to the presence of moisture shortens the life of the electrolyte, presumably because of current concentration effects at the edge of non-wetted areas, and their sodium/beta alumina/sodium tests confirm this. Like Gibson, they did not observe any noticeable dewetting when the temperature was decreased.

Although the early observations had shown that incomplete wetting of the beta alumina surface by sodium and consequent polarization problems were likely in sodium sulfur cells, little was reported for a number of years.

Work on the impedance of the interface between beta alumina and sodium and other electrodes continued (Armstrong *et al.*, 1974; Armstrong and Burnham, 1976; Bukun *et al.*, 1973, 1974; Tkacheva *et al.*, 1978). In a study at ambient temperature of the transfer of Na^+ across the interface between beta alumina and propylene carbonate containing Na^+ in solution, Voinov and Tannenberger (1974) proposed that the presence of water was the cause of time dependent and asymmetric resistances. Farrington (1976 and 1977), in more extensive investigations with the same system, found large increases in interfacial impedance for Na^+ entry and exit as a result of exposure of the beta alumina to water vapour.

Problems with the sulfur electrode and corrosion dominated sodium sulfur development work and were frequently blamed for high and increasing cell resistances. It is also probable that improvements in ceramic electrolyte fabrication techniques and changes of composition fortuitously minimized interfacial polarization losses.

As a result of improvements in sulfur electrode and current collector design and beta alumina fabrication, sodium sulfur cell resistances became lower and more stable. Consequently any polarization problems at the sodium/beta alumina interface became more obvious.

In 1980 workers at British Rail reported that very severe rates of cell resistance increase could occur which were attributable to the sodium/beta″ alumina interface (Demott, 1980). The rate of increase was very variable but in some cases the cell resistance doubled in less than ten cycles. Other effects were found to be associated with the increase in resistance: the resistance was asymmetric (i.e. lower on charge than on discharge); an even higher transient resistance occurred at the start of discharge or after a period on open circuit; the minimum discharge resistance decreased with increasing current (i.e. it was non-ohmic). These effects are very apparent in discharge and charge curves (Fig. 7.2)

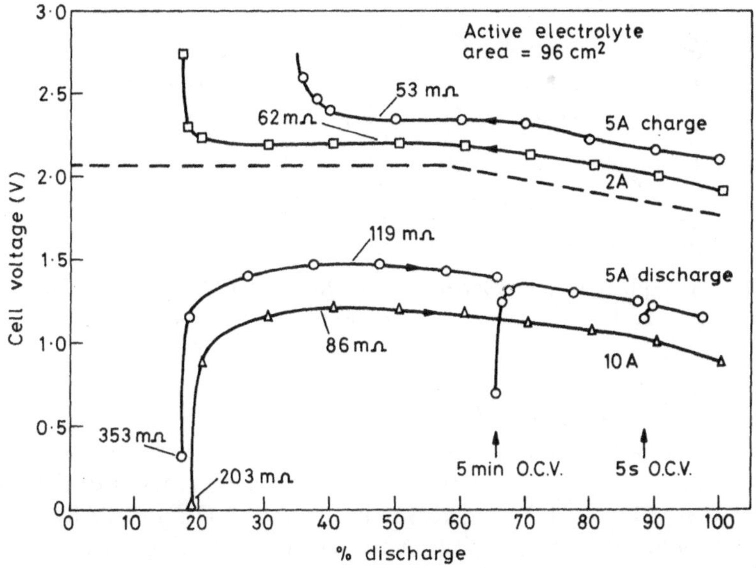

Fig. 7.2 Charge/discharge behaviour of a typical high resistance cell (Demott, 1980). Reproduced with permission of the Electrochemical Society.

and polarization curves (Fig. 7.3). The composition of this beta″ alumina was 0.7% Li_2O, 8.9% Na_2O, balance Al_2O_3. This composition had been adopted at a much earlier date because it had a much longer life than electrolyte doped with both magnesia and lithia and also, over a period of several years, seemed to be much less prone to interfacial polarization problems.

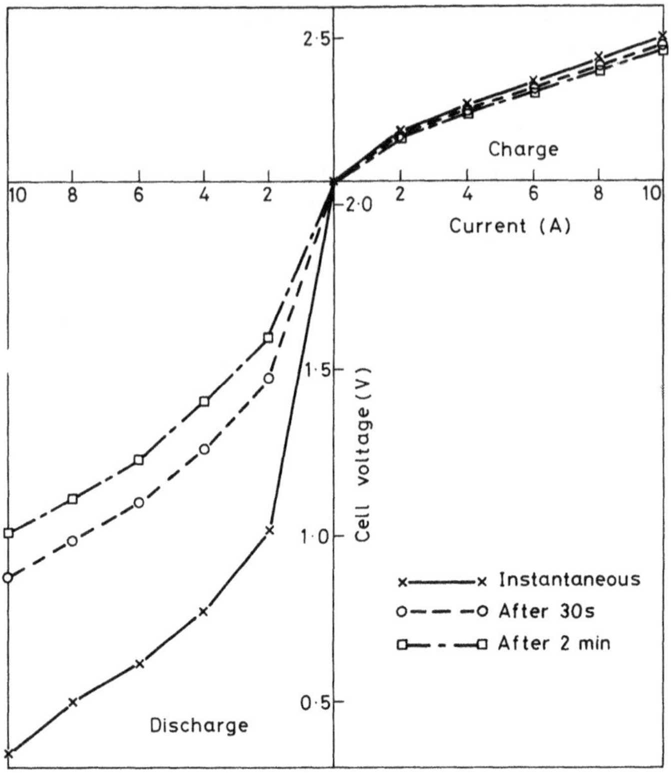

Fig. 7.3 Two phase region polarization data for the cell shown in Fig. 7.2 (Demott, 1980). Reproduced with permission of the Electrochemical Society.

7.2.2 Sodium electrode renewal

Conclusive proof that the resistance and associated effects were attributable to the sodium/beta″ alumina interface was obtained by renewal of the sodium electrode in cells which had increased in resistance (Demott, 1980). This was done by cooling the cell in the fully charged state and cutting away the cell case. Sodium was removed from the outside of the beta″ alumina tube by scraping and by reaction with ethanol; the complete beta″ alumina tube and sulfur electrode assembly could then be reassembled into a new cell case with fresh sodium. It was found that this procedure resulted in a reduced cell resistance, both for cells which had shown a rapid resistance increase over a few cycles (Fig. 7.4) and for others which had increased in resistance relatively slowly over a large number of cycles. Not only was the subsequent discharge resistance usually lower than the previous minimum value, but also it normally remained constant over a large number of cycles and the time dependent and non-ohmic effects were eliminated. An exception to this general behaviour was found when

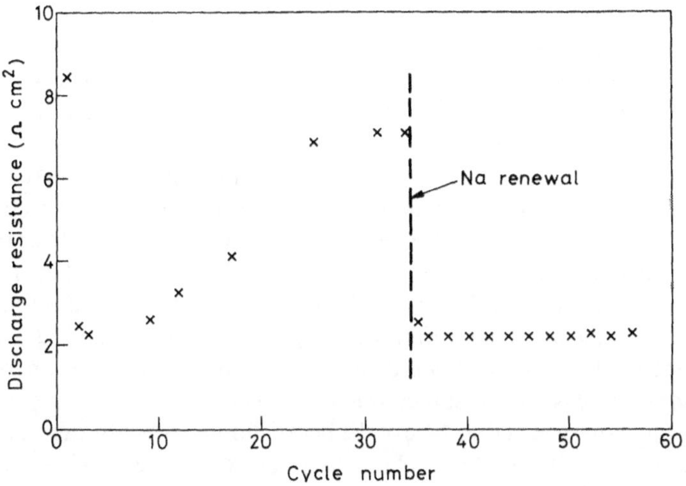

Fig. 7.4 Resistance behaviour of a typical cell before and after sodium renewal (Demott, 1980). Reproduced with permission of the Electrochemical Society.

the sodium electrode was renewed after only a few (< 30) cycles; although the resistance was initially reduced, it could again increase during further cycling. An example is shown in Fig. 7.5; this also shows that a stable resistance could be achieved after a second renewal of the sodium electrode. This suggests that material is accumulating or reacting at the electrolyte surface during operation

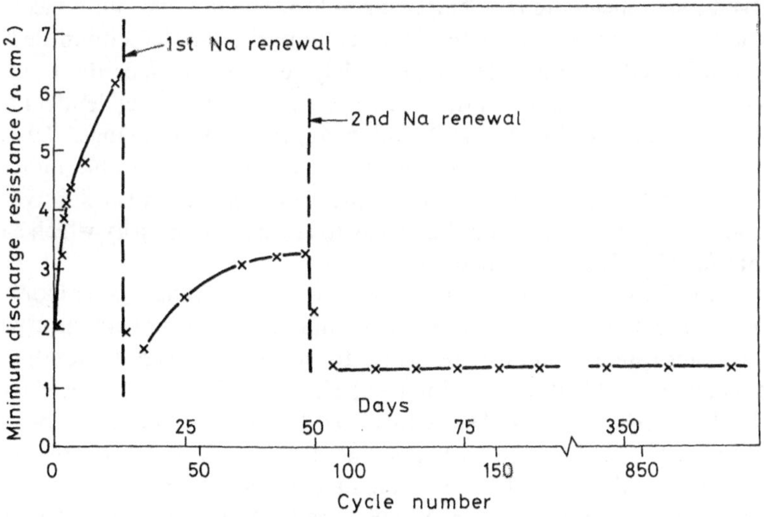

Fig. 7.5 Cell resistance behaviour before and after two sodium electrode renewals (Demott *et al.*, 1981). Reproduced with permission of British Rail Research.

of the cell and is removed during the sodium renewal process, and that not until all such material is removed is a stable resistance obtained. Clearly, renewal of the sodium electrode is not a practical solution to the problem and further experiments were performed (Demott *et al.*, 1981; Bugden *et al.*, 1981) to clarify the source of the problem and to find methods of overcoming it.

The effect of moisture was investigated by storing the electrolyte tubes under both dry and 100% relative humidity conditions for various periods before assembly into cells. No correlation was found between storage time or humidity and the rate of resistance increase. Also, cells which had exhibited the resistance increase phenomena and which had had their sodium electrodes renewed to eliminate interfacial polarization were again dismantled and the external surfaces of the tubes were exposed to water vapour for extended periods; on retesting no increase in resistance was observed.

Removal of the external surface of the beta″ alumina tubes by diamond grinding did not prevent the build-up of the interfacial phenomena. Also, some tubes removed from high resistance cells had a fresh surface exposed by grinding; when re-tested, these behaved in the same manner as previous cells after sodium renewal. That is, cells containing tubes which had been subjected to an extended period of cycling ($\geqslant 30$ cycles) had a low, stable resistance whereas those containing tubes which had received only a few cycles again rose in resistance.

These results suggested that the interfacial effects are not dependent on the properties of the as-fired surface (or material absorbed in the surface layers), but support the hypothesis that a species is being eluted from the bulk of the ceramic electrolyte.

Sources of impurities within the cell which could be responsible for the observed phenomena are the components of the sodium and sulfur electrodes and the beta″ alumina electrolyte. Although impurities in the sodium electrode may be involved, they are clearly not solely responsible since the resistance problems are eliminated when the electrode is replaced by one identical to the original. It was also found that if, following a sodium exchange, the sulfur electrode is also renewed, the cell resistance remains low and stable, indicating that neither is the sulfur electrode to blame. Thus, in the British Rail cells the ceramic electrolyte is clearly implicated as the source of impurity which causes the interfacial resistance to increase.

The impurity most likely to be responsible is calcium. It has been reported by various workers that calcium, occurring as an impurity in the ceramic powders used to fabricate the electrolyte, leads to resistivity increases in the beta alumina (Jones and Miles, 1971; Imai and Harata, 1972; and Buechele and De Jonghe, 1979). Also, Yasui and Doremus (1978) showed that additions of CaS to the sulfur electrode can cause very rapid increases in cell resistance; they found high levels of calcium within the bulk of the beta″ alumina in these cells, concentrated in grain boundaries or in the smaller grains, even in the absence of an applied electric field.

The British Rail workers confirmed the mobility of calcium by adding radioactive Ca-45 to the sulfur electrode (Demott *et al.*, 1981). After the cells had been cycled, or left on open circuit, calcium was found in both electrodes and on both beta″ alumina surfaces. Particularly high levels of activity were found on the beta″ alumina surface adjacent to the sodium electrode. When a thin layer was removed from this surface by grinding, the measured activity was greatly reduced, confirming a calcium concentration at the surface.

By analysis of ceramic samples from archive cells and by a parametric study using three Li_2O and four Na_2O doping levels, they found that the Li_2O content was critical, and that in combination with slight changes of the Na_2O content, widely differing rates of rise were possible, as shown in Fig. 7.6 (Bugden *et al.*, 1981). To show that Li_2O itself was not causing the resistance rise, cells containing electrolyte tubes which had been heated in molten $LiNO_3$ to raise the Li_2O content from 0.7 to 1.0% were constructed and tested: there was no significant difference in the behaviour of these cells and control cells.

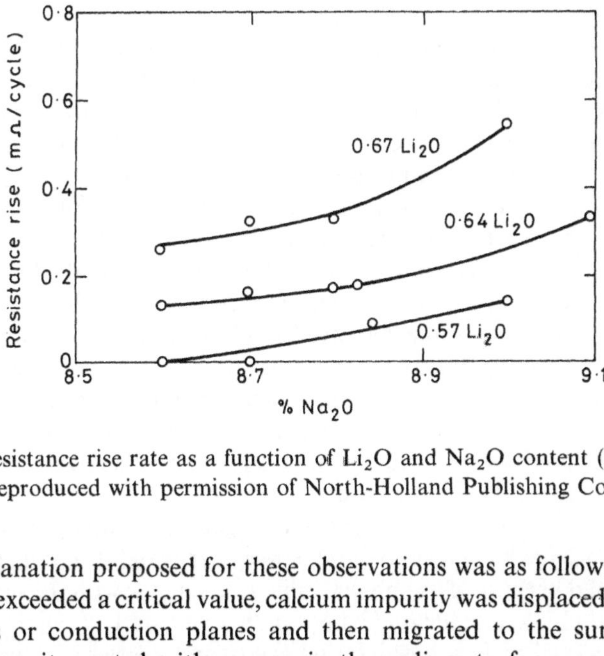

Fig. 7.6 Resistance rise rate as a function of Li_2O and Na_2O content (Bugden *et al.*, 1981). Reproduced with permission of North-Holland Publishing Company.

The explanation proposed for these observations was as follows. When the Li_2O level exceeded a critical value, calcium impurity was displaced to the grain boundaries or conduction planes and then migrated to the surface during cycling where it reacted with oxygen in the sodium to form an oxide. This hypothesis was tested by adding 0.22% CaO to a ceramic batch known to exhibit good resistance stability; tubes treated in this way caused resistance rise rates at least an order of magnitude greater than the control cells.

7.2.3 Treatments to cells

The experimental evidence strongly implies that impurities in the ceramic electrolyte can migrate to the surface during cycling and modify the sodium/beta″ alumina interface so that anodic sodium dissolution and, to a lesser extent, cathodic sodium deposition are impeded. Thus two approaches to overcoming the problems suggest themselves: changing the ceramic electrolyte formulation, and treating the sodium/beta″ alumina interface.

The work of Bugden *et al.* (1981) showed that at low dopant levels resistance rise can be minimized (Fig. 7.6); it is also clear that use of an alpha alumina precursor containing less calcium impurity will be beneficial.

Considerable success was achieved by treatment of both the ceramic surface and the liquid sodium. Treatment of the surface was aimed at improving its initial wettability by sodium (Wright, 1981; Demott *et al.*, 1981). Various surface treatments were investigated by observing the contact angle between a drop of liquid sodium and the surface of a beta″ alumina disc in a nitrogen atmosphere. The lowest contact angle was obtained with a surface which had been rubbed with metallic lead. Rubbing was not a practical method of coating the surface of electrolyte tubes with lead so an alternative procedure was adopted. The electrolyte surface was coated with lead acetate using a saturated aqueous solution, and this decomposed to metallic lead during thermal activation of the cell. Results from testing more than 100 cells containing electrolyte treated in this way showed it to be only partially successful: about half the cells had a constant resistance with no asymmetric or transient effects, whereas the remainder rose in resistance at various rates.

Levine and Anand (1977) had reported the addition of a titanium/zirconium alloy to sodium to act as an oxygen getter in order to prevent corrosion of the glass capillary electrolyte in the Dow cell design (as described in Chapter 4); they also found that this led to enhanced wetting of the glass by the sodium, although a reduction of the dissolved oxide level in the sodium would be expected to impair the wetting of ceramics (Bradhurst and Buchanan, 1961). Since oxygen appeared to be implicated in the interfacial polarization phenomena described earlier and wetting was also seen to be important, it was considered worth while to investigate the effect of oxygen getters. Thermochemical data for the oxides (Glassner, 1957) indicates that several metals should be capable of acting as oxygen getters in liquid sodium at 350 °C. Although CaO has the most favourable free energy of formation, calcium was not considered because of its known ability to exchange with the sodium ions in the beta″ alumina and degrade its conductivity and strength (Yasui and Doremus, 1978). Further consideration of cost and safety led to the selection of titanium and aluminium for addition to the sodium electrode to reduce its oxide content. Both titanium (sponge) and aluminium (flake) additions (approximately 3% by weight of the sodium) were found to reduce significantly the rate of resistance increase although they did not completely eliminate the problem (Demott *et al.*, 1981; Wright and Hames, 1982).

Since both the application of lead acetate to the electrolyte surface to improve wetting and the addition of titanium or aluminium to the sodium to act as an oxygen getter were found, separately, to be beneficial, a combination of both

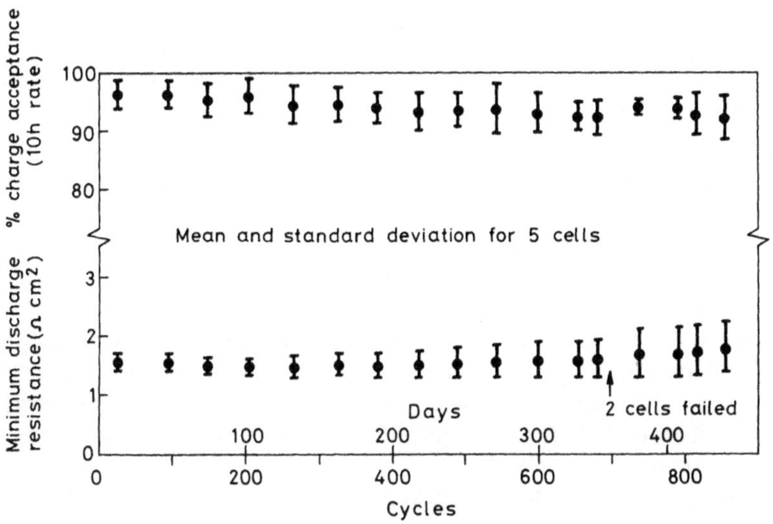

Fig. 7.7 Long term performance of cells containing beta″ alumina tubes treated with lead acetate and sodium treated with titanium powder (Demott *et al.*, 1981). Reproduced with permission of British Rail Research.

treatments was investigated. Fig. 7.7 shows that this was completely successful; cell resistances remained stable, symmetric and ohmic for hundreds of cycles.

Although these treatments have been found to be consistently effective, explanation of the mechanisms involved is somewhat tenatative. It is proposed that improved wetting, as a result of the treatment with lead acetate, has the effect of dispersing the calcium oxide layer over an increased surface area, thereby lessening its interference with sodium ion transport. The use of an oxygen getter minimizes the amount of calcium which is oxidized at the sodium/beta″ alumina interface; presumably most of the calcium impurity then remains mobile or possibly forms a titanate or aluminate at the surface of the getters. It is thought that it is only when calcium oxide is formed at the beta″ alumina surface that the problems arise. Accumulation of calcium at the beta alumina surface has also been observed by Lehnert and Hartmann (1982) at Brown Boveri. They found that asymmetric resistance behaviour occurred in sodium/beta″ alumina/sodium tests even without current flow. Surface analysis showed that this was due to a layer of $Ca(OH)_2$ and CaO. They have also confirmed that aluminium powder, acting as an oxygen getter, can entirely eliminate the asymmetry over extended periods of cycling. The transient and non-ohmic effects on discharge may result from localized heating of the electrolyte owing to current concentration caused by the surface film.

Workers at General Electric have also found similar asymmetric, non-ohmic and time dependent polarization effects at the sodium/beta alumina interface and have performed a variety of experiments to study them.

Breiter *et al.* (1980) found the asymmetric resistance behaviour to be very dependent on ceramic composition. Compositions containing 9.6% Na_2O with various Li_2O and MgO doping levels all showed substantial asymmetry; at lower (9.1%) soda contents there was a composition range in which either asymmetric or symmetric behaviour could occur and the composition 8.6% Na_2O, 0.5% Li_2O, 1% MgO was found to give symmetric behaviour. By using a sodium reference electrode in a thick walled tube they found the resistances at both interfaces to be larger during discharge than charge; the largest resistance contribution was at the sodium/beta alumina interface. Their suggested explanation for these observations is that a thin passivating film of Na_2O is formed at the interface owing to interaction of sodium with water, but they offer no explanation as to why such a film causes the asymmetric resistance behaviour. They also note that cell life is shorter when asymmetry exists, possibly because of current channelling through flaws in the film giving rise to high local current densities.

Breiter and Dunn (1981a) studied further the interfacial behaviour of various liquid electrodes with a beta″ alumina composition of 9.6% Na_2O, 0.75% Li_2O using cyclic voltammetry. In contrast to their previous paper, no polarization was found at the $NaNO_3$/beta″ alumina interface. With a sodium amalgam electrode the results were rather ambiguous owing to changes of wetting during sodium deposition. However, when sufficient sodium had been deposited into the mercury, ohmic and symmetric resistances were measured, although the values were somewhat larger than the computed bulk resistance of the beta″ alumina tube, suggesting that the oxidation properties of the amalgam either prevent a Na_2O interfacial film from developing or greatly inhibit its formation.

Dunn (1981) studied the effect of exposure to air on the resistivity, at room temperature, of both beta alumina (9.6% Na_2O, 0.25% Li_2O, balance Al_2O_3) and beta″ alumina (9.6% Na_2O, 0.75% Li_2O, balance Al_2O_3). Radial and axial resistivity measurements on samples of tubes showed that beta″ alumina was affected more than beta alumina and only in the radial direction, indicating a surface effect. He found that normal ohmic behaviour could be restored by heat treatment; considerable recovery occurred at 90 °C, but full restoration was not achieved until samples were heated to 600 °C. Thermogravimetry of powdered samples showed that the main weight loss occurred at ~100 °C followed by a slower loss up to ~600 °C (Fig. 7.8). The species evolved were identified by mass spectroscopy. Both H_2O and CO_2 were evolved; the H_2O was removed below 200 °C but CO_2 accounts for the weight loss at higher temperatures. He suggested that the excess resistance not removed below 300 °C is significant for sodium sulfur cell performance because it will be retained during normal cell operation.

Breiter and Dunn (1981b) made a more detailed study of the time dependence of the asymmetric behaviour using voltammetric and pulse techniques on a sodium/beta″ alumina/$NaNO_3$ system with a sodium reference probe within the thick wall of the electrolyte tube. Voltammetric sweeps suggest that at

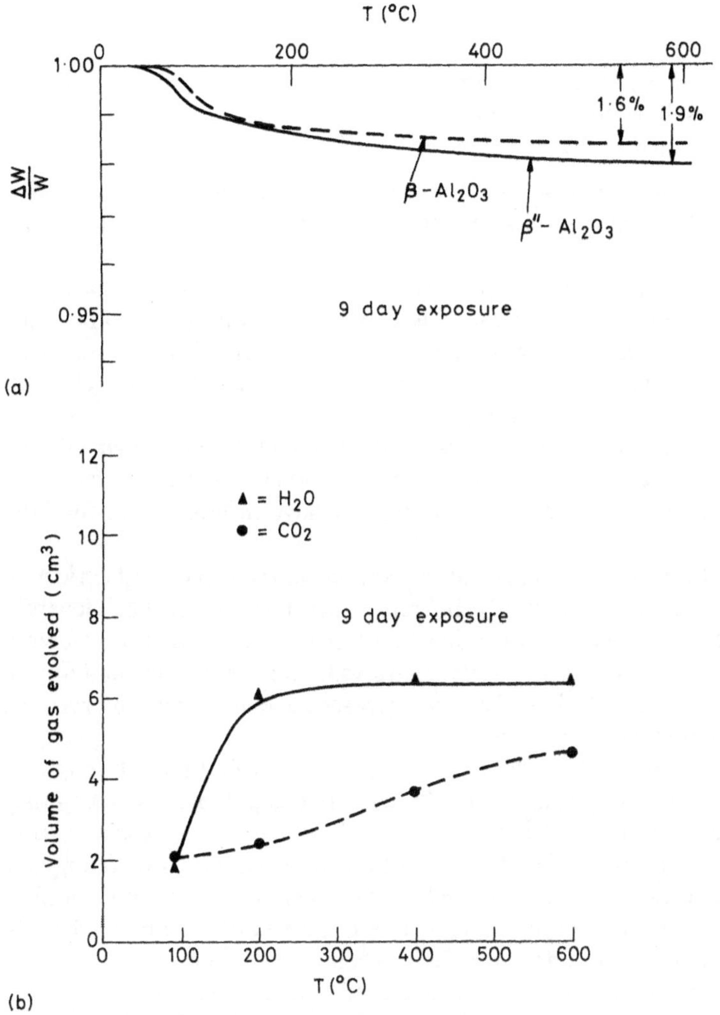

Fig. 7.8 Characterization of powdered beta and beta″ alumina tubes after exposure to air for nine days. (a) Thermogravimetric results (scan rate 1 °C min^{-1}); (b) cumulative volume of CO_2 and H_2O released (beta alumina only) (Dunn, 1981). Reproduced with permission of the American Ceramic Society.

anodic potentials a surface film becomes more passivating and that at sufficiently high cathodic potentials it is converted back to its less passivating state. It is not clear whether the film is inhomogeneous, containing small regions of metallic sodium, or discontinuous but still impeding sodium ion transport to the conduction planes. Calculation of the charge passed during film modification suggests that only a fraction of the film is modified during the sweeps.

Also, the extent of asymmetry was found to be very different for the three compositions studied.

Park *et al.*, (1981) described the resisitivity ageing characteristics in their sodium sulfur cells of ceramic of composition 9.6% Na_2O, 0.25% Li_2O, balance Al_2O_3. Three stages of resistance behaviour were observed:

(1) An initial decrease, or little change, of resistance. This was attributed to gradual removal of a moisture induced surface film leading to improved wetting.
(2) A resistance increase, which was shown by four point axial measurements to be due to increase of the bulk resistivity. It was proposed that this results from plugging of conducting channels in grain boundaries by a rapidly diffusing species such as water. It could be removed by high temperature (700 °C) annealing in air.
(3) A severe resistance increase which could be accompanied by surface degradation (spalling) of the ceramic. Other, non-ceramic, factors such as a poorly made sulfur electrode may also have an effect during this latter stage.

Thus the observations and interpretations of the General Electric and British Rail workers are manifestly different in several respects. The General Electric workers have clearly shown that a surface film, resulting from interaction with moisture, can seriously interfere with sodium ion transport and appears to be present initially. Bulk resistivity increases have also been observed which are also attributed to the presence of water.

The British Rail workers, however, frequently find that although the initial cell resistance is low, indicating little interfacial polarization, it can rise rapidly with cell cycling, and the sodium renewal results and exposure to moisture observations indicate that this is not due to moisture but to an impurity, most probably calcium, being eluted to the interface. The British Rail workers' observations that sodium sulfur cell resistances can be maintained at a low level over extended periods, either as a result of sodium electrode renewal, or by treatment of the interface with lead acetate to improve wetting combined with an oxygen getter in the sodium, indicate that interfacial polarization and associated cell resistance increase is not an inherent feature of the beta″ alumina/sodium interface.

7.3 SODIUM ELECTRODE CONSTRUCTION

There are three methods of maintaining sodium in contact with the full beta alumina surface. These are:

(1) feeding of sodium by gravity from a reservoir;
(2) wicking sodium to the beta alumina surface;
(3) forcing sodium from a reservoir by gas pressure.

7.3.1 Gravity fed sodium electrodes

In the early work on the sodium sulfur system, most groups used tubular cells in which the sodium was fed by gravity to the beta alumina tube from a sodium reservoir. A central sodium cell of this type is shown in Fig. 7.9. Central sulfur cells using similar gravity feed systems have also been described (Dell *et al.*, 1976). As ceramic development progressed, larger diameter beta alumina tubes became available and it became possible to build central sodium tube cells in which all the sodium could be contained within the beta alumina tube.

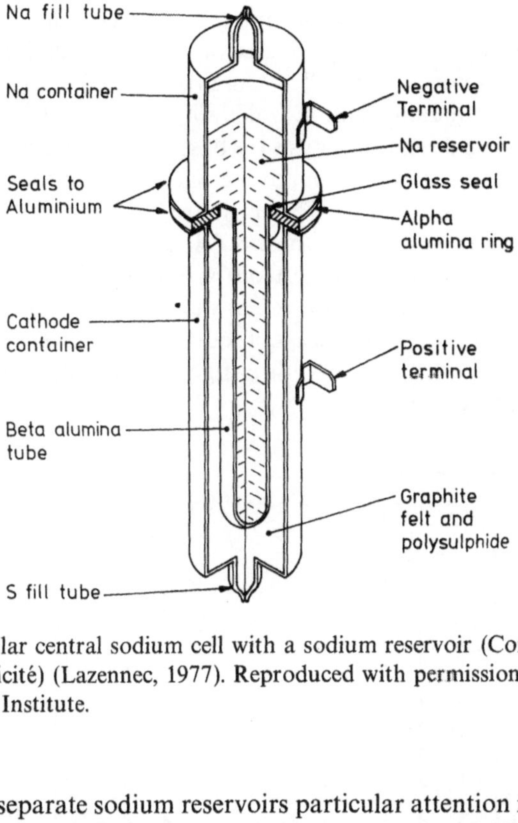

Fig. 7.9 A tubular central sodium cell with a sodium reservoir (Compagnie Général d'Électricité) (Lazennec, 1977). Reproduced with permission of the Electric Power Research Institute.

In cells with separate sodium reservoirs particular attention must be paid to the design of the cell seals. Failure of seals located in the middle of the cell will have more serious consequences than would be the case with vertically orientated cells which have the seals uppermost, and it is advisable to incorporate some sort of mechanical constraint to maintain the integrity of a cell should the primary seals fail. Aluminium thermocompression bond seals are particularly vulnerable owing to the relatively low melting point of aluminium. For load levelling applications, however, large cells with sodium reservoirs have been considered, as high capacity, low rate cells are required with no particular restriction on cell height (Minck, 1980). A gravity sodium feed system is also very appropriate for the Dow cell with its glass capillary electrolyte tubes which contain only a small volume of sodium.

7.3.2 Sodium capillary wicks

There are several advantages in eliminating the sodium reservoir and using a wick to maintain the sodium in contact with the beta alumina:

(1) cost reduction;
(2) the sealing problem may be simplified since the sodium electrode only requires a metal cap, and if the cell is operated vertically with the seals uppermost then the seals do not come into contact with the liquid reactants;
(3) it is possible to achieve a high degree of sodium utilization. In a cell which uses a sodium reservoir, the full internal volume of the electrolyte tube is still filled with redundant sodium when the cell is fully discharged. In a reservoirless cell design, as long as the base of the capillary wick is immersed in sodium at the end of discharge the full surface of the electrolyte is still wetted by capillary action. This is shown in Fig. 7.10. The quantity of sodium required in the reservoirless cell design is therefore only slightly greater than the stoichiometric quantity, only a slight excess being required to maintain contact with the wick at the end of discharge;
(4) the elimination of the large excess of sodium and the sodium reservoir gives a more compact cell design with a higher energy density which is more suited to electric vehicle use.

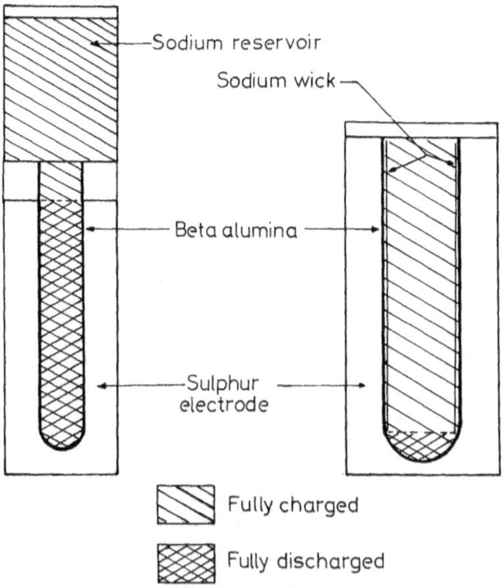

Fig. 7.10 Schematic comparison of a cell design with a sodium reservoir and a cell design that incorporates a sodium capillary wick.

The main requirements of a sodium wick material are:

(1) The wick material must be fully wetted by sodium at cell operating temperatures and preferably also at the sodium filling temperature, which is usually about 150 °C;
(2) the wick material should be resistant to chemical attack by liquid sodium, and it should not contaminate the sodium electrode with any harmful impurities;
(3) it should be cheap, light and easily incorporated into a cell;
(4) the wick material should not react exothermally with sulfur in the event of electrolyte failure;
(5) the wick should not outgas either at sodium filling temperatures or during warm-up of the cell to operating temperature.

It is relatively easy to obtain these requirements with metal wicks such as shims, gauzes and metal foams, but loose fill powder wicks, which in central sulfur cells perform a dual role of sodium displacement and sodium wicking, have given problems with outgassing (Stackpool, 1982), and fine powders of aluminium and iron which may be excellent wicks (Wright, 1982) can react exothermically with sulfur.

The three factors that govern the excess height of a column of sodium when using wicks are: the wetting behaviour of the materials in contact with the sodium; the surface tension of sodium; and the type of wick used.

(a) *The wetting behaviour of sodium*

The wetting behaviour of sodium metal on the beta alumina surface has been discussed in Section 7.2.1. The wetting behaviour of liquid sodium on various metals and non-metals has been studied fairly extensively, as it is of considerable interest for nuclear reactors of the fast breeder type where liquid sodium and sodium/potassium alloys are used as heat transfer media. Addison and his co-workers have studied the wetting of sodium as a function of time and temperature on nickel, cobalt, iron, molybdenum, tungsten, vanadium, titanium, zirconium and chromium (Addison *et al.*, 1962, 1968; Addison and Iberson, 1965). The receding contact angle was measured using a vertical plate technique, and all the metals had wetting temperatures below 200 °C. Addison concluded that the rate of change of contact angle was a measure of the rate at which sodium reduced the surface metal oxides. The reduction rate increased with temperature and depended upon the particular oxide present on the metal surface. Wetting was achieved once liquid sodium was in direct contact with the metallic surface. Additions of small quantities of calcium and barium to the sodium were shown to improve the wetting considerably, so helping to confirm that the wetting process involved oxide film reduction.

In similar work, however, Longson and Prescott (1973) showed that pure nickel and iron exhibited a minimum contact angle of 30° with pure sodium at 250 °C, and it was inferred that the reduction of the surface oxides was not taking place. On raising the temperature above 250 °C, zero contact angle was obtained as the oxide films were reduced. An increased wetting rate was also obtained by adding a trace of oxygen to the argon cover gas. The wetting behaviour and effect of oxygen impurity are shown in Fig. 7.11. With stainless steel samples, Longson and Prescott found that the contact angle did not fall below 30° as only wetting of the surface oxide took place, and this surface oxide was not removed by prolonged exposure to sodium even at higher temperatures.

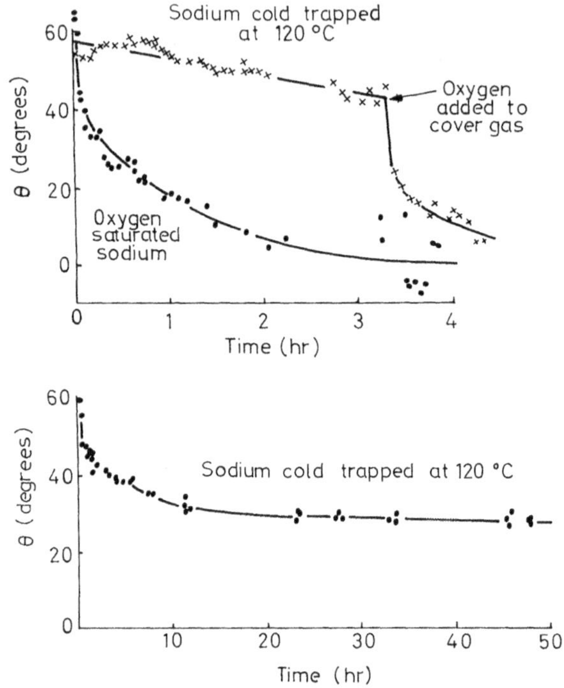

Fig. 7.11 Wetting curves obtained on abraded nickel at 250 °C (Longson and Prescott, 1973). Reproduced with permission of the British Nuclear Energy Society.

Sessile drop techniques have yielded different results, and in work by Hodkin *et al.* (1973) and Hodkin and Nicholas (1976) contact angles of 20° were not reached until 300 °C and above with pure metals such as iron, molybdenum, chromium and nickel, and stainless steels were not wetted at temperatures below 500 °C. Sessile drop experiments yield information about liquids advancing over a previously uncontacted surface, while vertical plate techniques characterize a liquid receding from a surface with which it has been in contact, which may explain the discrepancy between the values obtained by the two techniques. Hodkin and Nicholas found that surface preparation

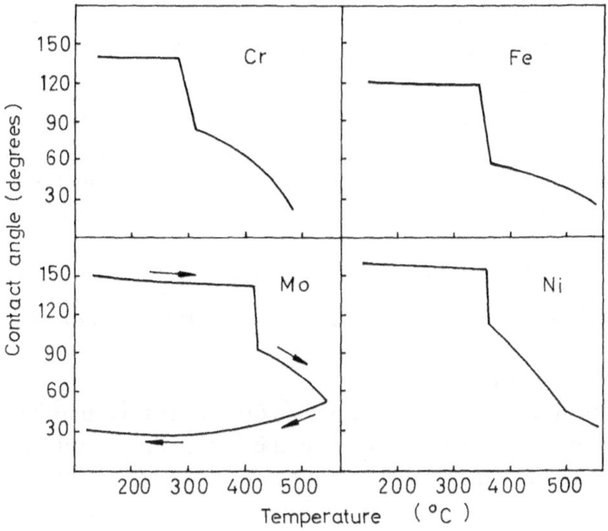

Fig. 7.12 The effect of temperature on the wetting of four pure metals by sodium (Hodkin and Nicholas, 1976). Reproduced with permission of the Atomic Energy Research Establishment, Harwell.

techniques had various effects on wetting behaviour; for instance, electrolytic polishing and chemical polishing had a detrimental effect; mechanical polishing had relatively little effect and ion bombardment significantly improved the wettability, as had been shown previously by Barlow and Planting (1969). Some of the results obtained by Hodkin and Nicholas are shown in Fig. 7.12.

Although the wetting behaviour of sodium on most metals has been characterized, the work on non-metals has been more limited. Hodkin and Nicholas (1979) have investigated the wetting characteristics of sodium on several oxides, carbides and nitrides. Virtually all the non-radioactive oxides studied (Al_2O_3, ZrO_2, Y_2O_3 and MgO) did not wet below 400 °C. In general, the carbides were more readily wetted than the corresponding oxides, and several nitrides wetted well below 400 °C. Earlier work by Stewart (1969) also showed that Al_2O_3 was not wetted below approximately 400 °C.

A large number of materials, then, are wetted at cell operating temperatures but not many are fully wetted at sodium filling temperatures. This non-wetting at filling temperatures has caused some problems (Stackpool, 1982), and sodium filling rigs which use a positive pressure to force the sodium into the various wick structures have been developed (Markin *et al.*, 1978; Riley and McGreavy, 1981). Methods of improving wetting at lower temperatures have been demonstrated including the use of treatments such as the lead acetate coating (Wright, 1981) discussed in Section 7.2.3. This and similar methods can be used as effectively on the various wick materials.

(b) *The surface tension of liquid sodium*

The surface tension of liquid sodium has been determined for relatively pure sodium and for a range of sodium oxide concentrations (Longson and Thorley, 1967; Taylor, 1954; Addison *et al.*, 1954, 1955). In the work by Longson and Thorley, the surface tension of sodium with a low oxygen content (approximately 3 ppm) and oxygen saturated sodium was determined over a temperature range from 98 to 202 °C. They obtained the following relationships:

$$\gamma = 203 - 0.08(T - 98) \quad \text{for pure sodium}$$
$$\gamma = 203 - 0.11(T - 98) \quad \text{for oxide saturated sodium}$$

where T is in degrees centrigrade.

The variation in surface tension with oxide content is therefore quite small ($<5\%$), and even the variation with temperature is not significant enough to affect the wicked heights of sodium in cell conditions.

(c) *Sodium wick construction*

Metal foil and woven metal cloths have been the most popular types of sodium wick. The capillary rise can be easily calculated for a simple foil wick applied to the beta alumina surface using the following equation:

$$h = \frac{2T \cos \theta}{\rho g r}$$

where ρ is the density of sodium at 350 °C, g is the acceleration due to gravity, θ is the contact angle, T is the surface energy of liquid sodium and r is the capillary gap.

In Fig. 7.13 the calculated capillary rise is plotted against the thickness of the sodium annulus between the foil and the ceramic electrolyte, assuming a contact angle of 20° at 350 °C (Lomax, 1977). Because beta alumina tubes are not perfectly cylindrical it is difficult to maintain very small capillary gaps, and this has resulted in the development of methods of maintaining foils in close contact with the beta alumina surface. In a patent by Lomax (1979) an iron foil wick is constrained against the beta alumina tube by aluminium mesh and aluminium liners. By means such as these the capillary rise required for horizontally operated tube cells and vertically operated plate cells can easily be achieved using metal foil wicks.

For a woven metal cloth wick, the approximate wicking height can be calculated using the equation for the wicked height of a bundle of vertical wires:

$$L = \frac{T \cos \theta}{\rho g} \left(\frac{2\pi n r}{1 - n\pi r^2} \right)$$

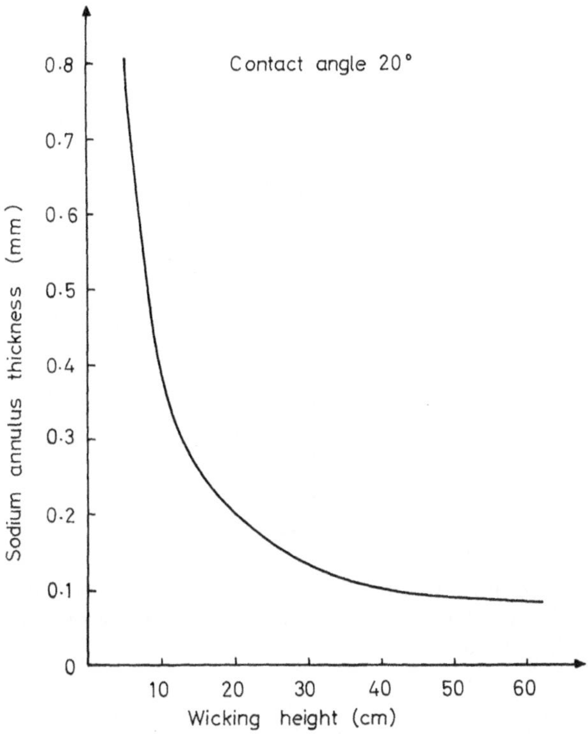

Fig. 7.13 The variation of wicking height with annular gap (Lomax, 1977). Reproduced with permission of the Society of Automotive Engineers Inc., USA.

where r is the radius of the wires, ρ is the density of liquid sodium at 350 °C, g is the acceleration due to gravity, T is the surface energy of the liquid, θ is the contact angle and n is the number of wires per unit area of cross-section. A typical cloth is 300 mesh stainless steel (Robinson and Jones, 1977). This gives a calculated wicked height of approximately 300 mm.

In the patent by Robinson and Jones, describing a sodium wick for a central sodium tube cell, four layers of 300 mesh stainless steel woven cloth are wrapped around a cylinder formed from a sheet of 0.05 mm stainless foil. The foil tends to unroll and presses the cloth layers into close contact with the inner wall of the electrolyte tube. This type of construction is used in the General Electric C45 cell design (a 578 Ah sodium reservoirless cell for load levelling application). In this design the sodium wick has to lift sodium to a height of 600 mm at a flow rate corresponding to 80 $cm^3 h^{-1}$ (Bast, 1982). The General Electric wick system is shown in Fig. 7.14 and a typical performance of a stainless mesh, capillary fed sodium electrode is shown in Fig. 7.15 (Jones, 1977). The foil and mesh wicks have to allow sodium to flow to the beta alumina surface at the required maximum discharge rate. If the beta alumina tube fractures the wicks

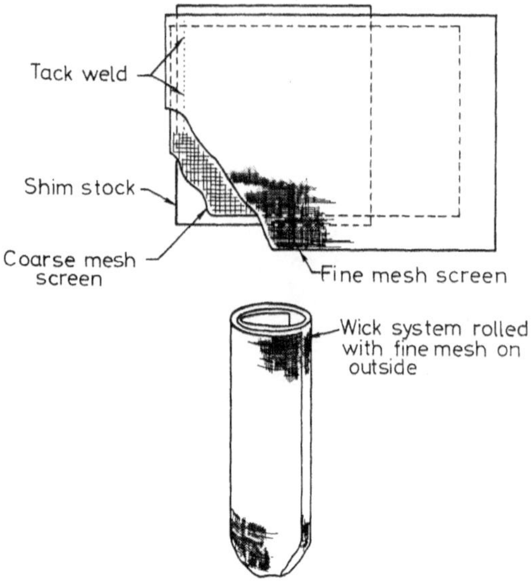

Fig. 7.14 Foil and mesh wick construction for a central sodium tubular cell (General Electric Co.) (Bast, 1982). Reproduced with permission of the Electric Power Research Institute.

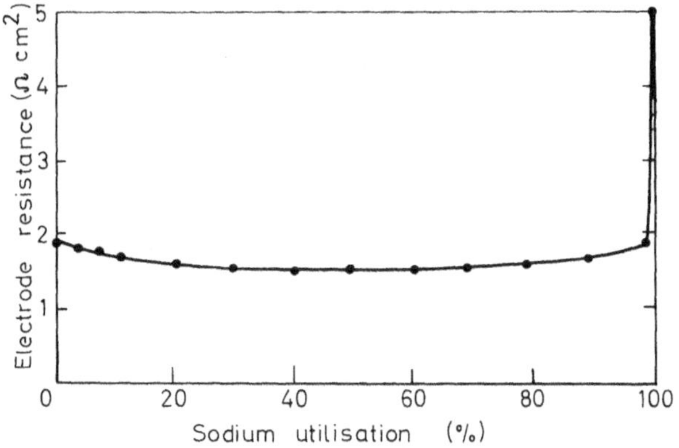

Fig. 7.15 Typical resistance and sodium utilization behaviour of a capillary wick fed sodium electrode (Jones, 1977). Reproduced with permission of Pergamon Press.

can form a safety feature by restricting sodium flow. The amount of sodium contained in a wick is also very small compared with the total Faradaic requirement, so the sodium immediately available for reaction on beta alumina tube failure is limited. A composite wick, constructed of two layers of iron foil separated by graphite foil, has been described by Lomax (1980) and appears to combine all the necessary safety features. The use of wicks as safety features is discussed more fully in Chapter 12.

Other materials that have been used as sodium wicks are porous metals and powders. The electrochemical behaviour of an electrode consisting of sodium contained in porous stainless steel felt has been investigated by Knoedler *et al.* (1977). In their work, the optimum pore size was found to be 80 μm at an optimum porosity of about 90%. To obtain a high utilization of sodium it was found that it was essential to achieve a close contact between the beta alumina electrolyte and the porous felt. Close contact was achieved by either an intermediate layer of wire wool or by accurate spark machining. A utilization of 90% was achieved with the correct pore size and good felt contact. A multitubular cell has also been described using this concept of completely absorbing sodium metal in fine pored metal felt (Baukal *et al.*, 1977). Apart from the safety advantage of immobilizing the liquid sodium, another claimed advantage of this system was that the sharp resistance rise occurring just prior to complete discharge could be used as a method of battery regulation. A similar sharp resistance rise is seen with other wick systems as shown in Fig. 7.15. The use of porous metals in sodium electrode construction has been mentioned in other patents concerned mainly with the safety of cells (Montgomery, 1978; Tilley *et al.*, 1978). The latter patent describes a method of forming thin porous coatings of materials such as nickel and aluminium by flame spraying or plasma spraying metal powder on to a beta alumina tube. Sodium wicking can be achieved by the capillary action of the pores in the metal coating, and possibly by the capillary action of the small gap which can be formed between the beta alumina and the metal coating by the differential thermal expansions of the materials when the cells is heated to its operating temperature.

The safety of central sulfur tube cells can be considerably enhanced by filling the annular gap between the electrolyte tube and the cell wall with inert material. These 'fillers' reduce the amount of sodium immediately available for reaction and reduce the flow of sodium from the main sodium reservoir in the event of beta alumina failure. (These aspects are discussed more fully in Chapter 12.) With a suitable choice of powder fillers, some measure of sodium wicking can be achieved. Materials suggested as fillers in the original patent were carbon spheres and powder, glass beads and ceramic powder (Tilley *et al.*, 1978). In the development of the tubular central sulfur cell, the British Rail workers found that carbon powders and carbon spheres, when used in conjunction with a pressurized sodium reservoir, were the most suitable materials for ensuring cell safety and performance (Tilley, 1980). Problems with outgassing and the non-wettability of those materials with sodium at the temperatures at which cells are normally filled led Stackpool (1982) to consider other materials. Zircon sand was found to meet most of the requirements; it was readily wetted by sodium at the temperature used for sodium filling and it was easy to outgas. In the Chloride Silent Power Ltd central sulfur design, which is operated horizontally, the sodium reservoir is located at the end of the cell. The thin sodium annulus around the length of the beta alumina tube is packed with zircon sand and this forms a capillary system, which keeps the electrolyte surface covered with sodium. Aluminium in the form of flake is also claimed to act as a sodium wick (Wright, 1982). Aluminium flake has not been used on its own, but it has been successfully used in combination with carbon fillers as an oxygen getter (see Section 7.2.3).

The use of capillary action has the advantage that no excess sodium is needed, which results in improved energy density, safety and cost. Metal foil wicks are only acceptable in horizontal tube cells and small plate cells where the wicking height is small. Powder fillers can also be used for horizontal tube cells. Porous metals are attractive on safety grounds, but they are considerably more expensive than foils or fillers and they require to be fitted tightly to the beta alumina surface. Thin, porous metal layers and metal cloths can support a considerable head of sodium and are most suitable for vertically operated cells.

7.3.3 Gas pressure sodium level control

The third method of maintaining the sodium level at the correct height is to pressurize the sodium reservoir with an inert gas. Multitubular and single tube central sodium cells have been described in patents (Jones and Miles, 1974). However, the first practical cell design utilizing this concept was developed at the British Rail laboratories. These cells incorporated a small amount of carbon felt in the sodium electrode reservoir. As the carbon felt is not wetted by sodium when the container is filled, nitrogen gas is trapped in the pores. The pressure of this gas doubles when the cell is heated to the operating temperature. The head of sodium which can be maintained during discharge depends on the volume of gas trapped in the felt and the difference between the filling and operating temperature. The cell designs that used this concept are described in a patent by Gibson and Tilley (1978) and one of these designs is shown in Fig. 7.16.

A variation in the cell assembly method has been described by Sudworth *et al.* (1977) which avoided the handling of liquid sodium during cell assembly. In this method the sodium containers still contained a small amount of carbon felt to define a gas space, but the containers were filled with liquid sodium prior to cell assembly and the sodium was allowed to solidify. The full 'pressure cans', as they were termed, were then inserted into cells. After the cell end caps had been welded into place, the sodium compartment was evacuated via a small pipe which was then crimp sealed. The cell was then heated to operating temperature. With this method of assembly, the seals had to be vacuum tight and the pressure can had to contain enough sodium to fill the annular gap between the electrolyte tube and the cell wall as well as the sodium required for cell discharge. This volume was reduced considerably by using powder fillers as described previously. This method of cell assembly was further improved by Verity (1979) who incorporated in the pressure can a substance that decomposed on heating to operating temperature and generated an inert gas. This procedure was advantageous because no space was required in the

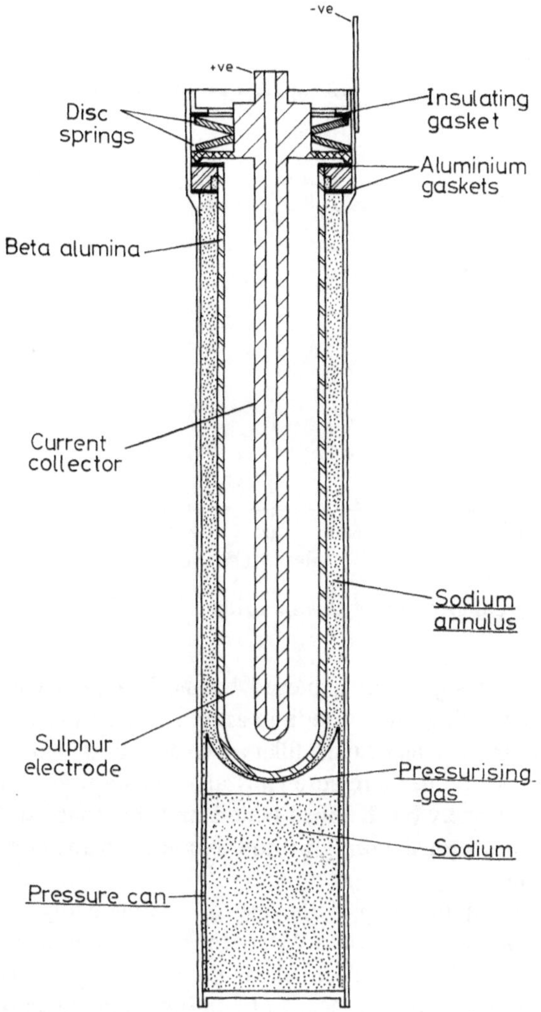

Fig. 7.16 A central sulfur cell design utilizing gas pressure sodium level control (Tilley, 1980). Reproduced with permission of British Rail Research.

pressure can to trap gas on sodium filling. Evacuation of the cell after insertion of the filled pressure cans was also found unnecessary, as the necessary gas pressure could easily be obtained. These higher pressures, coupled with the wicking action of the carbon annulus fillers, ensured that a 300 mm high beta alumina tube was completely covered with sodium during cell discharge. Several pressurizing compounds were tried, but the most convenient was found to be sodium azide. The decomposition of sodium azide is illustrated in Fig. 7.17. Incorporation of the correct amount of sodium azide for the two

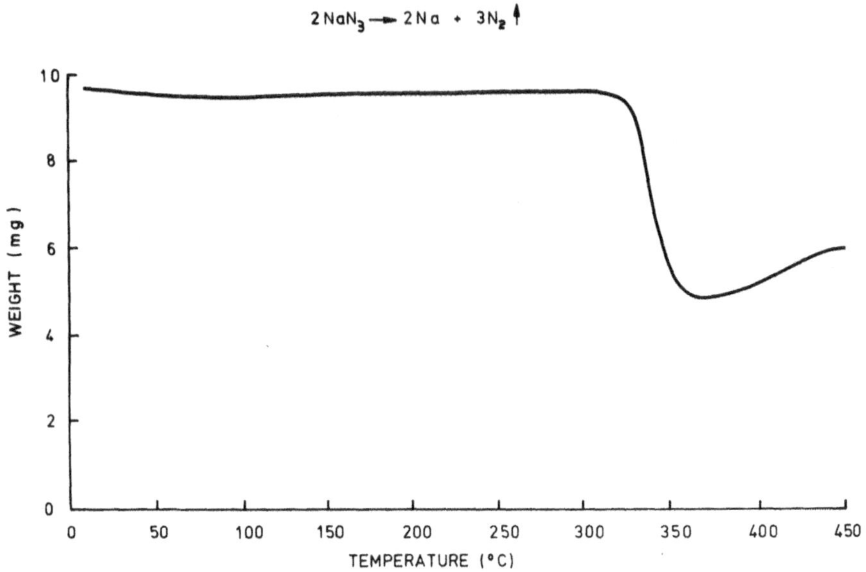

Fig. 7.17 The decomposition of sodium azide.

British Rail cell designs was achieved by using a pharmaceutical tableting method. The few milligrams of sodium azide required were pressed into small 5 mm diameter tablets, using inert fillers such as sodium chloride and talc. The use of a stock of prefilled pressure cans allowed sodium filling of cells to be carried out quite safely on the open bench, and this enabled the cell assembly rate to be increased considerably without recourse to any complicated sodium filling apparatus.

A further simplifying step is the use of commercially available prefilled sodium cans, similar to those sold to the foundry industry as NAVAC® (Sudworth, 1981). These sodium filled aluminium cans have gained international acceptance as a safe means of handling and transporting sodium.

For horizontally operated cells the use of a pressure system is more difficult. A method of correctly orientating the cell and pressure can during assembly and operation is needed to prevent loss of the pressurizing gas. The combination of a wick and a pressure can system would eliminate the need for orientating the cell once the wick had been filled with sodium, and thus a pressurizing sodium filling apparatus to wet difficult wick materials at lower temperatures would not be required (Verity, 1979).

The main disadvantage of a pressure can system is the need to provide a slight excess of sodium in the pressure can to fill any free space in the sodium compartment during heat-up to operating temperature. Excess sodium is required to allow for overdischarge, but this is common to capillary wick cells also.

In central sulfur cells the free space in the sodium annulus can be considerably reduced by using inert fillers, and then the overall reduction in cell energy density need be no more than 2–3% for a typical central sulfur cell of 100 Ah capacity. This is a small penalty to pay for a simple, quick method of cell assembly.

7.4 SUMMARY

The sodium electrode at first sight appears relatively simple compared with the sulfur electrode. This is far from being correct; as we have seen, the problems of interfacial polarization can severely affect cell performance, and the many possible electrode constructions can radically affect the cell design.

The degree of polarization at the sodium/beta alumina interface appears to be very dependent on the ceramic composition. In the worst cases it manifests itself in the form of a very rapid increase in cell resistance, which is also often asymmetric, non-ohmic and time dependent. There are two schools of thought on the cause of this behaviour. Breiter and Dunn believe that sodium reacts with moisture to form a surface film of sodium oxide. The observations of Demott, Bugden and Wright do not support this interpretation; their explanation is that an impurity, probably calcium, arising from the ceramic electrolyte is eluted to the electrolyte surface during cell cycling where it can form a surface film of calcium oxide. Fortunately, two means by which the problems can be completely eliminated have been identified (Demott, 1981; Wright and Hames, 1982) even for material which normally shows the most severe effects: either renewal of the sodium electrode, or treatment of the ceramic surface to improve wetting combined with use of an oxygen getter in the sodium. These, to some extent conflicting, observations illustrate the complexity of the interfacial phenomena. Clearly much work remains to be done before the sodium electrode is fully understood. However, this lack of understanding has not prevented the identification of solutions to the problems associated with interfacial polarization, and by employing them it is possible to eliminate these undesirable effects in practical cells.

Physically maintaining liquid sodium in contact with the whole surface of the beta alumina can be most easily achieved by using a sodium reservoir and gravity feed of sodium. This method, however, is really only applicable to load levelling cells, where the increased height and slightly inferior safety and energy density are not critical. For electric vehicle applications, capillary wick sodium electrodes are more suitable, especially for horizontal operation. A capillary wick system maximizes the specific energy as only a small amount of excess sodium is required, and there can also be safety advantages when using tight fitting capillary wicks. The pressure can system is able to give an extremely simple method of cell assembly with a small reduction in specific energy. For large vertical cells, a pressure system may be preferable to a capillary wick which may limit the sodium wicked height.

7.5 REFERENCES

Addison, C. C., Addison, W. E. and Kerridge, D. H. (1955) *J. Chem. Soc.*, 3047.

Addison, C. C. and Iberson, E. (1965) *J. Chem. Soc.*, , 1437.

Addison, C. C., Iberson, E. and Manning, J. A. (1962) *J. Chem. Soc.*, 2699.

Addison, C. C., Iberson, E. and Pulham, R. J. (1968) Soc. Chem. Ind. Monograph 28, 246.

Addison, C. C., Kerridge, D. H. and Lewis, J. (1954) *J. Chem. Soc.*, 2861.

Armstrong, R. D. and Burnham, R. A. (1976) *J. Electroanal. Chem.*, **72**, 257.

Armstrong, R. D., Dickinson, T. and Turner, J. (1973) *J. Electroanal. Chem.*, **44**, 157.

Armstrong, R. D., Dickinson, T. and Willis, P. M. (1974) *J. Electroanal. Chem.*, **53**, 389.

Barlow, M. and Planting, P. J. (1969) *Zeitschrift für Metalkunde*, **60**, 719.

Bast, J. (1982) EPRI Report EM-2579, Project 128–6.

Baukal, W., Knoedler, R. and Kuhn, W. H. (1977) US Patent 4 038 464.

Bradhurst, D. A. and Buchanan, A. S. (1961) *Aust. J. Chem.*, **14**, 397.

Breiter, M. W. and Dunn, B. (1981a) *J. Appl. Electrochem.*, **11**, 685.

Breiter, M. W. and Dunn, B. (1981b) *Electrochim. Acta*, **26**, 1247.

Breiter, M. W., Dunn, B. and Powers, R. W. (1980) *Electrochim. Acta*, **25**, 613.

Buechele, A. C. and De Jonghe, L. C. (1979) *Ceram. Bull.*, **58**, 861.

Bugden, W. G., Barrow, P. and Duncan, J. H. (1981) *Solid State Ionics*, **5**, 275.

Bukun, N. G., Evtushenko, V. V. and Ukshe, E. A. (1974) *Elektrokhimiya*, **10**, 677.

Bukun, N. G., Ukshe, E. A. and Evtushenko, V. V. (1973) *Elektrokhimiya*, **9**, 406.

Dell, R. M., Sudworth, J. L. and Jones, I. W. (1976) Proceedings 11th IECEC Conference, Nevada.

Demott, D. S. (1980) *J. Electrochem. Soc.*, **127**, 2312.

Demott, D. S., Wright, M. L. and Hames, M. D. (1981) Extended Abstracts 159th Meeting Electrochemical Society, Minneapolis, Abstract 469.

Dunn, B. (1981) *J. Am. Ceram. Soc.*, **64**, 125.

Farrington, G. C. (1976) *J. Electrochem. Soc.*, **123**, 591.

Farrington, G. C. (1977) *J. Electroanal. Chem.*, **76**, 165.

Gibson, A. (1977) *in Power Sources*, vol. 6 (ed. D. H. Collins), Academic Press, London, 673.

Gibson, J. G. and Tilley, A. R. (1978) US Patent 4 104 448.

Glassner, A. (1957) Argonne National Laboratory Report ANL-5750.

Hodkin, E. N., Mortimer, D. A. and Nicholas, M. G. (1973) Liquid Alkali Metals Conference, British Nuclear Energy Society, Nottingham, p. 167.

Hodkin, E. N. and Nicholas, M. (1976) UKAEA Harwell Report AERE R7406.

Hodkin, E. N. and Nicholas, M. (1979) *Science of Ceramics*, **10**, 395.

Imai, A. and Harata, M. (1972) *Jpn J. Appl. Phys.*, **11**, 180.

Jones, I. W. (1977) *Electrochim. Acta*, **22**, 681.

Jones, I. W. and Miles, L. J. (1971) *Proc. Br. Ceram. Soc.*, **19**, 161.

Jones, I. W. and Miles, L. J. (1974) UK Patent 1 356 576.

Knoedler, R., Baukal, W. and Kuhn, W. H. (1977) *J. Electrochem. Soc.*, **124**, 236.

Lazannec, Y. (1977) EPRI Report EM-413.

Lehnert, G. and Hartmann, B. (1982) Extended Abstracts 162nd Meeting Electrochemical Society, Detroit, abstract 335.

Levine, C. and Anand, J. (1977) Proc. of the Load Levelling Symposium, vol. PV 77–4, p. 292, The Electrochemical Society.

Lomax, G. R. (1977) Proceedings SAE Congress, Detroit, paper 770280.

Lomax, G. R. (1979) UK Patent 1 548 263.
Lomax, G. R. (1980) *Electric Vehicle News*, **9**, 20.
Longson, B. and Prescott, J. (1973) Liquid Alkali Metals Conference, British Nuclear Energy Society, Nottingham, England, p. 171.
Longson, B. and Thorley, A. W. (1967) *The Alkali Metals*, The Chemical Society, p. 153.
Markin, T. L., Junkinson, A. R. and Warwick, M. A. (1978) UK Patent 1 519 451.
Minck, R. W. (1980) Sodium/Sulphur Battery Review, Ford presentation to DoE, Newport Beach, Ca.
Montgomery, A. G. (1978) UK Patent 1 530 274.
Park, D. S., Powers, R. W. and Breiter, M. W. (1981) *Solid State Ionics*, **5**, 271.
Riley, D. J. and McGreavy, G. (1981) UK Patent Application 2 061 598.
Robinson, G. and Jones, I. W. (1977) UK Patents 1 461 071 and 1 461 072.
Stackpool, F. M. (1982) UK Patent Application 2 089 559.
Stewart, D. (1969) Fulmer Research Institute Report, R316/1.
Sudworth, J. L. (1981) United Nations Seminar on Energy Sources, Madras.
Sudworth, J. L., Tilley, A. R. and Bird, J. M. (1977) Extended Abstracts 152nd Meeting Electrochemical Society, Atlanta, vol. 77–2, Abstract 88.
Sudworth, J. L., Tilley, A. R. and South, K. D. (1973) in *Fast ion transport in solids* (ed. W. van Gool), North-Holland, p. 581.
Taylor, J. W. (1954) *J. Inst. Metals*, **83**, 143.
Tilley, A. R. (1980) 4th DoE/EPRI Sodium Sulfur Battery Workshop, Derby.
Tilley, A. R., Hames, M. D., Sudworth, J. L. and Bird, J. M. (1978) UK Patent 1 511 152.
Tkacheva, N. S., Ukshe, E. A. and Leonova, L. S. (1978) *Elektrokhimiya*, **14**, 1405.
Verity, C. (1979) UK Patent 1 558 241.
Viswanathan, L. and Virkar, A. V. (1982) *J. Mater. Sci.*, **17**, 753.
Voinov, M. and Tannenberger, H. (1974) *Electrochim. Acta*, **19**, 959.
Wright, M. L. (1981) UK Patent Application 2 067 005.
Wright, M. L. (1982) UK Patent Application 2 080 607.
Wright, M. L. and Hames, M. D. (1982) UK Patent Application 2 080 608.
Yasui, I. and Doremus, R. H. (1978) *J. Electrochem. Soc.*, **125**, 1007.

8 Cell construction

A. R. TILLEY

8.1 INTRODUCTION

In Chapter 4 the various cell designs were briefly introduced. The central sodium and central sulfur tube cell configurations are the two main cell designs which use beta alumina electrolyte. The construction and assembly of the various developers' cell designs varies, not only because of these two distinct design configurations, but because of other features such as orientation in operation, sodium electrode construction, seals type etc. The developers' emphasis on the various cell designs has changed over the years, primarily because of the various cell and battery technical problems but also sometimes for marketing or funding reasons.

The various designs that have been considered are described briefly in Tables 8.1 and 8.2. Source references are given (even if unpublished) and the figure number is shown where a design has been discussed in other sections. The data on specific energies is given only for approximate comparison purposes; they are subject to continuous updating which makes them virtually obsolete at the time of writing. These figures, however, are considered to be realistic and not

Table 8.1 Load levelling cell design details

Developer	Cell configuration	Electrolyte size (mm)	Orientation in operation	Sodium electrode configuration	Capacity (Ah)	Specific energy and rate ($Whkg^{-1}$)	($kWhm^{-3}$)	(h)	Figure no.	Reference
Ford	Central sodium	34 OD × 250	Vertical	Reservoir	150	125	272	5		Minck, 1983
General Electric (GE)	Central sodium	40 OD × 600	Vertical	Wick No reservoir	405				Fig. 4.4	Bast, 1982
General Electric/ Chloride Silent Power Ltd (GE/CSPL)	Central sodium	30 OD × 600	Vertical	Reservoir						
Compagnie Générale d'Electricité (CGE)	Central sodium		Vertical							
Yuasa	Central sodium	36 OD × 300	Vertical	Wick	150	235	420	8		Iwabuchi, 1983
Dow	Central sodium	Glass fibres 50 × 80 × 10	Vertical	Reservoir	12	196	258	5	Fig. 4.1	Levine, 1983
British Rail	Central sulfur	33 OD × 300	Vertical	Pressure system Annular sodium storage	100	149	292	5	Fig. 12.5	Bull, 1983
British Rail	Central sulfur	33 OD × 300	Vertical	Pressure system	100	180	281	5	Fig. 4.5	Tilley, 1980

Table 8.2 Electric vehicle cell design details

Developer	Cell configuration	Electrolyte size (mm)	Orientation in operation	Sodium electrode configuration	Capacity (Ah)	Specific energy and rate ($Whkg^{-1}$)	($kWhm^{-3}$)	(h)	Figure no.	Reference
Brown Boveri (BBC)	Central sodium	25 OD × 210	Vertical	Wick No reservoir	40	190		2		Fischer, 1983
Chloride Silent Power Ltd (CSPL)	Central sulfur	33 OD × 600	Horizontal	Wick and reservoir	185	160	360	?		Lomax, 1979
Ford	Central sodium	25 OD × 220	Horizontal	Inner reservoir	42	145	290	3		Minck, 1983
British Rail	Central sulfur	33 OD × 300	Vertical	Pressure can reservoir	100	160	250	2	Fig. 4.5	Tilley, 1980
British Rail	Central sulfur	33 OD × 300	Vertical	Pressure system Annular sodium storage	100	130	256	2	Fig. 12.5	Bull, 1983
Yuasa	Central sulfur	36 OD × 300	Vertical	Wick	150	235	420	8		Iwabuchi, 1983

projected claims. For the construction of batteries, where the degradation of specific energy due to the incorporation of insulation, outer battery casing, cooling etc. can be as much as 30–40%, these actually obtained cell specific energy figures should be borne in mind.

The various seals designs and production processes for the cell components are discussed in the following sections. Early laboratory test cells and production cells are very different, and many elegant but expensive solutions to cell technical problems have been discarded because they were commercially unacceptable. The mass production cost and ease of incorporation of the technical solutions to cell problems have to be considered at an early stage in cell development.

Quality control procedures and adoption of simple cell assembly methods are also more important with large batch production processes than one-off laboratory cell construction, and these two factors are discussed briefly at the end of this chapter.

8.2 SEALS

The most vital component of sodium sulfur cells is the thin walled beta alumina electrolyte. The selection of the optimum ceramic in terms of composition, fabrication etc. is largely based on its electrical and durability properties in cell operation. This defines its physical properties – any cell sealing method must then be compatible with the selected electrolyte (and not vice versa).

Hostile reactants – liquid sodium and liquid sulfur/polysulfide – are in compartments either side of the beta alumina electrolyte. These compartments must be sealed both from each other and from the atmosphere; in addition they must be electrically insulated from each other throughout the cell's life.

The electrode current collectors in most cases are metallic, doubling as the cell case on the outside of the cell. These metal components must ultimately be sealed to the beta alumina.

It has become common practice to accomplish this in more than one stage. Most cell designs employ an alpha alumina ceramic collar to provide electronic and ionic insulation between the electrodes. Generally, the solid electrolyte tube or plate is bonded to this alpha alumina collar by means of a glass. The seals between the metal components and the ceramic components are then made by bonding the metals to the collar. The alpha alumina collar also serves the useful purpose of strengthening the beta alumina at the point of attachment to the remainder of the cell.

The seal construction has a strong influence on the cell design and the overall cell performance. The seal can affect the orientation of a cell and limit the maximum temperature and/or pressure that a cell can attain. The effect of the seal on the cell gravimetric energy density and the proportionate cost of the seal assembly can be reduced in tubular cells by increasing the length of the ceramic electrolyte. The seal type can also limit the overall diameter of a cell, and this can be a severe limitation with the plate cell design. Thermocompression and compression seals which are greater than 80 mm in diameter are difficult to construct, owing to the inherent weakness of thin section alpha alumina rings. At a maximum diameter of 80 mm the plate cell is inferior to the tubular cell designs. The size of the seal assembly used in tubular cells markedly affects the cell volumetric energy density. This is not necessarily reflected in the overall battery energy density, as a finite intercell spacing is probably required in a battery for safety, cooling and ease of maintenance.

The main types of metal/alpha alumina seals that have been used are gasket compression, glass, brazed and diffusion bond. The basic requirements of these seals are:

(1) The seals must contain sodium and sulfur/sodium polysulfides during cell operation;
(2) the cost of the seal component and manufacture must be acceptably low;
(3) the seal must have a minimal effect on cell specific energy.

To achieve a final low cost in mass production, the seal not only has to be capable of being easily manufactured by mass production machinery, but also must use lightweight, low cost, simply shaped components. Another factor that affects the ultimate production cost is the number of stages involved. If one stage in the total seal manufacture is the glassing of the alpha alumina/beta alumina seal, then, unless the metal ceramic seal is made at the same time, a minimum of two stages is needed. This being the case, the process requiring the highest temperature must obviously be carried out first. There is a clear distinction here between the ambient temperature assembled compression seals and the three chemically bonded seals which all require to be made at elevated temperatures.

If the manufacture of the glass/alpha alumina/beta alumina seal, which will be discussed first, can be carried out at the same time as a glass/metal sealing operation or braze seal production, then a single stage heating cycle production process which completes all the hermetic seals becomes possible, with consequent cost savings.

8.2.1 The alpha alumina/beta alumina glass seal

An enormous amount of effort has been devoted to the task of manufacturing an acceptable glass seal. As mentioned earlier, the beta alumina component is selected first. The selection of the glass and alpha alumina component is then made with regard to both physical and chemical properties.

The sealing glass must be capable of producing a hermetic seal with a lifetime of up to five years at 330 °C and should be capable of withstanding the occasional thermal cycle to room temperature during this time. The elevated working temperature of the cell, coupled with the chemically hostile environment within it, requires careful selection of a durable glass sealing material. The glass must not contain leachable impurities which will harm the beta alumina if they enter the liquid sodium, e.g. calcium and potassium. Normally the seal has

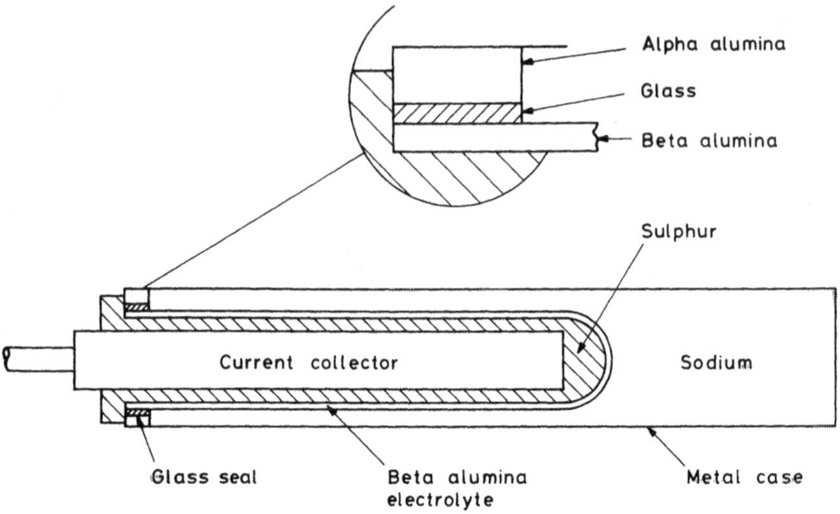

Fig. 8.1 A typical alpha alumina/beta alumina glass seal arrangement for a tubular cell (Barry *et al.*, 1980). Reproduced with permission of Deutscher Verlag für Schweisstechnik (DVS) Gmbh, Dusseldorf.

a concentric geometry consisting of an alpha alumina ring surrounding the beta alumina tube and bonded to it by means of an intermediate layer of the sealing glass, as shown in Fig. 8.1. The detailed shape of the alpha alumina component does vary slightly between different developers. The whole assembly is jigged into position during the glass sealing operation, which can be carried out in air or a controlled atmosphere.

Ideally, the selection of the alpha/beta alumina components for this geometry of seal would result in a seal having a net radial compressive stress at room temperature but a stress-free seal at operating temperature. This can be achieved by selecting an alpha alumina material with a slightly higher coefficient of thermal expansion than the inner beta alumina.

The expansion coefficient of sintered alpha alumina is essentially linear over the temperature range of interest (i.e. from ambient to the setting point of the glass, generally 600 °C). The expansion coefficient of pure alumina is approximately $8.0 \times 10^{-6}\ K^{-1}$, compared with $\sim 7.1 \times 10^{-6}\ K^{-1}$ for beta alumina over the same temperature range. This expansion coefficient is too high to allow pure alumina to be employed in such seals, and a debased alumina with a lower coefficient of expansion is preferred. There are a number of commercially available debased aluminas which give a better match to the expansion of beta alumina, the additives generally being calcium and silica. The final selection of the debased alpha alumina will be determined by its coefficient of thermal expansion, and more importantly by its resistance to attack by liquid sodium. (High silica content materials are vulnerable to such attack at 350 °C (Brinker and Klein, 1980)).

The instability of silica containing glasses to liquid sodium has led to extensive research programmes for developing more suitable glasses.

A group of glasses based on the alkaline earth aluminoborates (Hirayama, 1961) shows good resistance to attack by liquid sodium (Elyard and Rawson, 1962). As a consequence of the non-linear expansion of these glasses, however, seals between alpha alumina and beta alumina would normally be in tension at 350 °C; also these glasses are dimensionally unstable at 350 °C as a result of molecular reorganization, which leads to large tensile stresses in the sealed alpha/beta alumina assembly (Jones, 1981). A large research programme was carried out by Chloride Silent Power Ltd (CSPL) in association with the National Physical Laboratory (NPL) to develop aluminoborate glasses which overcame these limitations. This work was reported by Barry *et al.* (1980) and outlined by Jones (1981). Photoelastic stress analysis was carried out on alpha alumina/glass/beta alumina sandwiches using a large range of glass compositions, and it was found to be necessary to add dopants to the aluminoborate glasses in order for the seal to operate in compression between room temperature and 350 °C, the operating temperature of the battery. The dopants could be SiO_2 or mixed alkaline earth oxides and SiO_2. The relevant stress/temperature curves are shown in Figs 8.2 and 8.3. These glasses are not easy to handle, but Barry (NPL) and Stackpool (CSPL) have developed a

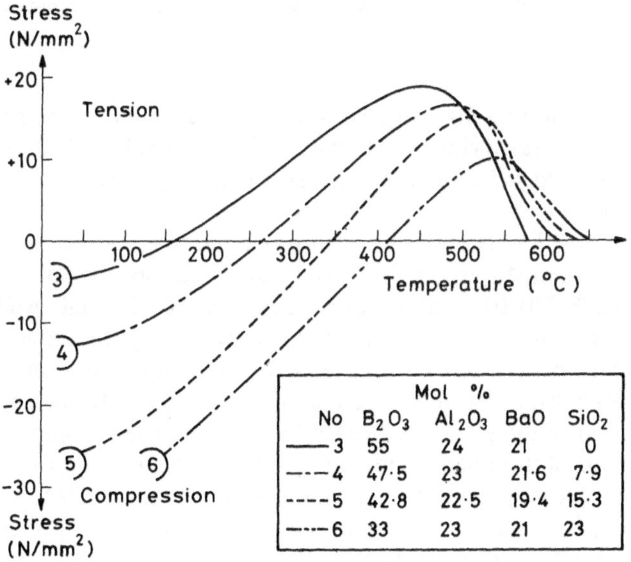

Fig. 8.2 The influence of silica addition to the stress characteristics of aluminoborate glasses (Barry *et al.*, 1980). Reproduced with permission of Deutscher Verlag für Schweisstechnik (DVS) Gmbh, Dusseldorf.

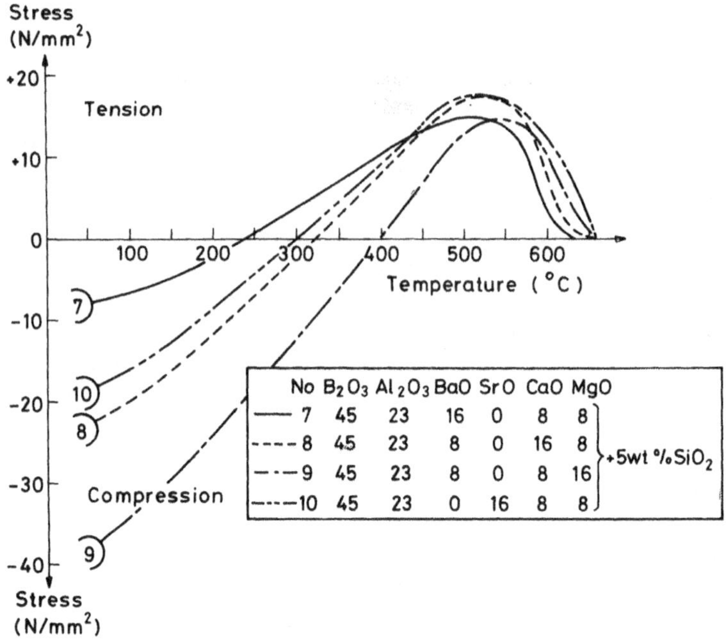

Fig. 8.3 The influence of mixed alkaline earth oxides and silica on the stress characteristics of aluminoborate glasses (Barry *et al.*, 1980). Reproduced with permission of Deutscher Verlag für Schweisstechnik (DVS) Gmbh, Dusseldorf.

process for drawing long, thin rods of the glass which are subsequently converted into spiral springs, as shown in Fig. 8.4. Single coils of the prepared glass are detached for assembly between the alpha alumina collar and beta alumina tube.

Glass compositional development along similar lines has been carried out at General Electric (Mitoff, 1978). Early seals used borosilicate glasses; one composition, S 750–02, led to poor resistance to thermal cycling of the alpha/beta alumina seal. This was found to be due to the growth of crystal dendrites in the beta alumina; the absence of sodium and silicon in such crystals led to the conclusion that the alpha alumina crystals had formed as a result of the interaction of the glass and the beta alumina at the glassing temperature. Further details of this phenomenon are given by Chatterji (1976), in which it is shown that interdiffusion of barium and silicon had taken place into the beta alumina. No such reaction occurs at the glass/alpha alumina interface.

Crystal growth at the glass/beta alumina interface was also a problem with early aluminoborate compositions (e.g. GE 2087) employed at General Electric, but in this case the crystallization occurred in the glass phase. In an improved aluminoborate glass, GE 2093, this crystallization was suppressed by reducing

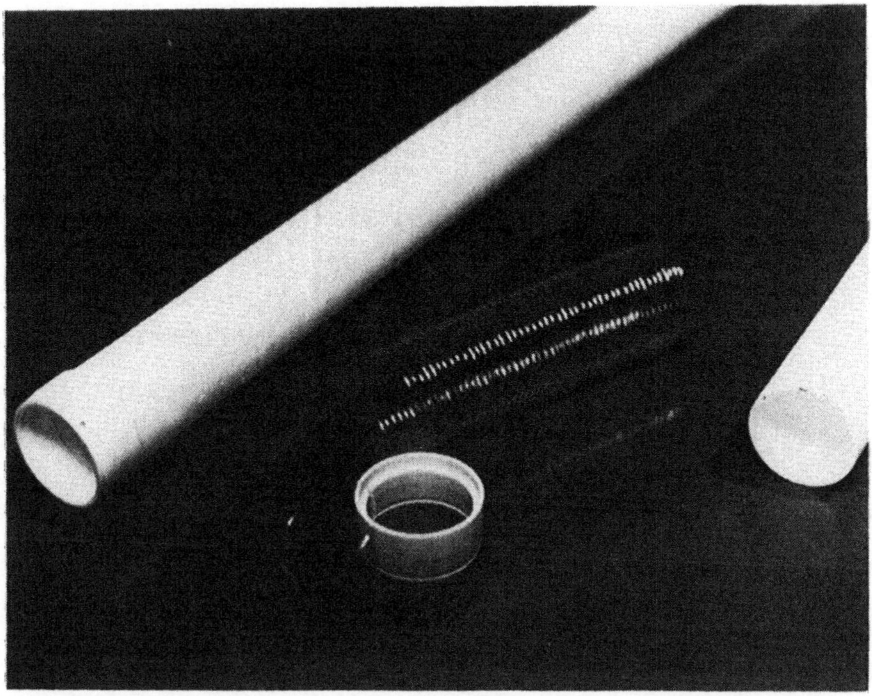

Fig. 8.4 The components of an alpha alumina to beta alumina glass seal. Photograph kindly provided by Wynn Jones (CSPL).

the concentration of modifiers and adjusting the glass compositions to be well within the glass forming regions.

In general, care must be taken in the actual seal fabrication stage not to use excessive temperatures because of the potential problem of interdiffusion between the glass and the ceramics; an annealing stage should also be used when cooling.

Attempts have been made to quantify the stresses that can be developed in a fabricated seal using theoretical finite element stress analysis; the stresses are dependent on both the materials and the seal geometry. An example was given by Barrow (1980). In this example, shown in Fig. 8.5, a glass (ME1) was used which because of its low thermal expansion ($\sim 5 \times 10^{-6}\ K^{-1}$) causes a large compressive stress to be developed within the glass when cooled to room temperature. Despite this, as Fig. 8.5 shows, both upper and lower surfaces of the glass are under tension ($\sim 10\ N\ mm^{-2}$). This is undesirable and can lead to the propagation of cracking from these surfaces into the glass.

The alpha/beta alumina glass seal cannot be designed in isolation from the rigours of cell operation. For example, due consideration must be given to the question of sodium ion current flow through the beta alumina tube wall

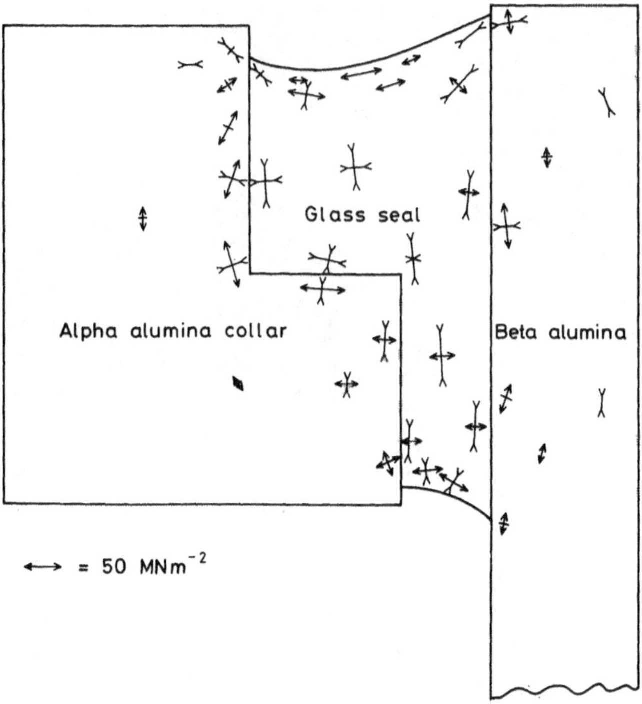

Fig. 8.5 A simplified finite element stress analysis plot for a MEI glass alpha to beta alumina seal (after cooling from 550 to 15 °C) (Barrow, 1980). Reproduced with permission of British Rail Research.

adjacent to the glass seal. If the beta alumina is masked on the sodium side of the cell over a greater area than the active sulfur electrode, it is possible to cause extensive current intensification in the electrolyte at the edge of the glass. In the charging mode this can lead ultimately to beta alumina breakdown in this region.

The glass seal between the alpha and beta alumina components of the cell is almost fully developed now in a satisfactory form, owing largely to the massive efforts of the workers referenced in this section. Any further tuning of the state of the art will only be necessary if the results of long term cell operation demand it.

8.2.2 Compression seals

Virtually all the early development work on tube and plate cells utilized simple bolted flange compression seals. In these early cell designs, sealing gaskets were axially compressed against an alpha alumina insulating flange which had been previously glassed to the beta alumina disc or tube. Typical of these designs were the TRW boiler plate cell (Silverman *et al.*, 1976) and the GE 16 Ah laboratory test cell (Chatterji, 1976) shown in Figs 8.6 and 8.7. This type of

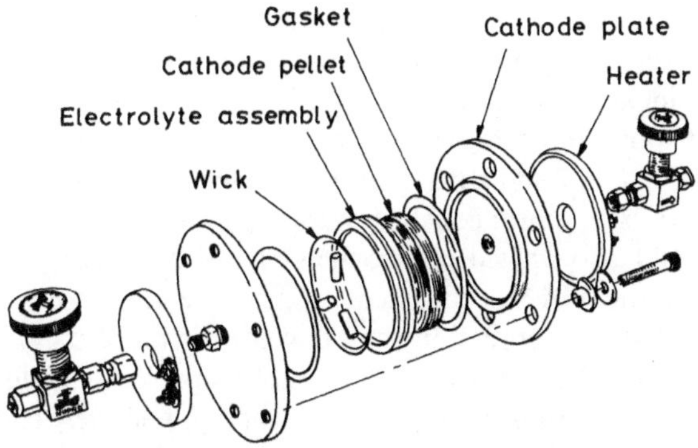

Fig. 8.6 A plate cell design utilizing bolted flange compression seals (TRW) (Silverman *et al.*, 1976). Reproduced with permission of the Electric Power Research Institute.

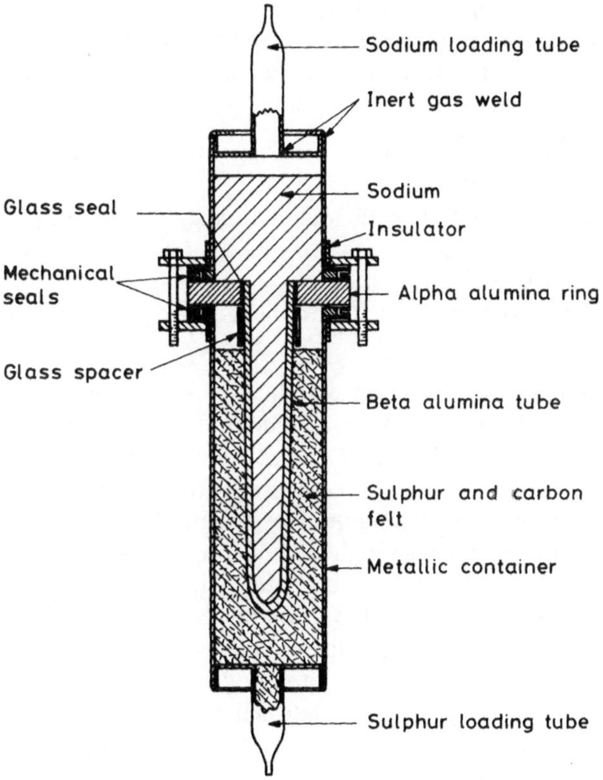

Fig. 8.7 A central sodium tube cell utilizing bolted flange compression seals (General Electric Co.) (Chatterji, 1976). Reproduced with permission of the Electric Power Research Institute.

design which used bulky bolted flanges was extremely useful for development work, as cells were easily assembled on the bench and incorporation of new cell development ideas was relatively easy. The components were robust and reusable and post mortem work was made very simple. Several of the developers – Ford, GE and BBC – continued with this type of construction for several years whilst simultaneously developing more refined sealing techniques. British Rail (BR) also started cell work using bolted flange compression seals, but as part of the UK coordinated programme they concentrated on the development of the compression seal, developing more compact lightweight seal assemblies. The British Rail seal probably represents the furthest development of the axial compression seal which, although relatively cheap and easy to assemble in large numbers, ultimately cannot really compete with the high temperature seals that are used by the other developers.

A further development of the compression seal is the Ford radial compression ring seal. This seal is as compact as the high temperature seals and has the advantages of the axial compression seal, i.e. low cost assembly at ambient temperature using simple components.

The various compression seals that have been developed have used a variety of gasket materials and clamping methods. The gasket material has to have the following properties:

(1) It has to withstand continuous operation at 350 °C and possible excursions to higher temperatures;
(2) it has to be compatible with sodium and sulfur/sodium polysulfides;
(3) it should be flexible and compressible to take up any uneveness in the flanges at acceptable loads;
(4) it has to be easy to manufacture to the required shapes and tolerances, and low in cost.

Several materials have been used as sodium compartment sealing gaskets: flat gaskets such as flexible graphite, aluminium, nickel and Rodar alloy, and shaped gaskets such as Inconel C rings, stainless steel hollow O rings and aluminium O rings. The only gasket material that can meet the requirements listed earlier however, is aluminium. Flexible graphite gaskets have been used and they can form excellent gas tight seals at low sealing stresses, but they have a limited life as a sodium seal because unfortunately sodium forms an intercalation compound with the graphite (Silverman, 1976; Tilley, 1976). Similarly, for the sulfur compartment seal only aluminium appears to be suitable. TRW found the flexible graphite gaskets satisfactory for over 3000 hours' operation with their boiler plate cell design, but flexible graphite seals performed far worse than aluminium seals in the more lightweight British Rail seals assemblies, especially when cells were thermally cycled (Tilley, 1976). The British Rail cells are operated vertically, seals uppermost, so gasket contact with the liquid sulfur/polysulfide melt is unlikely. Aluminium seals, however, have performed satisfactorily when immersed in the melt, even though certain grades of aluminium have been known to be attacked by sodium polysulfides under polarized conditions (Miles and Jones, 1972; Tilley, 1976).

The performance of a compression seal is, not surprisingly, affected by the quality of the sealing faces and their alignment. Careful quality control of the surface finish, flatness of the sealing components and checking of the seal loading and welding jig were essential in achieving reliable seals at British Rail. The British Rail cells still began to leak sodium, however, after approximately 200 days, even though sulfur loss tests had shown exceptionally good seal performance for over 600 days (Tilley, 1980). The leakage was at first thought to be similar to previously reported problems with aluminium diffusion bond seals. CGE had seen sodium leakage due to a ternary sodium/aluminium/silicon compound (Lazennec, 1977), and GE had reported problems with diffusion bonds owing to the formation of a porous, cracked, mild steel/aluminium intermetallic when using mild steel components instead of stainless or chromized steels (Mitoff, 1978). The aluminium gaskets used in the BR seal were found to contain extremely low levels of silicon ($<0.01\%$), and intermetallic growth was only detected in seals from extremely long life tests that had been operated at temperatures higher than 400 °C. The leakage problem was, in fact, thought to be due to simple corrosion of the mild steel, and cells with protected mild steel seal faces showed a greatly improved seal life of >400 days (Tilley, 1980).

Most of the developers have published papers and patents which show bolted axial compression seal constructions for tube and plate cell similar to the TRW design shown in Fig. 8.6. The main problem with any sealing assembly of this type is to maintain adequate stress on the gaskets during cell operation. Seal load loss can occur due to the differential expansion of the alpha alumina collar (coefficient of thermal expansion (CTE)$\sim 6.8 \times 10^{-6}\ K^{-1}$) and the steel components (CTE $\sim 12 \times 10^{-6}\ K^{-1}$). With the early bolted assemblies, the tension of the bolts was sufficient to maintain the integrity of the seals at operating temperature. With lightweight clamped assemblies the problem of maintaining a fairly constant sealing load from assembly to operating temperature is more difficult. The initial British Rail lightweight clamping system is shown in Fig. 8.8 (Sudworth and Tilley, 1976). The main reasons for the change in the seal assembly, apart from the large reduction in size and weight, were that cell assembly was made much easier, and a much more even and accurate sealing load could be obtained using a hydraulic compression jig and welding technique compared with the original hand tightening of a ring of flange bolts. A reduction in clamping force occurred with the new lightweight assemblies, not only owing to the differences in the thermal expansion of the components but also owing to stretch, compression and bending of the components after assembly. The original attempts to reduce this load loss resulted in the incorporation of compensating materials having a high coefficient of thermal expansion (Gibson and Duncan, 1974). Mild steel was used for the low expansion components and initially aluminium

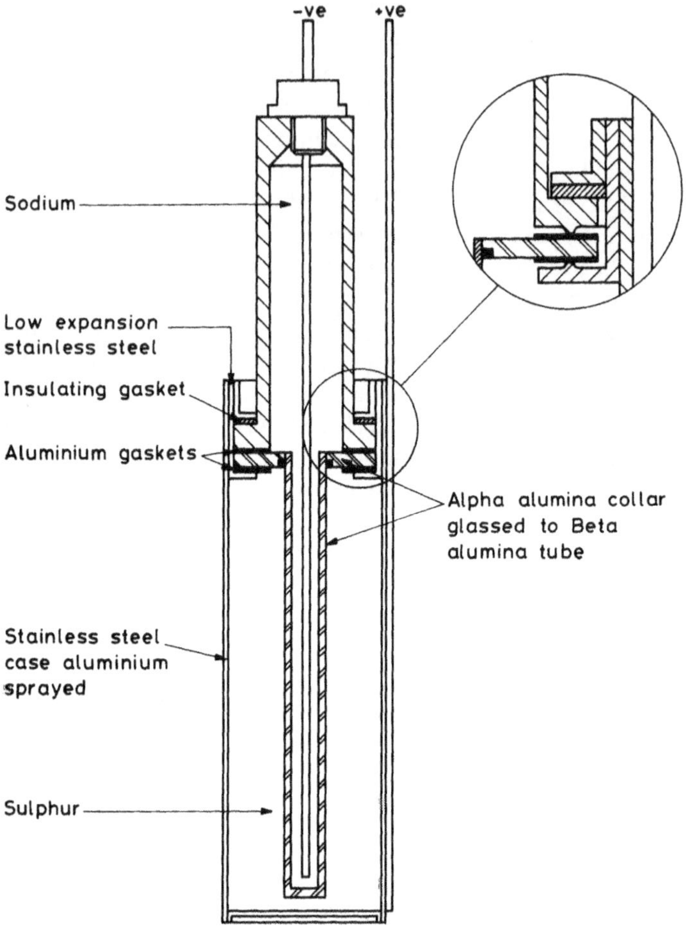

Fig. 8.8 A central sodium tube cell incorporating a lightweight compression seal assembly (British Rail) (Tilley, 1976). Reproduced with permission of the Electrochemical Society Inc., and Argonne National Laboratory.

(CTE $\sim 23 \times 10^{-6}$ K^{-1}) was used for the high expansion material. It was found that at 350 °C the stresses resulted in creep of the aluminium and austenitic stainless steel was substituted. This method, however, required stringent control of material specifications and dimensional tolerances to achieve reproducible seal stresses at 350 °C, and it did not overcome initial loss in clamping force after assembly.

The change in seal stress was reduced to a small amount by the use of disc springs of low spring rate compared with the other components of the seal assembly (Tilley, 1976). The disc springs were easily incorporated into the seal assemblies as shown in the central sodium cell design in Fig. 8.9, and central

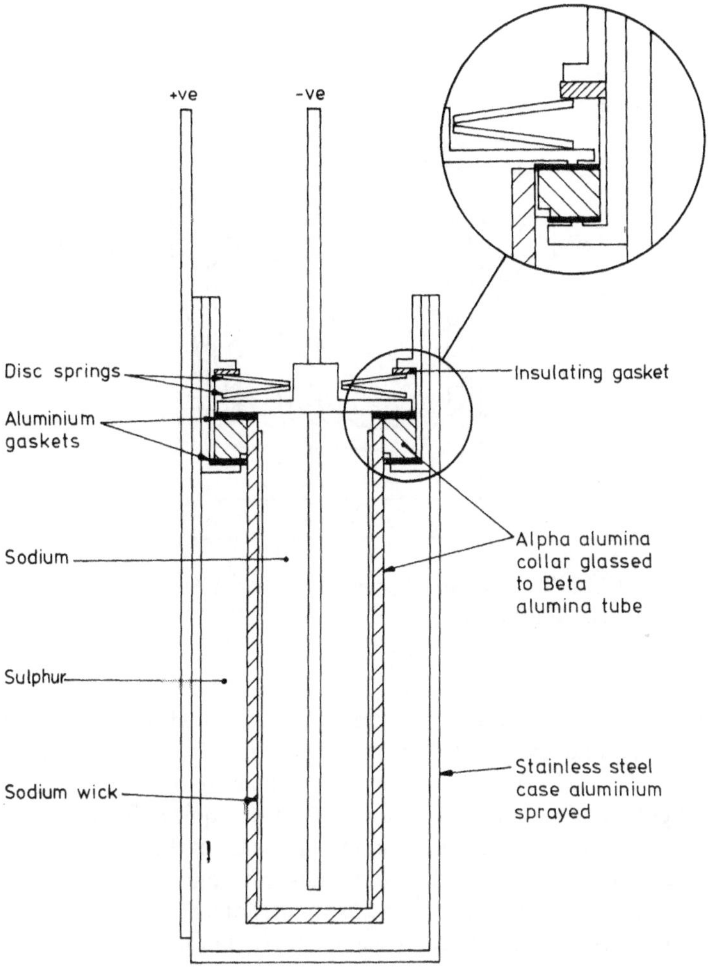

Fig. 8.9 A central sodium tube cell incorporating a disc spring compression seal assembly (British Rail) (Tilley, 1976). Reproduced with permission of the Electrochemical Society Inc. and Argonne National Laboratory.

sulfur tube cell designs incorporating disc springs have also been described (Tilley, 1978). The British Rail seal was refined using finite element stress analysis to evaluate and optimize various designs. The design programme concentrated on using large batch produced, easily formed components, and the seal assembly was optimized for minimum load loss commensurate with minimum size, weight and ease of assembly. The optimized seal design was used in the BR16 cell which formed the basis of all the central sulfur cell development at British Rail, and more than 6000 of these seal assemblies were constructed and used successfully. The BR16 seal assembly is shown in Fig. 8.10. The main

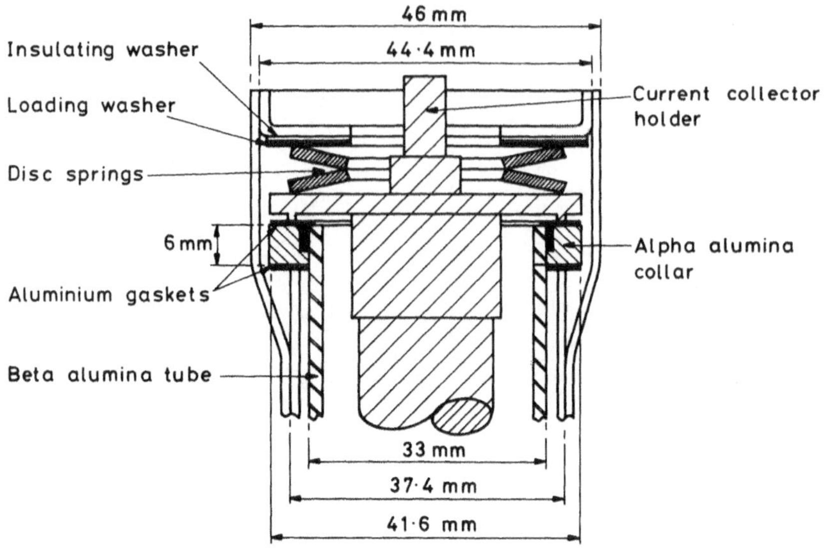

Fig. 8.10 The British Rail BR16 seal assembly (Tilley, 1980). Reproduced with permission of British Rail Research.

disadvantage of this seal was its bulkiness; it had a seal offset of 5 mm which reduced the cell and battery volumetric energy density considerably. For the BR30 cell design, which BR used in their vehicle battery, the seal size was reduced slightly and an offset of only 2.5 mm was required for this design. The reduction in size also reduced the bending moments of the components and so reduced the load loss on assembly and heat-up to only 5%. The disc springs were relatively expensive, and even in this smaller diameter seal they accounted for approximately 50% of the total seal cost. No suitable cheap material was found, however, that could operate at the fairly high stress levels and at 350–400 °C, and the only suitable material used by the British Rail workers was a FV520 turbine steel (Tilley, 1980).

The two disadvantages were not thought to be significant, as space was needed between cells in a battery for safety, cooling and ease of maintenance. Although the cost of the disc springs was relatively high compared with the other seal components, it was compensated by the fact that the alpha alumina collar was a fairly simple shape compared with some of the more complicated shapes required for the high temperature seals, and also the disc springs were reusable. The seal cost, in fact, only represented approximately 10% of the total BR16 cell cost for small laboratory batch quantities, and for the larger, high capacity BR30 cell design the seal cost was approximately 4% of the total cell cost for large production batch quantities (Tilley, 1980). It was considered that any disadvantages of this type of compression seal were far outweighed by the advantages: ambient temperature assembly was easy; simple, easily formed, low dimensional tolerance metal components were used; no expensive assembly equipment was required to assemble seals at the rate of ~100 per day. A possible advantage over hermetic high temperature seals was that disc spring assemblies could act as a pressure release valve if high pressures occurred owing to cell failure.

There are several patents on radial compression seals by CSPL and Ford. The patent by Jones and Miles (1974) describes the use of an elastomeric seal which seals directly on to the beta alumina tube. Cells using silicone rubber O rings were used at CSPL for routine laboratory tests. The cells were tested in a temperature gradient to prevent degradation of the silicone rubber seals (Brennan, 1977). The other patented design by Jones *et al.* (1977) describes a similar radial seal in which the seal to the beta alumina tube is made by a packing of caesium or rubidium chloride and graphite foil. These two types of seal have been superseded by a diffusion bond seal which is described in Section 8.2.5.

The Ford patents describe two types of radial seal. The designs described by Topouzian (1979) and Mikkor (1981) use a C shaped band to apply pressure on to soft, deformable rings which are separated by an electrically insulating collar. The deformable rings are forced against the beta alumina tube to form radial seals for the sodium and sulfur compartments. The later patent by Mikkor is a modification of Topouzian's original design and it uses thin aluminium coatings on the sealing areas to obtain hermetic seals more easily. Chase *et al.* (1979) describe a very simple radial seal which uses a glassed on alpha alumina collar, similar to previous axial compression seals, which acts as a load bearing component. The ceramic collar is contoured to accept the bell shaped ends of the sodium and sulfur compartment casings. Each compartment casing has a sealing band which engages on the open end of the casing and compresses the casing on to the ceramic collar to form a seal. This simple seal has been described in more detail by Elkins (1980) and is shown in Fig. 8.11. It seems to be able to compete with more sophisticated high temperature seals; cells made

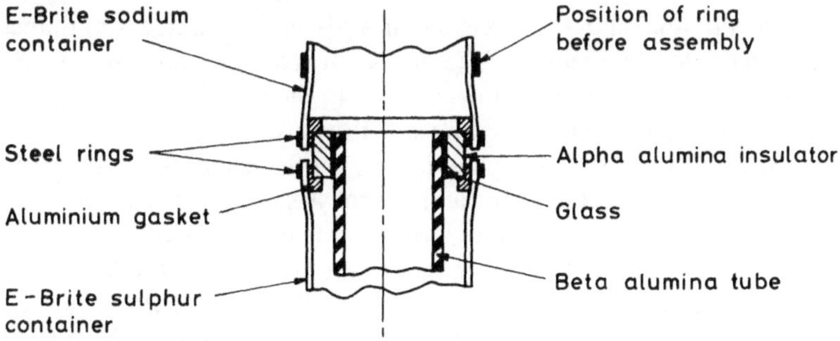

Fig. 8.11 The Ford Radial compression seal (Elkins, 1980). Reproduced with permission of the Ford Aerospace and Communications Corporation.

with these seals have been on test for three years without leaking and are resistant to freeze/thaw cycling. The advantages claimed by Ford are high volumetric efficiency, lower potential cost, freeze/thaw capability and improved reliability.

8.2.3 Glass/metal seals

It has been established that it is feasible to bond a glass to both alpha and beta alumina. The technical process of hermetically bonding glass to metals is a well established art (Hull and Burger, 1934), used extensively in the lamp and electronics industries.

Consideration of these two facts leads to the obvious thought that a simple solution to the problem of bonding metal components to ceramic components in beta (sodium sulfur cells) would be to use a glass as the 'adhesive'. Most groups have at least dabbled in this approach for cells. Basically there are two distinct types of glass/metal seals, namely matched and compression seals. In the former, the component materials are specially selected to minimize any difference in thermal expansion, thereby minimizing residual stress in a joint. In the latter, the metal and glass are deliberately selected to have considerably different coefficients of expansion, such that when the seal cools below the set point of the glass, one of the components is permanently under compression. A general review of this subject has been published by Buckley (1979).

Whilst compression seals are most suitable for wire lead-throughs etc., matched seals are desirable for the larger components used in tubular cells; all developments have concentrated on matched seals. The general principle of successful glass/metal bonding in matched seals is discussed by Hull and Burger (1934). The metal surface is preoxidized in a controlled manner; this oxide film must be wetted by the liquid glass forming the complex oxide bond; the seal must be annealed while cooling.

Superimposed on these general principles are further restrictions dictated by the specific requirements of the sodium sulfur cell, such as geometric constraints and chemical compatibility with the liquid sodium and sulfur/polysulfide melts.

The manufacturing process is particularly simple. By suitable jigging, the alpha/beta alumina and the glass/metal seals can be made simultaneously owing to their close proximity. Pressure is not required and as-fired ceramics have adequate dimensional tolerance owing to the mobile nature of the liquid glass: expensive machining of ceramics is thus not required.

The geometric constraints depend largely on the cell design – in particular, whether the cell is of the central sodium or central sulfur type. The former design of cell facilitates easier seal design. Two groups have carried out most of the research and development of the glass/metal seal option – the Yuasa Battery Co. Ltd, Japan, and Harwell (as part of the UK National Programme). The former group was concerned with the central sodium design whilst the latter group concentrated on central sulfur cell designs. The various seal designs evolving from these quite separate studies are shown in Figs 8.12 and 8.13,

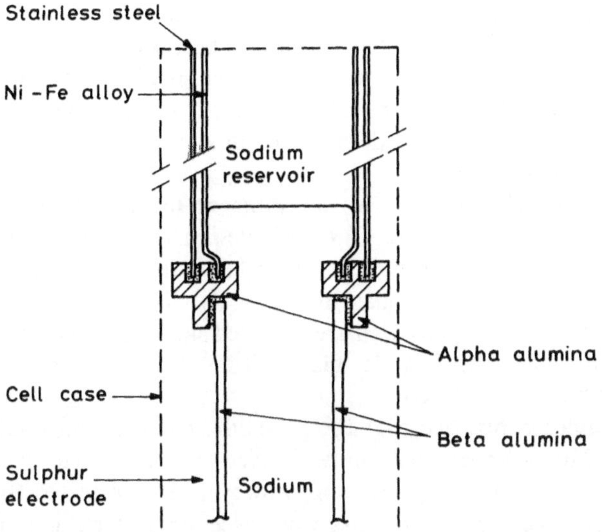

Fig. 8.12 A central sodium cell glass/metal seal arrangement (Yuasa Battery Co. Ltd).

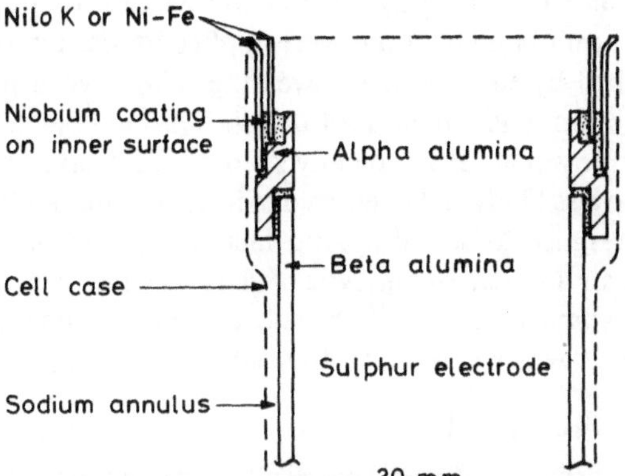

Fig. 8.13 A central sulfur cell glass/metal seal arrangement (UKAEA Harwell).

drawn to the same scale. The differences are immediately apparent in that the central sodium design cell (Yuasa) is based on a ~15 mm diameter beta tube. Additional restrictions on the UK team were that the open end of the tube must remain completely open at all times, facilitating insertion of the solid cathode preform, and the seal annular thickness should not exceed 4 mm in order to maximize volumetric energy density of the batteries of cells.

Little published information is available about the details of the latest Yuasa seal in terms of glass composition or additional corrosion protection, but its durability in operation is reported to be very good (Hattori *et al.*, 1977). Five cells were subjected to thirty thermal cycles between room temperature and 350 °C at rates of up to 150 °C h^{-1} without failure. The incorporation of the sodium reservoir at the top of the cell enables long metal cylinders to be used for the glass seals: this is a considerable advantage for the subsequent welding operation to the cell case.

Details of the glass/metal seal development in the UK are given in the literature (Baker *et al.*, 1980). One such design is shown in Fig. 8.13; both consecutive metal ring components were fabricated from nickel iron based alloys commonly used in glass/metal seals. The expansion coefficients (250–600 °C) of the materials used are: debased alpha alumina $7.07 \times 10^{-6}\ K^{-1}$; beta alumina $7.1 \times 10^{-6}\ K^{-1}$; glass $5.91 \times 10^{-6}\ K^{-1}$; and alloy rings $8.00 \times 10^{-6}\ K^{-1}$ (20–500 °C).

It was found in practice that the preoxidized inner ring used to seal the cathode (sulfur) compartment showed adequate resistance to corrosion by sulfur/polysulfide. The early test cells, however, demonstrated that the oxide bond between the metal and glass was susceptible to reduction, and hence penetration by the sodium (vapour and liquid) in less than three months. This problem was overcome by the application of a thin (8–15 micron) niobium coating or a niobium foil on selected portions of the alloy rings which, when subsequently bonded to the glass, were exposed to sodium vapour or liquid (Bones, 1978). The niobium coatings were applied by sputter ion plating and the foils were attached by electron beam welding. The oxygen partial pressure at 350 °C over niobium oxide in equilibrium with niobium metal is less than that over sodium employed in cells, typically 2–10 ppm dissolved oxygen. The oxide bond is therefore less likely to be reduced. All seal components, metal, glass and ceramic, were designed to be self-jigging, facilitating easy production. Fig. 8.14 shows components and seal of one type of glass/metal seal. The fragile nature of such seals led to some difficulties in the subsequent welding operation when the tube/seal assembly was welded into the cell case.

8.2.4 Brazed seals

The technique of using a soft metal braze alloy to bond metals to ceramic is also well established. It is common to use a debased alpha alumina and a premetallized layer on the surface of this ceramic.

Seals of a design similar to those for the glass/metal seals of the central sulfur cell were developed at Harwell. One such seal is shown in Fig. 8.15. A β bronze alloy, 97 wt% copper, 2.95 wt% nickel and 0.05 wt% boron, was used for the sodium seal. A stainless steel component plus an alumina support ring replaced the nickel alloy rings on the cathode side. This was bonded to the alpha alumina using either gold or an 80 wt% gold, 20 wt% nickel alloy.

Strong, hermetic seals were invariably obtained using this system. The main differences between the brazed seals and the glass/metal seals can be appreciated by reference to the summary Table 8.3. Machined components are required,

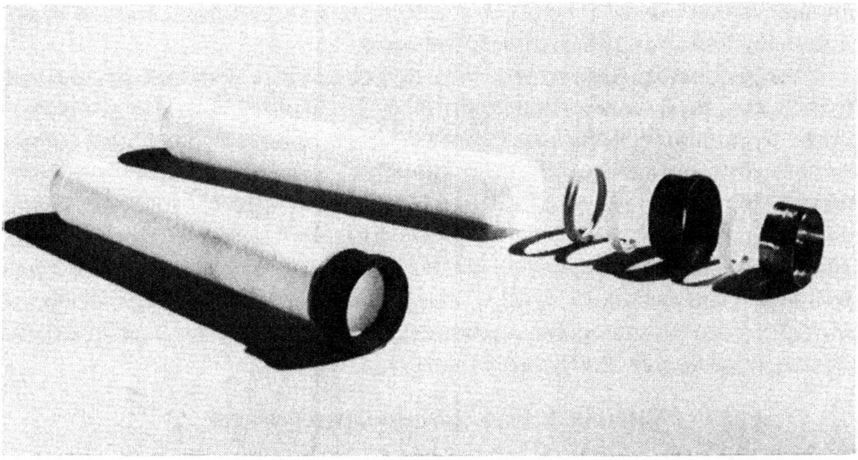

Fig. 8.14 Components of a central sulfur glass/metal seal assembly (Baker *et al.*, 1980). Reproduced with permission of Deutscher Verlag für Schweisstechnik (DVS) Gmbh, Dusseldorf.

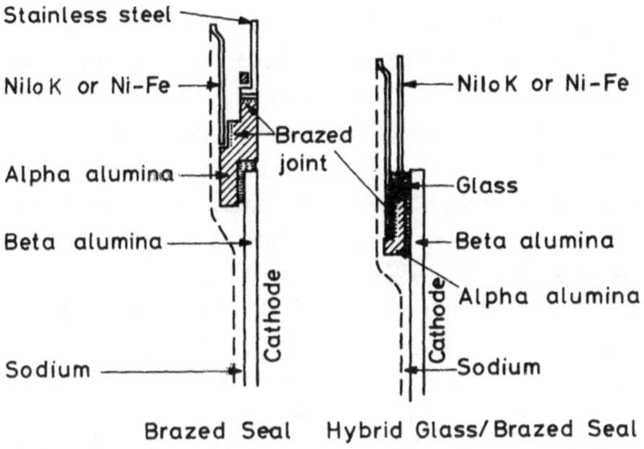

Fig. 8.15 A brazed seal and a hybrid braze and glass/metal seal arrangement.

more stages are involved and the processing conditions are more severe. However, the finished brazed seals are less brittle, giving no difficulty at the cell assembly stage.

The brazed seal is more desirable on the sodium side of the cell, but the glass/metal seal is preferred for the sulfur side. These two facts led to the development of a hybrid braze and glass/metal seal (Baker and Bones, 1979; Fig. 8.15) in which the outer metal ring was brazed to the alpha alumina. The inner alloy ring was then glassed in position at the same time as the alpha/beta alumina seal was made. A twenty cell battery had been operated using a mixture of the glass/metal and brazed seals discussed.

Investigations into the use of the brazing technique for bonding stainless steel to alpha alumina have also been reported by GE (Mitoff, 1978). They found that brazes containing noble metals (e.g. silver copper, silver palladium, gold nickel) were rapidly attacked by sodium. On the other hand, brazes composed of iron, nickel or chromium were stable in liquid sodium, but suffered some passivating corrosion in contact with sodium polysulfide. For the low vitreous content alpha alumina, a premetallization of the ceramic was required, the active metal technique using titanium hydride being preferred. Tensile tests on bonds simulating those used for cell seals demonstrated that this type of metal/ceramic joint was adequately mechanically strong.

8.2.5 Diffusion bond or thermocompression seal

The solution to the metal/ceramic bonding problem currently favoured in most laboratories is the diffusion bond. In this process, simultaneous application of heat and pressure leads to a bond being formed by atomic interdiffusion of a metal and a ceramic. Pioneering work on this type of direct bond was carried out by Dawihl and Klinger (1969), Klomp (1971, 1972) and at the Harwell Laboratory (UK) by Nicholas and Crispin (1982a, 1982b), where it was found to be advantageous to use a soft metal interlayer between the metal and ceramic components. The most convenient interlayer is aluminium; fortunately this metal is also passive to corrosion by both the sodium and sulfur/polysulfide. The soft metal interlayer enables the accommodation of metal and ceramic components with substantially different expansion coefficients. Recent comprehensive general reviews by the original developers discuss detailed conditions, with respect to time at temperature and pressure for the optimum seals (Nicholas and Crispin, 1982a, 1982b; Mortimer, 1979). Other factors of importance are surface preparations of all components and seal geometry, including the thickness of the aluminium interlayer. It is important that the protective oxide film on the aluminium layer is disturbed so that fresh aluminium surfaces contact the metal and ceramic surfaces. A key parameter for the quality control monitoring of such seals is the thickness of the intermetallic layer formed between the aluminium and the metal component: the aluminium diffusion into the high alumina ceramic is more difficult to quantify. Nicholas and Crispin (1982a) report that the bonding operation is best carried out in a vacuum: in some cases an additional annealing heat treatment, without external pressure, is advantageous in terms of bond strength. Two distinctly different interfaces must be considered in this type of seal. The first is the bond between the aluminium and the alpha alumina which accommodates a large difference in thermal expansion coefficients. The bond is a surface phenomenon: Klomp (1971) showed that beyond a 50 Å distance from the interface no chemical reaction occurred. The second bond to be considered is that between the metal component and the aluminium. In this case an intermetallic layer does form, e.g.

between aluminium and mild or stainless steel. There is no difficulty in producing hermetic seals reproducibly by the diffusion bond method.

In practical cells, two distinctly different geometries have been developed, one for the central sulfur cell and one primarily for the central sodium cell. These two types are illustrated schematically in Figs 8.16 and 8.17. As with the glass/metal seals, the central sulfur cell design, which employs a narrow sodium annulus, imposes severe geometric restrictions in the seal annular thickness.

Extensive development work by Bindin at CSPL has provided a unique solution to this problem. In his design, approximate concentric bonds are made

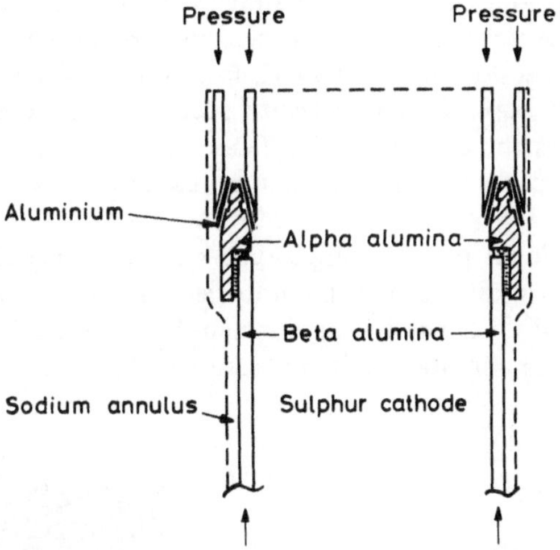

Fig. 8.16 A central sulfur cell diffusion bond seal (Chloride Silent Power Ltd).

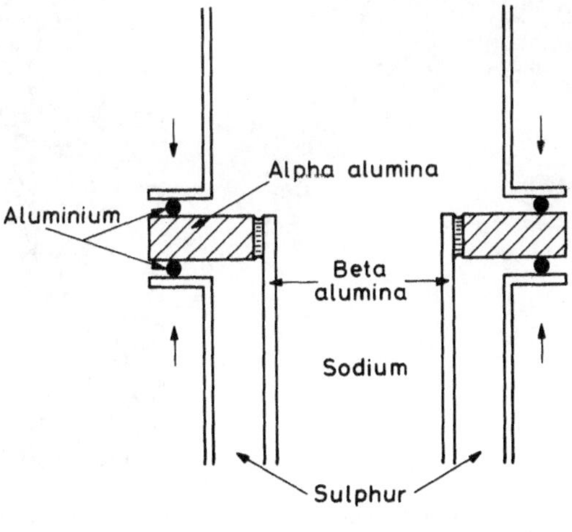

Fig. 8.17 A central sodium cell diffusion bond seal (General Electric Co.).

simultaneously on a conical shaped ceramic, as shown in Figs 8.16 and 8.18. A groove is provided on each ceramic tapered surface to accommodate surplus, displaced aluminium: this also leads to a corresponding higher pressure on the remaining reduced conical surface area. Details of the variety of conditions under which successful bonds can be made are given by Bindin (1977) in his patent. The aluminium can be employed either as a thin foil or as a coating on the components parts. Accurate alignment of the apparatus is needed to produce excellent seals. Because the alpha/beta alumina seal has to be carried out first, the subsequent apparatus has to accommodate the beta alumina tube. It is preferable to use inconel or stainless steel as the outer annular metal ring components, and to use a nickel iron alloy as the inner metal ring (for central sulfur cells). A typical example of processing conditions given in Bindin's patent is: a temperature of 500–650 °C, from 0.1 to 10 minutes in argon or vacuum, and an applied axial pressure such as to produce a pressure of 2 N mm^{-2} across the conical sealing faces. On cooling, radial compressive forces are created by the differential thermal expansion coefficients of the components.

More conventional parallel flange type seals have been used for central sodium cells by workers at other laboratories, e.g. CGE, BBC, GE and Ford (Fig. 8.17). Basically, a flat alpha alumina annular disc is used and the bonds are made under similar conditions of temperature and pressure. Both flanged and

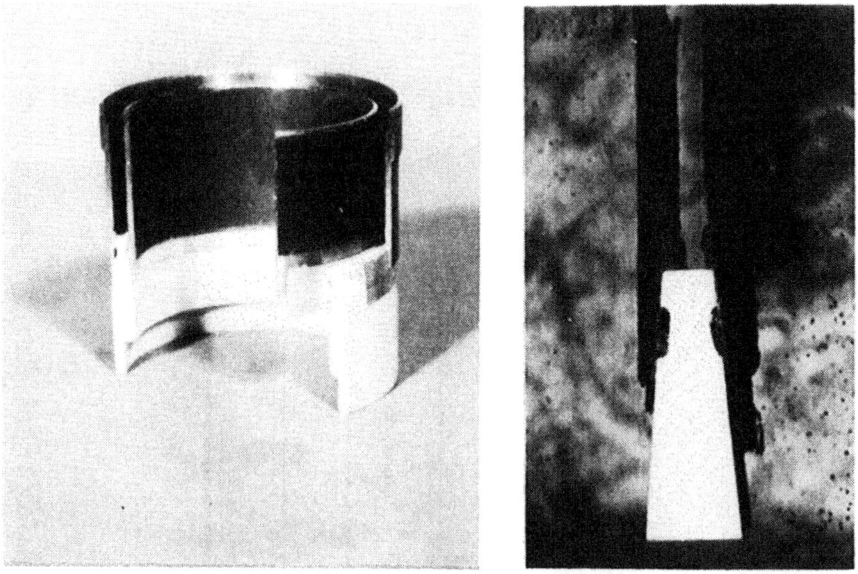

Fig. 8.18 Sections of a Chloride Silent Power tapered diffusion bond. Photographs kindly provided by Wynn Jones (CSPL).

non-flanged tubular metal components have been used (Desplanches *et al.*, 1977).

An early common problem was identified with long term use of this type of bond at 350 ± 50 °C, as might be experienced in a battery. At these temperatures there was a tendency for the intermetallic layer formed at the aluminium/base metal interface to continue to 'grow' slowly owing to further interdiffusion. This leads eventually to a weakening of the bond. This effect can be minimized by the use of a high chromium containing layer on the surface of the base metal.

Joint GE/CSPL work has also resulted in the use of the annular, conical diffusion bond in a central sodium cell (Bast, 1982). There is no doubt that the diffusion bond is one of the best options for the sodium sulfur cell at present and for the foreseeable future.

8.2.6 Field assisted bonding

A related but distinctly different means of bonding aluminium to a ceramic has been described by Mitoff (1978). In this case the ceramic has to be a good ionic conductor: thus the bonding is directly on to the beta alumina. The aluminium component is made the anode and a minimum of 1×10^{-5} equivalents per cm^2 must be transported in order to form a bond. The sodium ions in the beta alumina become polarized towards the cathode and depleted at the anode. At the anode, aluminium ions partially replace the sodium ions. The added advantage of the loss of the sodium ions is that a surface layer of alpha alumina insulator is formed at the interface of these vacuum tight bonds – a vital requirement if this type of seal is to be used to enclose a sodium sulfur cell.

8.2.7 Summary of sodium sulfur cell seals

The main methods for joining the metal and ceramic components are summarized in Table 8.3. There is no really outstanding seal arrangement, as all the seals have some disadvantages. The option preferred by the majority of the developers is the aluminium diffusion bond, but this seal, in the central cathode arrangement, can be expensive owing to the complex alpha alumina collar shape that is required. Fragility and cost will perhaps rule out glass and brazed seals, although these two types have been extensively developed in applications other than sodium sulfur. All the seals listed have been demonstrated to have excellent operational lives in cells, and have been developed to a semi mass production stage.

The most promising seal in terms of cost and ease of assembly appears to be the Ford radial compression seal.

8.3 MANUFACTURING PROCESSES

8.3.1 The sodium electrode

Early laboratory test cells were all filled with liquid sodium metal in inert gas glove boxes, but as larger numbers of cells were required for large batteries,

Table 8.3 Summary of the various seal configurations

Seal type	*Bonding conditions: Atmosphere*	*Temp. (°C)*	*Elevated pressure*	*Ground surfaces required*	*Number of stages*	*Hermetic*	*Durability in cell assembly and use*	*Comments*
Mechanical (+gaskets)	Air	Ambient	Yes	Yes	2	Yes (?)	Difficult to make hermetic. Some designs very robust but bulky.	Used successfully in large numbers of test cells. Alpha alumina/beta alumina seal carried out first.
Glass/metal	Inert e.g. argon	800–1000	No	No	1	Yes	Some designs are fragile causing problems on cell assembly. Corrosion protection required on the sodium side.	Used extensively by Yuasa (Japan) in central sodium cells. Development for central sodium cells in UK. Alpha alumina seal carried out simultaneously with the glass/metal seal. The alpha alumina component could be eliminated. Attractive long term option for providing simplest low cost seal.

Conventional braze	H_2	1100	No	Yes + metallizing	3	Yes	Strong durable bonds. Noble metals required to resist corrosion. Extra protection on the sulfur side desirable.	Expensive and little used in cell tests. Alpha alumina seal carried out after brazing.
Aluminium diffusion bond or thermo-compression seal	Vacuum or inert	650	Yes	Yes	2	Yes	Strong durable bonds that are corrosion resistant. Tendency for intermetallic layer to continue growing during cell operation at 350 °C.	Used extensively in UK, Europe and USA in many differing geometric designs. Currently the preferred option. Can be rather expensive if complicated ceramic shapes are employed. Alpha alumina/beta alumina seal carried out before diffusion bonding.

reliability, safety tests etc. the developers used simpler sodium filling methods which followed two distinct routes. Either a sodium liquid filling apparatus was used which injected sodium directly into a cell, or cells were assembled with solid sodium using prefilled containers.

(a) *Liquid sodium filling*

Filling with liquid sodium requires a machine that will evacuate the sodium compartment, inject a metered amount of liquid sodium and then seal off the compartment. In addition to the normal hazards of handling liquid sodium, there are other disadvantages in using this method. If the sulfur electrode were in place during sodium filling, a broken beta alumina tube would result in reaction of sodium and sulfur with potentially hazardous results. In addition, unless the cell is preheated, the thermal shock caused by contact with liquid sodium at around 130 °C may damage certain cell components. Liquid filling, however, has many advantages. It ensures that any sodium wicks and the beta alumina surface are in physical contact with sodium before heating to the operating temperature, and this encourages better wetting of these two surfaces (Markin *et al.*, 1978). It is also easier, with liquid filling, to reduce the level of oxide impurity and thus reduce the possibility of any sodium flow restriction devices becoming blocked with sodium oxide. The utilization of the available volume in the sodium compartment is also higher if liquid filling is used, as the tolerances to which containers of solid sodium and the cell can be made inevitably lead to wasted space. Also, it is not possible to fill small annular gaps in cells using solid filling methods.

Several developers have used liquid sodium filling machines, and some of these have been automated to allow continuous production of sodium filled cells. Fig. 8.19 shows schematically the Ford filling station (Ritchie, 1980). Similar machines have been described in the literature by Harwell, CSPL and GE.

The Harwell machine was not automatic and was used only for small batches of cells. In their method the cells, after filling, were temporarily sealed with taper pins and were finally sealed permanently by welding the pins into place after removal from the filling apparatus (Markin *et al.*, 1978). An automated sodium filling machine has been described in detail by Riley and McGreavy (1981) for use in pilot plant scale cell production. In this method a fill tube is incorporated into the cell end cap and a temporary seal is formed around this tube, so that the cell sodium compartment can be flushed with argon gas and evacuated prior to filling. A tube of smaller diameter than the fill tube is used to feed sodium into the cell. (This feed tube is used to prevent contamination of the fill tube with sodium, as this would hamper the final welding and sealing operation.) A cell is filled with liquid sodium at 150 °C; it is then evacuated, and the feed tube is partially withdrawn from the fill tube which is then sealed by crimping and welding. To effect a slow cell heat-up and cool-down before and after filling, cells

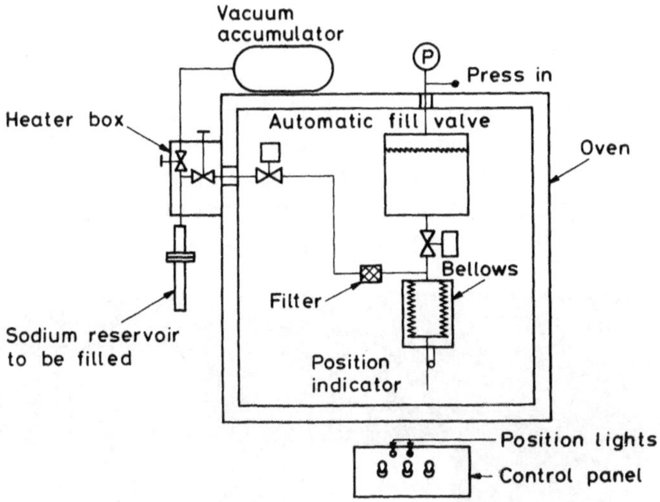

Fig. 8.19 A schematic of the Ford automatic sodium filling station (Ritchie, 1980). Reproduced with permission of the Ford Aerospace and Communications Corporation.

are traversed through a multistage oven and cooling zone; typically heating and cooling rates of 10 °C min^{-1} are not exceeded.

(b) *Prefilled sodium containers*

The use of prefilled sodium containers is particularly suited to cells in which the sodium level is maintained by gas pressure. In these systems a cell is activated as the sodium melts on heating and flows out of the container into contact with the beta alumina. The use of prefilled containers enables cells to be assembled easily and safely. The British Rail cells which used the pressure can system could be assembled on the open bench quickly and easily without the need for complex equipment.

Sodium filling has been used in other industries for many years. The foundry industry uses large quantities of sodium for the modification of aluminium silicon alloys. An addition of approximately 0.012% sodium to an alloy such as LM6 (10–12% silicon) improves the microstructure, giving an alloy with an increased strength and ductility. Sodium for this use is supplied by Foseco (FS) Ltd, Tamworth, Staffs. The sodium is sealed into small aluminium cans and sold worldwide under the trade name of NAVAC®. NAVAC® is accepted as a safe containment for sodium and no special handling techniques are used for its transport or storage. Because of its fairly extensive use in the foundry industry, large scale production of several sizes of NAVAC® cans is undertaken by Foseco in the UK and Europe.

The British Rail pressure cans and the Foseco NAVAC® cans are simple aluminium extrusions as used in vast quantities for small aerosol containers. The filling of the British Rail pressure cans by Foseco at rates in excess of 3000 per day presented no problems, and the need for handling liquid sodium for BR16 and BR30 cell assembly was therefore eliminated at British Rail. As the cans are sealed with aluminium lids they can be stored for several months before use; the removal of the lid and cutting the can to the correct length in one operation by a guillotine is all that is required prior to inserting the can into a cell. The British Rail pressure can components are shown in Fig. 8.20.

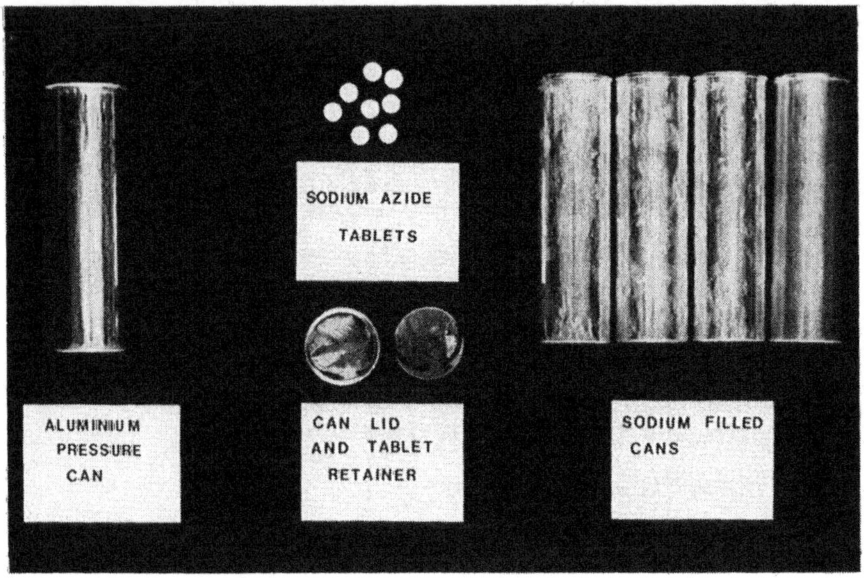

Fig. 8.20 The British Rail pressure can components. Reproduced with permission of British Rail Research.

Another large scale use of sodium, which could possibly be of some use in the introduction of sodium filling and handling for sodium sulfur cells, is the use of sodium filled valves in internal combustion engines for car and aircraft use. Plants for solid and liquid filling of valves have been operating for many years in Europe and the United States. The foundry and automotive uses for sodium have been established for many years, and if sodium handling can be accepted into these industries, sodium filling of sodium sulfur cells should not present any problems.

8.3.2 The sulfur electrode

Graphite or carbon fibres are now commonly used as the electronically conductive matrix material in the cathode compartment. To obtain good contact to the current collector the carbon or graphite fibres have to be under compression, and to obtain a low cathode resistance it is preferable to have the fibres parallel to current flow, i.e. normal to the current collector and beta alumina surfaces. Early cells were filled with molten sulfur after the carbon or graphite felt had been packed into the cells, and some developers initially used sulfur filling rigs similar to the sodium filling systems (Bush, 1975; Lewis, 1978).

The sulfur electrode manufacture now used by all the developers involves the use of preformed electrodes. The electrodes are made up from elements which are shaped in a mould whilst impregnated with liquid sulfur. While the felt is still under compression the sulfur is cooled and solidified. These solid preforms are inserted into cells during assembly, and when the cell is heated up the sulfur melts and allows the carbon or graphite matrix to expand and contact the beta alumina and current collector surfaces. The use of sulfur electrode preforms was first described by Storey in 1977. The advantages claimed were ease of cell assembly, rapid production by hot moulding or injection moulding using established thermoplastic manufacturing techniques, and a less uneven loading of felt into the cathode compartment. The latter problem had caused breakage of beta alumina tubes, as the packing of the cathode area with dry carbon felt had exerted excessive bending moments in long cells. This original preform idea was used at the British Rail laboratories for producing trough shaped electrodes by pressing resilient carbon felt soaked in liquid sulfur in hot or cold moulds. Two of the trough shaped preforms were fitted into the BR central sulfur cells to form a complete sulfur electrode (Piercy and Sudworth, 1977; Bird *et al.*, 1978). Similar systems have been developed at Ford (Ritchie, 1980), CGE (Bast, 1980), GE (Bast, 1982) and BBC. The concept has been developed using injection moulding techniques by British Rail and CSPL. The CSPL sulfur preform is multisegmented and is produced by injection moulding (Brennan, 1980; Jones, 1981). The CSPL preforms and the arrangement of the electrode in a cell are shown in Fig. 8.21. The early sulfur preforms, although enabling much quicker and easier cell assembly methods to be used, were not particularly satisfactory. The various forms of carbon or graphite felts that were available all had structures in which most of the fibres were aligned parallel to the plane of the felt sheets. The preformed electrodes therefore all had fibres normal to the direction of current flow. Ways of reorientating the fibres and then moulding the preforms have been described by Brennan (1980) and Riley *et al.* (1982). The requirement for a 'structured' high performance sulfur electrode as described in Chapter 5, however, meant that a modification of the fabrication methods was needed. The cost of the viscose rayon felt precursor materials (such as the Le Carbone RVC 4000 felt) was also prohibitive for large scale manufacture, and the introduction of the cheaper spun pitch materials (such as the Union Carbide Thornel VMA felt) was required. The handling of the spun pitch materials is not as easy as the viscose rayon felts; the fibres are more brittle and the felt is more easily damaged. These loose, brittle fibres are much more irritating to the skin.

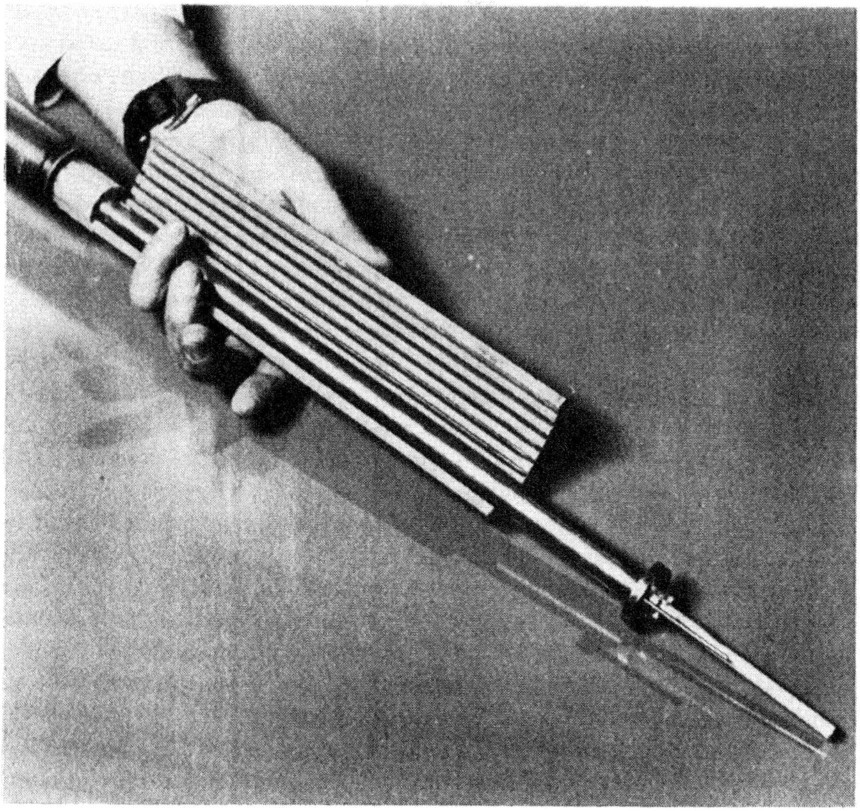

Fig. 8.21 The Chloride Silent Power multisegmented sulfur electrode preform. Photographs kindly provided by Wynn Jones, (CSPL).

Two methods which overcome these problems and enable structured alumina/carbon sulfur electrodes to be constructed have been disclosed. In the manufacturing method described by Robinson *et al.* (1982), a sheet of the carbon fibre material is impregnated with sulfur, whilst under compression, by injection moulding or by simply pressing sheets of the material that have been impregnated in liquid sulfur. After the sulfur has cooled and solidified, the sheets are then quite strong and rigid as the sulfur binds all the fibres together. Washers are then punched from the sulfur carbon composite sheets using conventional sheet metal punching systems. Stacks of these washers can be threaded on to the current collector pole of a central sulfur cell and the fibres are then oriented parallel to the direction of current flow. Incorporation of alumina structuring sheets can be done before the sulfur impregnation stage or at the current collector/sulfur electrode assembly stage.

In a more elegant production method described by Wright (1982), the problems of handling, structuring and orientation of the fibres are overcome and also more radial compression is introduced into the electrode, a feature that is not achieved in the method described by Robinson *et al.* (1982). In the process developed by Wright, a sheet of the carbonized spun pitch fibre mat is impregnated with sulfur and then pressed between two moulds. The fibres lie parallel to the plane of the sheet. Each small element formed by the mould is hexagonal and has a central aperture. The sheet is shown with each shaped element and with the stamping tool in Fig. 8.22(a) and (b). Spun pitch fibre material is preferred, because its brittleness enables the apertures to be formed easily, and after the sulfur has solidified the hexagonal elements are easily separated. Hexagonal shapes are preferable to circular ones because there is then less waste of material. The next stage in manufacture is to thread the hexagons on to a rod, as shown in Fig. 8.22(c). They are secured at each end by small retaining washers. The assembled electrode is then heated to 150 °C to melt the sulfur and then compressed in a cylindrical mould as shown in Fig. 8.22(d), any excess sulfur being squeezed out as the mould halves are closed. The moulds are cooled and after the sulfur has solidified the assembly rod is removed. The size of the moulds and assembly rod is chosen so that the electrode is a fairly close fit on to the cell current collector, but slightly smaller than the internal diameter of the beta alumina tube. When the cell is heated, the sulfur melts and the compressed felt expands, contacting tightly the current collector and electrolyte tube wall. The incorporation of alumina felt structuring is easy with this construction, as the original hexagonal mouldings can be made with alternate layers of carbon and alumina felt sheets.

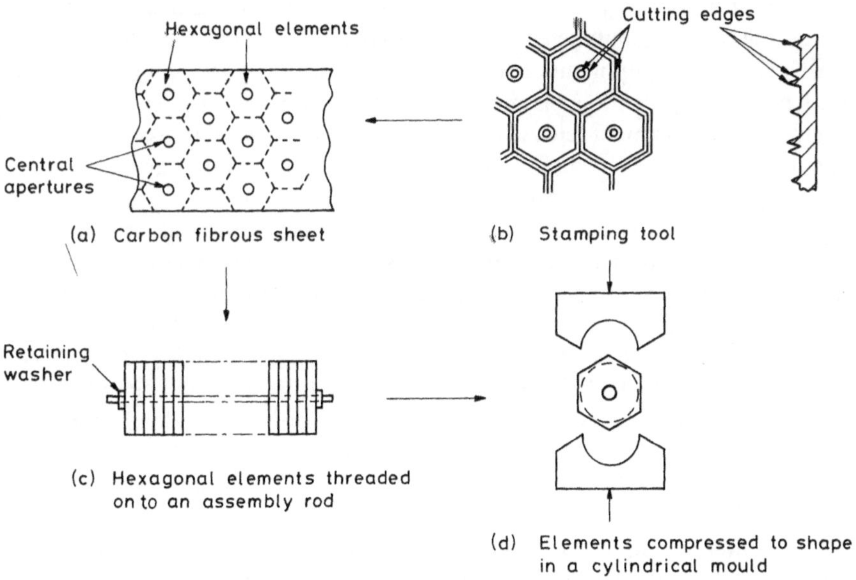

Fig. 8.22 Production of the British Rail structured sulfur electrode (Wright, 1982).

Wright also describes a simple way of adding a 'loosefil' layer to the outside surface of the electrode. The advantages of using a combined structured and loosefil layer electrode have been discussed in Chapter 5. The loosefil mix of carbon fibres, alumina fibres and sulfur is rolled out as a thin layer on to thin PTFE glass cloth. The cloth is heated to approximately 200 °C and then this thin layer of fibres and molten sulfur is wrapped around a cold premoulded sulfur electrode. The loosefil layer sticks to the electrode and the PTFE cloth is easily removed and reused. This type of sulfur electrode construction has an outstanding performance and can be manufactured with little waste using low cost spun pitch felt. The manufacturing method reduces the problems of handling the spun pitch felt, because after the sulfur impregnation stage contact with any free fibres is virtually eliminated. It is doubtful if an electrode of equivalent performance can be manufactured more easily and cheaply by any other method.

8.3.3 Metal component manufacture

A description of the various production processes of individual cell metal components is not really practicable, owing to the considerable variation in cell designs. The approach to large scale cell production by all the developers is, however, broadly similar. Changing to large scale production of cells from small scale laboratory prototype testing involves a complete change of approach to cell design and assembly. Features such as cost, ease of assembly and lightweight components are not paramount with development cells, but production cells have to be designed with these features in mind. The changes meant that early designs, which used accurately machined components that were often carefully matched and fitted, had to be discarded in favour of cheaper mass produced components using simple pressing and stamping techniques. With metal components such as cell case tubing, case end caps, restrictor plates, electrode compartment closures etc. the larger tolerances associated with quantity production processes could be easily accommodated. Some cell designs, however, still have areas such as seals that require a very tight tolerance control. The cost advantages in choosing a seal configuration that can utilize as-fired ceramic components and simple pressed metal components, instead of a seal that requires ceramic grinding or an accurate metal machining operation, are considerable. An indication of the cost reduction that is possible by changing to large scale production methods is shown in Fig. 8.23. A cost breakdown of a 100 Ah British Rail central sulfur tube cell is shown in Fig. 8.24, and as can be seen the metal components account for a fairly small proportion of the overall cell cost. The higher cost items of the cell have been given more attention and the beta alumina production methods and costings have been covered fairly extensively by all the developers (see Chapter 3). Not only is the beta alumina the highest cost item in any cell design, but also it is the only sodium sulfur cell material that is not already manufactured on a large scale or uses an already established large scale manufacturing process. Certain other

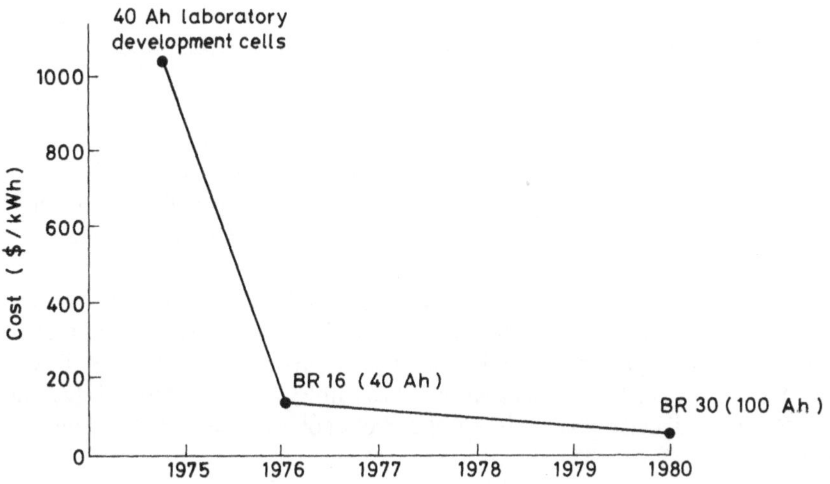

Fig. 8.23 The cost reduction achieved by changing to mass production methods for British Rail 40 Ah and 100 Ah cells (Tilley, 1980). Reproduced with permission of British Rail Research.

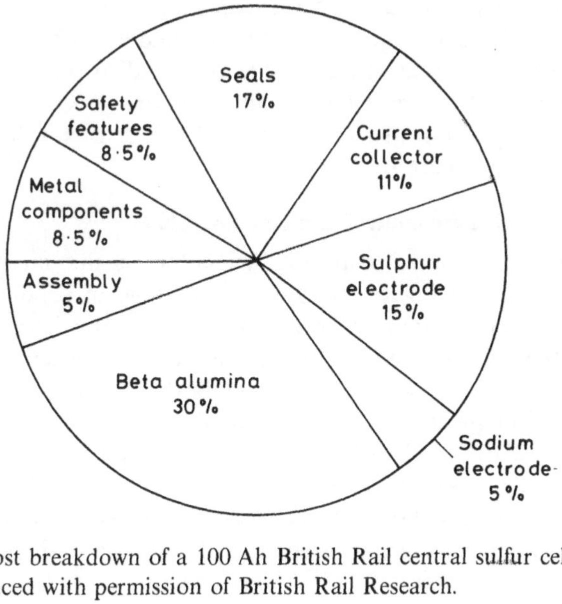

Fig. 8.24 A cost breakdown of a 100 Ah British Rail central sulfur cell (Tilley, 1980). Reproduced with permission of British Rail Research.

features, however, can sometimes appear to be prohibitively expensive, but with some cell designs it may be possible to use a relatively costly component because it is able to provide a dual function. For instance, in a costing of safety features for the British Rail 100 Ah cells, a cost of $10.8 per kWh was quoted for quantities of 500 000 (Tilley, 1980). This high cost was mainly due to the flexible graphite sheet that was required to protect the outer cell case. A central sodium cell design can use such a graphite layer as corrosion protection not only for safety but also for a suitable sulfur electrode current collector (Hartmann, 1980).

Large scale production of cells is required by all the developers as they reach the stage of battery trials. Brown Boveri and CSPL now have fully operational pilot production plants for producing large numbers of cells, but details of the production and cell assembly methods have not yet been published. Details of the British Rail cell construction and current collector construction have been reported.

The current collector assembly used in the BR cell design is quite simple. Two aluminium rods are friction welded to a stainless steel blank, a short 6 mm diameter rod for the positive terminal and a large 9 mm rod for the current collector. After the stainless steel blank has been machined to the correct dimensions, the aluminium current collector rod is prepared by grit blasting. A thin nichrome coating is then applied to this roughened surface by flame spraying, and a layer of flexible graphite paper is isostatically pressed on to this nichrome layer. The components for this production are shown in Fig. 8.25, and Table 8.4 lists the production costs (Tilley and Wright, 1981).

The metal components in a sodium sulfur cell do not appear to present any particular problem with regard to large scale manufacture. Standard processes that are easily automated can be used to form and assemble the components. Processes which have been used include tube end forming, metal pressing, injection moulding, friction welding, flame spraying, TIG and laser welding. Mass production processes such as these will be essential if the load levelling cost goals (Bast, 1982) are to be met.

8.3.4 Quality control and cell assembly

Commercialization of the sodium sulfur battery is not far away, and the developers are all moving to the pilot plant stage of production with the emphasis on the application and production engineering. It is not surprising that, as with the cell components, little published data exists on the present cell assembly methods. The assembly procedure for the Dow cell is slightly different from the beta alumina cells, because of the tube sheet glass fibre assembly and the need for liquid sulfur filling. The metal components, the liquid sodium filling and the metal/glass seals are, however, similar to the beta alumina electrolyte cells. The various stages of the Dow cell assembly are shown in Fig. 8.26, and it is claimed that all these stages are amenable to automation (Levine and Anand, 1977).

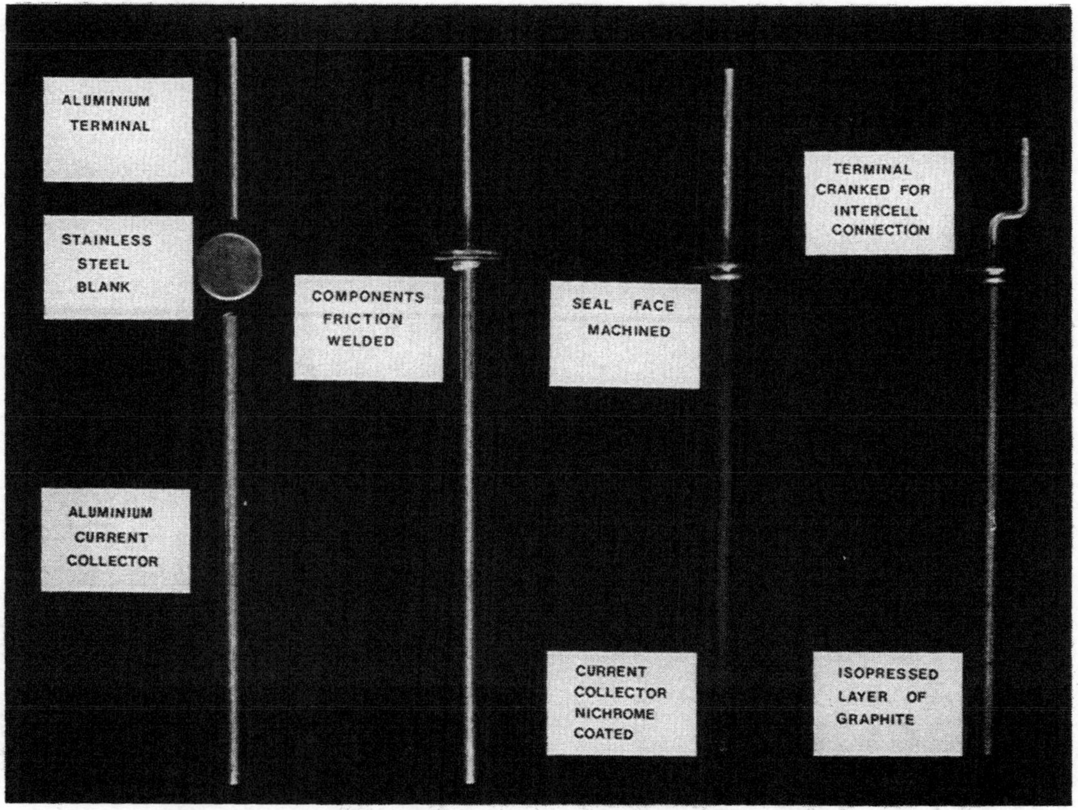

Fig. 8.25 Central sulfur cell current collector components (Tilley and Wright, 1981). Reproduced with permission of the American Society of Mechanical Engineers.

Table 8.4 Production costs for the BR30 current collector for quantities of 1000 (1981 prices)

	Cost (\$)	*Labour (man hours)*
Aluminium +ve terminal (6 mm diam × 100 mm)	0.15	
Aluminium current collector (9 mm diam × 300 mm)	0.35	
Stainless steel holder	1.1	
Friction welding (two welds)		0.09
Nichrome powder	0.17	
Gas (oxygen/hydrogen)	0.7	
Spraying and spraying preparation		0.11
Graphite paper	0.27	
Isostatic pressing		0.1
Bending of +ve terminal for intercell connection		0.03
Total	2.74	0.33
At \$14 per man hour, total cost per current collector is \$7.36		

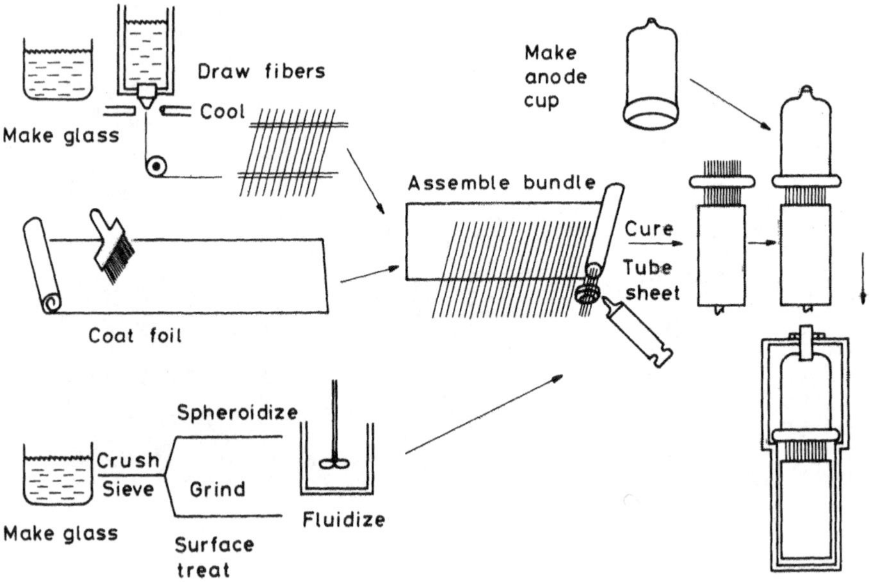

Fig. 8.26 The various stages of the Dow cell construction (Levine and Anand, 1977). Reproduced with permission of the Electrochemical Society, Inc.

The quality control procedures required for the raw materials, the finished components and intermediate stages of cell assembly are broadly similar for all cell designs. At the raw material procurement stage the materials have to be checked, metal components have to be ratified with the supplier certifications and the relevant standard specifications (AISI, DIN, BS etc.), and checked by chemical analysis, metallography and possibly welding trials. Materials such as the sulfur and sodium require batch checking by chemical analysis. At the component stage, dimensional checks are required by careful sampling inspection, and some components even require 100% checking using simple go/no go gauges. Most developers have tolerance checks on the alpha/beta alumina glassed assembly and the other close fitting metal components, such as the safety insert/sodium fill tube in the central sodium cell and the outer cell case in the central sulfur cell. The aim with all the quality control checking is to be able to build up a stock of components that fit together, with no problems on assembly and cell warm-up, and give a reproducible performance. The importance of quality control is not to be underestimated, and the change from building small numbers of prototype cells to constructing large numbers of cells from production components has not been without its problems for most of the developers. For the smaller quantity cell production, the Ford 'shop traveller' system ensures that the correct procedures from raw material to component stage are carefully adhered to. The inspections, any special procedures for cleaning or handling, quality control checks etc. are all recorded and any particular cell can be traced back to particular batches of materials or components (Ritchie, 1980).

The British Rail cell assembly of the BR16 and BR30 cells was a simple, straightforward operation. An extremely high success rate on cell assembly and cell warm-up was achieved, which was only possible because of the many quality control checks on the components (Sudworth, 1983). The most critical checks were the alignment of the beta alumina/alpha alumina assembly and the seal face alignment of the outer case and the current collector. The larger BR30 cells had obviously much higher component reject rates than the shorter BR16 cells. The checks were undertaken by unskilled staff using simple go/no go gauges and cells were assembled by hand on the open bench at rates of up to 50 per day (Tilley, 1980). The metal components, electrodes and alpha/beta alumina assemblies, which had previously been subjected to quality control checks, were taken from a large stock and laid out prior to assembly. This procedure was always used to ensure no vital components could be left out of a cell. Assembly of the parts was then carried out, the compression seal being formed first: a flat aluminium gasket was slipped over the beta alumina tube up to the alpha alumina collar, and the tube assembly was then inserted into the cell case so that the gasket was trapped between the mild steel sealing face and the alpha alumina collar. The preformed sulfur electrode was inserted into the beta alumina tube, followed by the current collector assembly complete with an aluminium gasket which again located between the sealing face on the current collector assembly and the alpha alumina collar. Two disc springs, a load spreading washer, an insulated washer and the inner clamping collar were then fitted around the central positive terminal post, and centralized by using a temporary locating collar. The whole assembly was then transferred to a hydraulic press and a load of approximately 200 kg was applied to the seal assembly. The inner clamping collar was then automatically TIG welded to the outer collar, the cell was removed from the loading/welding jig and the temporary locating collar was removed. The flexible graphite liner was fitted into the outer cell case and the annular gap between the electrolyte tube and the cell case was filled with carbon spheres to reduce the volume of sodium required to fill this annular gap. The final operation was the insertion of the pressure can full of sodium metal and final closure of the cell by automatically TIG welding a cell end cap into place.

The labour content of each cell using this hand assembly is relatively high, but although this is a disadvantage in the high wage economy countries of the major sodium sulfur developers, it could be a suitable system in developing countries where there is an abundance of cheap manpower.

8.4 SUMMARY

Large scale component production and cell assembly is one of the major tasks for the developers as the work progressively moves closer to commercialization. Production and costing have been largely focused on the beta alumina, as the reactants and other components are relatively cheap. (The key to low cost sodium sulfur batteries, especially for load levelling, is to produce large beta alumina tubes at low cost.) The sodium and sulfur in fact, as has been described, can be packaged and incorporated fairly easily and cheaply into cells using established mass production methods. Improvements in production and assembly costs can come from combining several assembly methods into a single process, such as a one shot heating process to produce all the necessary metal/alpha alumina and alpha/beta alumina seals.

For load levelling applications an extra effective cost reduction can be obtained with an increase in scale, and here a single cell capacity of 400–600 Ah is probable. For the traction applications, the sodium sulfur cell using the present semi-production methods has already reached cost competitiveness with lead acid batteries. There appear to be no major problems associated with the transition to full production and assembly of the various components of sodium sulfur cells.

8.5 REFERENCES

Baker, D. J. and Bones, D. J. (1979) UK Patent 1 538 110.

Baker, P. J., Bones, R. J. and Partridge, B. A. (1980) Deutscher Verlag fuer Schweisstechnik Conference, Baden-Baden, West Germany.

Barrow, P. (1980) 4th DoE/EPRI Sodium Sulfur Battery Workshop, Derby.
Barry, T. I., Schajer, G. S. and Stackpool, F. M. (1980) Deutscher Verlag fuer Schweisstechnik Conference, Baden-Baden, West Germany.
Bast, J. (1980) EPRI Report EM-Project 128-4, -5.
Bast, J. (1982) EPRI Report EM-2579, Project 128-6.
Bindin, P. (1977) US Patent 798 277.
Bird, J. M., Tilley, A. R. and Sudworth, J. L. (1978) 13th IECEC, San Diego paper 789201.
Bones, R. J. (1978) US Patent 4 074 026.
Brennan, M. P. (1980) UK Patent Application 2 042 244.
Brennan, M. P. J. (1977) *Proc. Electrochem. Soc.*, **77–6**, 963.
Brinker, C. J. and Klein, L. C. (1980) *Phys. & Chem. Glasses*, **21**, 141.
Buckley, R. G. (1979) *Ceram. Ind.*, **112**, 20.
Bull, R. N. (1983) Beta Research and Development Ltd Private Communication.
Bush, J. B. (1975) EPRI Report, Project 128-2.
Chase, G. G., Harlow, R. A. and Gupta, V. K. (1979) US Patent 4 170 694.
Chatterji, D. (1976) EPRI Report EM-266, Project 128-3.
Dawihl, W. and Klinger, E. (1969) *Ber. Deutsch. Keram. Ges.*, **46**, 12.
Desplanches, G., Lazennec, Y. and Wicker, A. (1977) US Patent 4 037 027.
Elkins, P. E. (1980) Sodium/Sulfur Battery Review, Ford presentation to DoE, Newport Beach, CA.
Elyard, C. A. and Rawson, H. (1962) *Advances in Glass Tech.*, Part 2. p. 270–286. Plenum Press, New York.
Fischer, W. (1983) Brown Boveri Co. Private communication.
Gibson, J. G. and Duncan, J. H. (1974) UK Patent 1 356 312.
Hartmann, B. (1980) 4th DoE/EPRI Sodium Sulfur Battery Workshop, Mannheim.
Hattori, S., Yamaura, M., Kimura, S. and Iwabuchi, S. Proceedings SAE Congress, Detroit, paper 770281.
Hirayama, C. (1961) *J. Am. Ceram. Soc.*, **44**, (12), 602.
Hull, A. W. and Burger, E. E. (1934) *Physics*, **5**, 384.
Iwabuchi, ? (1983) Yuasa Battery Co. Private communication.
Jones, I. W. (1981) *Philos. Trans. R. Soc. London*, **A302**, 339.
Jones, I. W., McNamee, M. and Bindin, P. J. (1977) UK Patent 1 496 407.
Jones, I. W. and Miles, L. J. (1974) UK Patent 1 356 576.
Klomp, J. T. (1971) *Sci. Ceram.*, **5**, 501.
Klomp, J. T. (1972) *Bull. Am. Ceram. Soc.*, **51**, 683.
Lazennec, Y. (1977) EPRI Report EM-413, Project 726-1.
Levine, C. (1983) Dow Chemical Co. Private communication.
Levine, C. and Anand, J. (1977) 152nd Meeting Electrochemical Society, Atlanta.
Lewis, J. K. (1978) Sodium/Sulfur Battery Review, Ford presentation to DoE, Newport Beach, Ca.
Lomax, G. R. (1979) Journal of Power Sources, **4**, 301.
Markin, T. L., Junkison, A. R. and Warwick, M. A. (1978) UK Patent 1 519 451.
Mikkor, M. (1981) US Patent 4 245 012.
Miles, L. G. and Jones, I. W. (1972) in *Power Sources*, vol. 3 (ed. D. H. Collins), Oriel Press.
Mink, R. W. (1983) Ford Aerospace Communications. Private communication.
Mitoff, S. P. (1978) EPRI Report EM-683, Project 128-4.

Mortimer, D. A. (1979) UKAEA Harwell Report AERE-R9533.
Nicholas, M. G. and Crispin, R. M. (1982a) *J. Mater. Sci.*, **17**, 3347.
Nicholas, M. G. and Crispin, R. M. (1982b) *Proc. Br. Ceram. Soc.*, **32**, 33.
Piercy, R. and Sudworth, J. L. (1977) 151st Meeting of the Electrochemical Society, Philadelphia, USA.
Riley, D. J. and McGreavy, G. (1981) UK Patent Application 2 061 598.
Riley, D. J., Ozanlar, V., McNamee, M. and Ansell, R. O. (1982) UK Patent 2 095 026.
Ritchie, E. E. (1980) Sodium/Sulfur Battery Review, Ford presentation to DoE, Newport Beach, Ca.
Robinson, G., Ansell, R. O. and McNamee, M. (1982) UK Patent Application 2 095 027.
Silverman, H. P., Marcoux, L. S. and Seo, E. T. (1976) EPRI Report EM-226, Project 127.
Storey, M. A. (1977) UK Patent 1 472 975.
Sudworth, J. L., (1983) 5th DoE/EPRI Sodium Sulfur Battery Workshop, Washington DC.
Sudworth, J. L. and Tilley, A. R. (1976) UK Patent 1 452 481.
Tilley, A. R. (1976) Proceedings of Symposium and Workshop on Advanced Battery Research and Design, Electrochemical Society and Argonne National Laboratory Report ANL 76-8 B231.
Tilley, A. R. (1978) UK Patent 1 506 505.
Tilley, A. R. (1980) 4th DoE/EPRI Sodium Sulfur Battery Workshop, Derby.
Tilley, A. R. and Wright, M. L. (1981) 16th IECEC, Atlanta, paper 819394.
Topouzian, A. (1979) US Patent 4 169 919.
Wright, M. L. (1982) UK Patent Application 2 097 992.

9 Cell optimization

J. L. SUDWORTH

LIST OF SYMBOLS

R	Cell area resistance
ρ_E	Electrolyte specific volume resistivity
t_E	Electrolyte thickness
t_s	Sulfur electrode thickness
ρ_c	Specific volume resistivity of the cell case material
ρ_{Na}	Specific volume resistivity of sodium metal
t_c	Cell case thickness
t_{Na}	Sodium wick thickness
d_c	Diameter of cell case
L	Length of cell case
d	Electrolyte tube diameter
η	Voltage efficiency
P	Cell power per unit area of electrolyte
V_0	Mean reversible EMF
E	Cell energy per unit area of electrolyte
T	Discharge time
K	A constant equal to one plus the volume fraction of carbon in the sulfur electrode
δ_s	Density of Na_2S_3
E_t	Theoretical specific energy of the cell
V_c	Volume of cell
L_t	Length of cell terminals
F	Faraday's constant
M_a	Mass of active material
M_a'	Mass of active material plus matrix material
δ_m	Density of porous matrix
$\delta_{Na_2S_3}$	Density of Na_2S_3

δ_{Na}	Density of sodium
D	Maximum diameter of cell
M_s	Mass of seal components per unit area of electrolyte
s	Mass of seal component per unit circumference of electrolyte tube
L_c	Total length of cell
M_c	Mass of cell case per unit area of electrolyte
M_e	Mass per unit area of electrolyte
M_r	Mass of sodium reservoir per unit area of electrolyte
M_i	Mass: power ratio of inert materials in the cell
M	Mass: power ratio for the whole cell
t_α	Width of alpha alumina collar in seal
ρ_{cc}	Specific volume resistivity of current collector material

9.1 INTRODUCTION

The dimensions of a sodium sulfur cell will be determined by two factors – the energy:power ratio, and the relative importance of the various cell characteristics such as efficiency, specific energy etc. The effect on these characteristics of varying the dimensions of the cell components can be calculated and the results of these calculations used to select the optimum cell dimensions.

The sodium sulfur cell lends itself to such calculations – its internal resistance is ohmic, the utilization of sodium and sulfur is independent of discharge rate and the cell operates at 100% current efficiency. The cell resistance and the cell EMF vary with depth of discharge, and when using the equations developed here a decision must be made as to which values to use – the average values or the values at a certain depth of discharge. The average cell voltage is 2.015 at 350 °C.

The most important parameter in cell design is the mass of sulfur per unit area of the electrolyte. This is proportional to the thickness of the sulfur electrode for a given cell geometry, and it is convenient to calculate the characteristics of interest – specific energy, efficiency, cost – as a function of sulfur electrode thickness.

Two cell designs will be considered – central sulfur and central sodium tubular cells – but the same procedure can be applied to any cell design with beta alumina solid electrolyte. For the glass electrolyte cell, the important dimension of tube diameter is fixed and these calculations are not relevant. For simplicity, it will be assumed that the cell capacity is equal to the theoretical capacity, but in practice it will always be less and will be a function of charge time. A second assumption will be that the cell is evacuated; this simplifies the calculation and is a desirable feature of cell design since it results in the maximum cell capacity.

An expression for the cell resistance can be derived for each cell configuration, and if the area resistance (units Ω cm^2) is used the quantity of sulfur required per unit area of electrolyte can be calculated for any given cell resistance, voltage efficiency and discharge rate. This then determines the weight and volume of the other cell components.

The cell resistance can be expressed in terms of the electrolyte resistivity and thickness and the sulfur electrode thickness:

$$R(\Omega\ \text{cm}^2)=\rho_E t_E+0.4t_s \tag{9.1}$$

where ρ_E is the resistivity of the electrolyte, t_E is the wall thickness of the electrolyte and t_s is the thickness of the sulfur electrode. The figure of 0.4 Ω cm^2 corresponds to the value obtained in one type of cell (see Section 5.4.4) and will vary according to the sulfur electrode structure.

It is implicit in Equation 9.1 that contributions to the cell resistance, from for example the cell case, can be neglected. This is not strictly correct and Equation 9.1 has to be modified to include these other resistances.

9.2 CENTRAL SODIUM CELL CALCULATIONS

An additional resistance will arise from the cell case which must conduct the current from the positive electrode to the terminal, and if the sodium contained within the electrolyte tube is utilized on discharge the resistance of the sodium wick must also be considered. In both cases, the current flowing up the cell is assumed to change linearly from zero at the bottom to I, the total cell current at the top of the cell. To correct for this the resistance of the two components will be taken as half the actual value. The resistance expressed in Ω cm^2 will be:

$$R_x=\frac{\rho_x L^2 d}{2t_x d_x} \tag{9.2}$$

where ρ_x is the resistivity of the material, d_x is the diameter of the component, L is the length of the cell case, d is the electrolyte diameter and t_x is the thickness of material.

Thus the total cell resistance is given by:

$$R(\Omega\ \text{cm}^2)=\rho_E t_E+0.4t_s+\frac{\rho_c L^2 d}{2t_c d_c}+\frac{\rho_{Na} L^2}{2t_{Na}} \tag{9.3}$$

where the subscripts c and Na refer to the case material and the sodium wick respectively.

From the cell resistance and the voltage efficiency (η) the cell power can be calculated using Ohm's law:

$$P(\text{W cm}^{-2})=\frac{V_0^2\eta(1-\eta)}{R} \tag{9.4}$$

where V_0 is the mean cell EMF and the area refers to the surface of the electrolyte in contact with the electrodes. The cell energy E is simply the product of the power and the discharge time T (hours):

$$E(\mathrm{Wh\ cm^{-2}}) = \frac{V_0^2\eta(1-\eta)}{R}T \tag{9.5}$$

Thus the total cell energy is given by:

$$E(\mathrm{Wh}) = \frac{V_0^2(1-\eta)T\pi dL}{R} \tag{9.6}$$

The volume of the sulfur electrode (V) which equals the volume of sodium trisulfide plus the volume of carbon in the electrode matrix is proportional to the total energy:

$$V = \frac{KE}{E_t\eta\delta_s} \tag{9.7}$$

where K is equal to one plus the volume fraction of carbon in the electrode, E_t is the theoretical specific energy of the cell (0.760 Wh g^{-1}) and δ_s is the density of sodium trisulfide at the cell operating temperature.

The volume of the sulfur electrode is obviously equal to the difference of the total cell volume and the volume of the electrolyte tube of external diameter d:

$$\frac{KE}{E_t\eta\delta_s} = \frac{\pi(d+2t_s^2)L}{4} - \frac{\pi d^2 L}{4} \tag{9.8}$$

Substituting Equation 9.6 and solving for η gives:

$$\eta = 1 - \frac{R(dt_s + t_s^2)E_t\delta_s}{KV_0^2Td} \tag{9.9}$$

By substituting for η in Equation 9.6, the cell energy can be calculated. The cell volume (V_c) is easily calculated if all the Faradaic sodium can be contained within the electrolyte tube:

$$V_c(\mathrm{cm^3}) = \frac{\pi D^2(L+L_t)}{4} \tag{9.10}$$

where D is the cell diameter $(d+2t_s)$ and L_t is the length of the terminals.

If some of the sodium has to be accommodated in a reservoir, the cell volume is given by:

$$V_c = \frac{\pi D^2}{4}\left\{L + L_t + \left[\frac{23 \times 3600\ E}{2\eta \mathrm{F}\delta_{\mathrm{Na}}} - \frac{\pi(d-2t_e)^2}{4}\right]\frac{4}{\pi(d+2t_s)^2}\right\} \tag{9.11}$$

In equation 9.11 the atomic weight of sodium is taken as 23; 3600 is the number of seconds in an hour. From these equations the volumetric specific energy can be calculated; to calculate the gravimetric specific energy, however, requires a more detailed knowledge of the cell design. The mass of the cell is made up of two components – the active materials, i.e. sodium and sulfur, and the inert materials which comprise the rest of the cell, i.e. cell case, seals, electrolyte etc.

The mass of the active material is a function of cell energy and can be expressed as the mass: power ratio (g kW^{-1}):

$$M_a = \frac{E}{10^3 E_t \eta P} = \frac{T}{10^3 E_t \eta} \tag{9.12}$$

The ratio of matrix material to active material will be fixed and can be included by modifying Equation 9.12:

$$M'_a = \frac{K'T}{10^3 E_t \eta} \tag{9.13}$$

where

$$K' = 1 + \frac{\delta_m}{\delta_{Na_2S_3}} \tag{9.14}$$

where δ_m and $\delta_{Na_2S_3}$ are the densities of the porous matrix and sodium trisulfide respectively.

The mass of the inert components can be expressed in terms of mass per unit area of electrolyte from which, using Equations 9.4 and 9.5, the power:weight ratio and the energy:weight ratio can be calculated.

For the seal components, the mass per unit area of electrolyte is given by:

$$M_s = \frac{(d+2t_s)\sum s}{dL} \tag{9.15}$$

where s is the mass per unit circumference for each seal component.

For the cell case the mass per unit area of electrolyte is given by:

$$M_c = \frac{cL_c}{dL}\left[(d+2t_s) + \frac{(d+2t_s)^2}{2L_c}\right] \tag{9.16}$$

where L_c is the total length of the cell (Equation 9.11) and c is the weight per unit area of the cell case.

For the electrolyte:

$$M_e = 3.2t_e \tag{9.17}$$

In cells of more complex design the additional components can be treated in the same manner. For example, if a sodium reservoir is required the mass per unit area is given by:

$$M_r = \frac{7.9t_c\pi D^2}{(d+2t_s)dL}\left[\frac{23\times 3600\ E}{2\eta F\delta_{Na}} - \frac{\pi(d-2t_e)^2L}{4}\right] \tag{9.18}$$

The mass:power ratio of the inert materials is given by:

$$M_i = \frac{R(M_s+M_c+M_e+M_r)}{10^3\ V_0^2\eta(1-\eta)} \tag{9.19}$$

The total mass:power ratio of the cell (kg W^{-1}) is therefore:

$$M = M_i + M'_a \tag{9.20}$$

The mass:energy ratio is obtained simply by dividing M by the discharge time.

By substituting the cost of the components for the weight, the same equations can be used to calculate the total cost of the cell components which when added to the cost of assembly will give the total cell costs.

9.3 CENTRAL SULFUR CELL CALCULATIONS

The cell resistance is given by:

$$R = \rho_e t_e + 0.4t_s + \frac{2\rho_{cc}dL^2}{d_{cc}^2} + \frac{\rho_{Na}L^2}{2t_{Na}} \tag{9.21}$$

where the subscript cc refers to the current collector rod and d is the internal diameter of the electrolyte tube.

The efficiency is given by (cf. Equation 9.9):

$$\eta = 1 - \frac{R(dt_s - t_s^2)E_t\delta_s}{KV_0^2Td} \tag{9.22}$$

If the Faradaic sodium is assumed to be contained in a reservoir at the end of the cell, the volume of the cell can be calculated. The maximum cell diameter is given by:

$$D = d + 2t_\alpha \tag{9.23}$$

where t_α is the width of the alpha alumina collar which is glass sealed to the open end of the beta alumina tube. The cell length is given by the following equation, where the third term is the length of the sodium reservoir calculated from the volume of sodium and the cross sectional area of the cell:

$$L_c = \left[L + L_t + \frac{0.858 V_0 (1-\eta) 4dLT}{RD^2} \right] \tag{9.24}$$

The cell volume is therefore:

$$V_c = \frac{\pi D^2 L_c}{4} \tag{9.25}$$

The volumetric specific energy can be calculated from Equations 9.6 and Equations 9.21 to 9.25. The gravimetric specific energy can be calculated in a similar manner to that of the central sodium cell. The mass per unit area of electrolyte of the seal components is again given by:

$$M_s = \frac{(d + 2t_e + 2t_{Na}) \sum s}{dL} \tag{9.26}$$

but the mass per unit area of electrolyte of the cell case becomes:

$$M_c = \frac{cL_c}{dL} \left[(d + 2t_e + 2t_{Na}) + \frac{(d + 2t_e + 2t_{Na})^2}{2L_c} \right] \tag{9.27}$$

There is also an additional component, the central current collector, whose mass per unit area is given by:

$$M_{cc} = \frac{(d - 2t_s)^2}{4d} \rho_{cc} \tag{9.28}$$

The mass:power ratio of the inert materials is given by:

$$M_i = \frac{R(M_s + M_c + M_{cc} + M_e)}{10^3 V_0^2 \eta (1-\eta)} \tag{9.29}$$

The mass:power ratio of the cell (kg W^{-1}) is therefore:

$$M = M_i + \frac{K'T}{10^3 E_t \eta} \tag{9.30}$$

The specific energy of the cell is therefore:

$$ED = \frac{T}{M} \tag{9.31}$$

The cost of the cell can be calculated in the same way as for the central sodium cell.

9.4 OPTIMIZATION OF CELL CHARACTERISTICS

In Fig. 9.1, the specific energy and other cell characteristics are plotted against sulfur electrode thickness for a central sodium cell having an electrolyte tube 2.5 cm OD by 30 cm long. The energy:power ratio of the cell is 2 hours.

It can be seen that the specific energy passes through a maximum at a sulfur electrode thickness of 8 mm. This is a consequence of the fall in the cell voltage with increasing current density. The choice of sulfur electrode thickness will be dictated by practical considerations and the value chosen may not coincide with that which results in maximum specific energy. Thus the practical optimum electrode thickness from Fig. 9.1 might be 6 mm: a cell having an electrode of this thickness would not need a sodium reservoir, would have a high efficiency and a relatively low current density.

This choice of electrode thickness might be further modified if cost is considered to be an important factor. The cost will pass through a minimum and this may or may not occur at an electrode thickness of 8 mm.

Fig. 9.2 shows the same plot for a central sulfur cell and clearly the shape of the curves is somewhat different. The sharp reduction in specific energy and efficiency which occurs as the thickness of the sulfur electrode approaches the radius of the electrolyte tube is a direct consequence of the increased cell resistance due to the reduced cross sectional area of the current collector.

The electrode thickness corresponding to maximum specific energy occurs at 12 mm, but again there are good reasons for choosing a value somewhat less than this.

It is noticeable that the volumetric specific energy of a central sulfur cell is considerably less than that of a central sodium cell. The gravimetric specific energy, however, is similar for both cell designs.

The other cell dimensions of interest are the diameter and length of the electrolyte tube. Fig. 9.3 shows the effect of tube diameter for a central sodium cell. Each point on this graph is the maximum specific energy for that electrolyte tube diameter; consequently the sulfur electrode thickness increases from 8 mm at 2.5 cm diameter to 10 mm at 7 cm diameter.

The optimum tube diameter will depend upon the relative importance of the gravimetric and volumetric specific energies, and lies in the range 2–4 cm. Only in the case of the 2.5 cm diameter tube is a sodium reservoir necessary. The corresponding plot for a central sulfur cell is shown in Fig. 9.4. In this case,

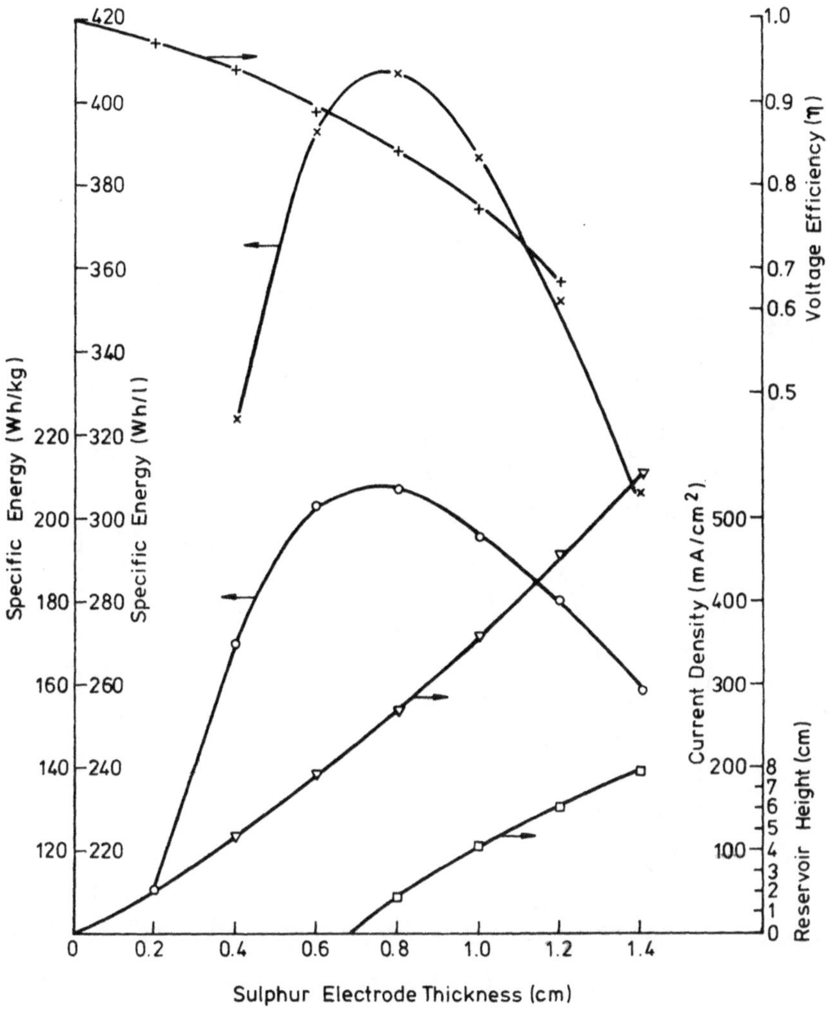

Fig. 9.1 Specific energy and efficiency of central sodium cells. Electrolyte tube diameter 2.5 cm; discharge time 2 hours, electrolyte tube length 30 cm. ○ Wh kg^{-1}; × Wh l^{-1}; + voltage efficiency; ▽ current density; □ sodium reservoir height.

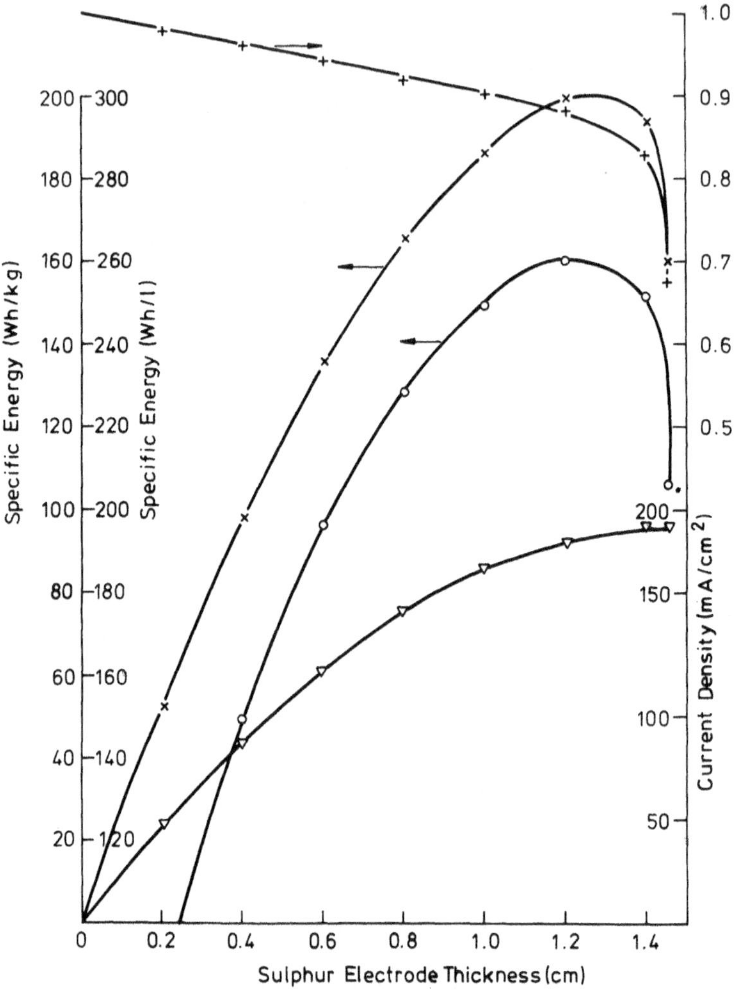

Fig. 9.2 Specific energy and efficiencies of central sulfur cells. Electrolyte diameter 3.0 cm; electrolyte length 30 cm; discharge time 2 hours. × Wh kg^{-1}; ○ Wh l^{-1}; + voltage efficiency; ▽ current density.

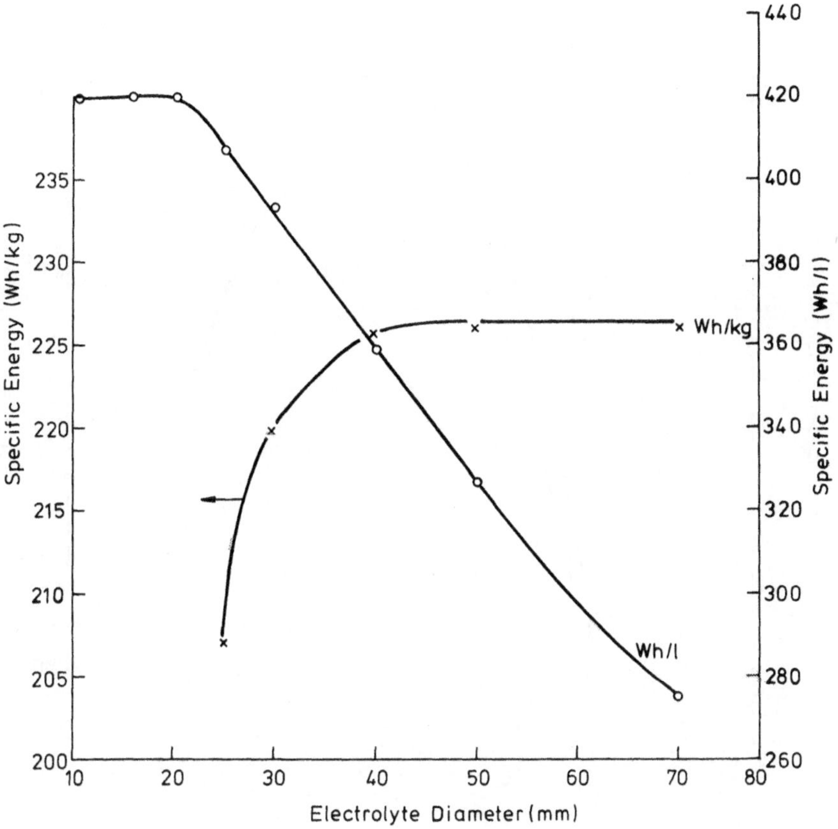

Fig. 9.3 Effect of electrolyte tube diameter on specific energy of central sodium cells. Electrolyte length 30 cm; discharge time 2 hours. × Wh kg^{-1}; ○ Wh l^{-1}.

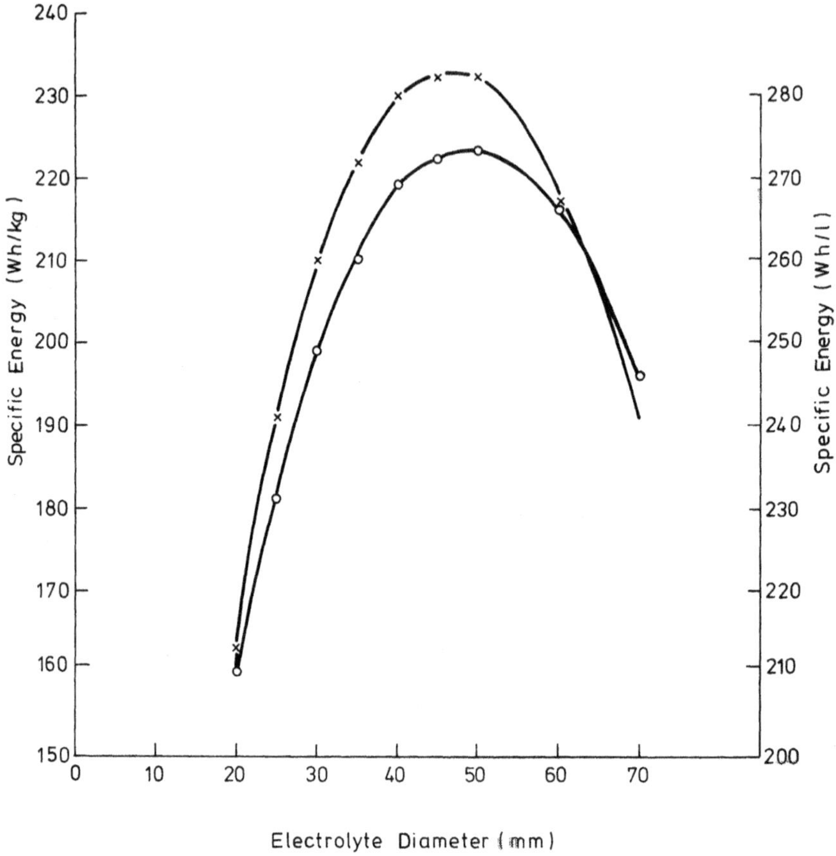

Fig. 9.4 Effect of electrolyte diameter on specific energy of central sulfur cells. Electrolyte length 30 cm; discharge time 2 hours. × Wh kg^{-1}; ○ Wh l^{-1}.

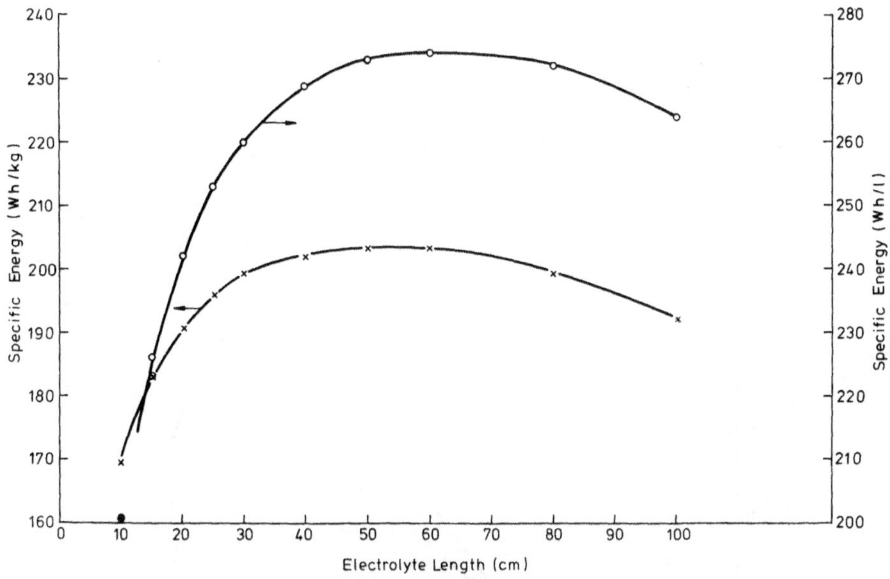

Fig. 9.5 Effect of electrolyte length on specific energy of central sulfur cells. Electrolyte diameter 3.0 cm; discharge time 2 hours, sulfur electrode thickness 1.2 cm. × Wh kg^{-1}; ○ Wh l^{-1}.

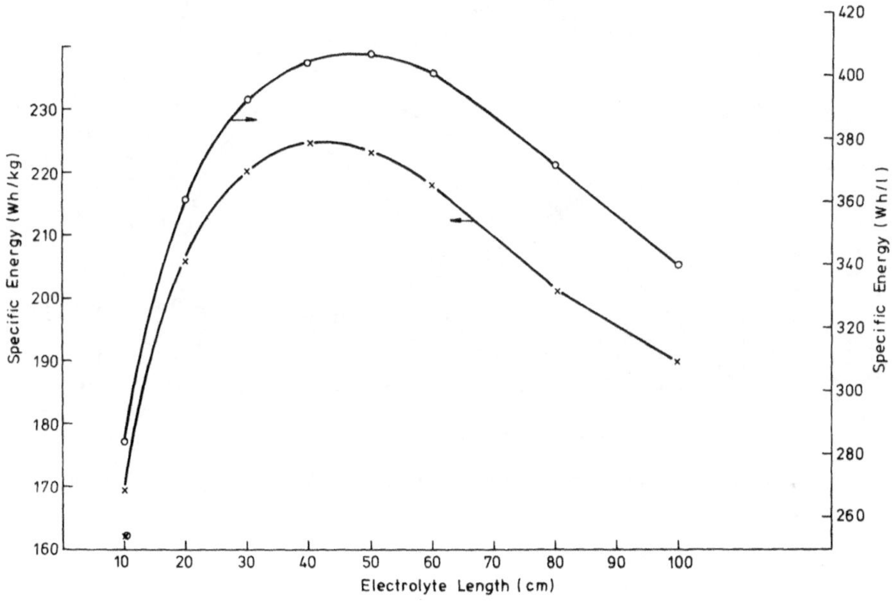

Fig. 9.6 Effect of electrolyte length on specific energy of central sodium cells. Diameter of electrolyte 3.0 cm; discharge time 2 hours, sulfur electrode thickness 1.2 cm. × Wh kg^{-1}; ○ Wh l^{-1}.

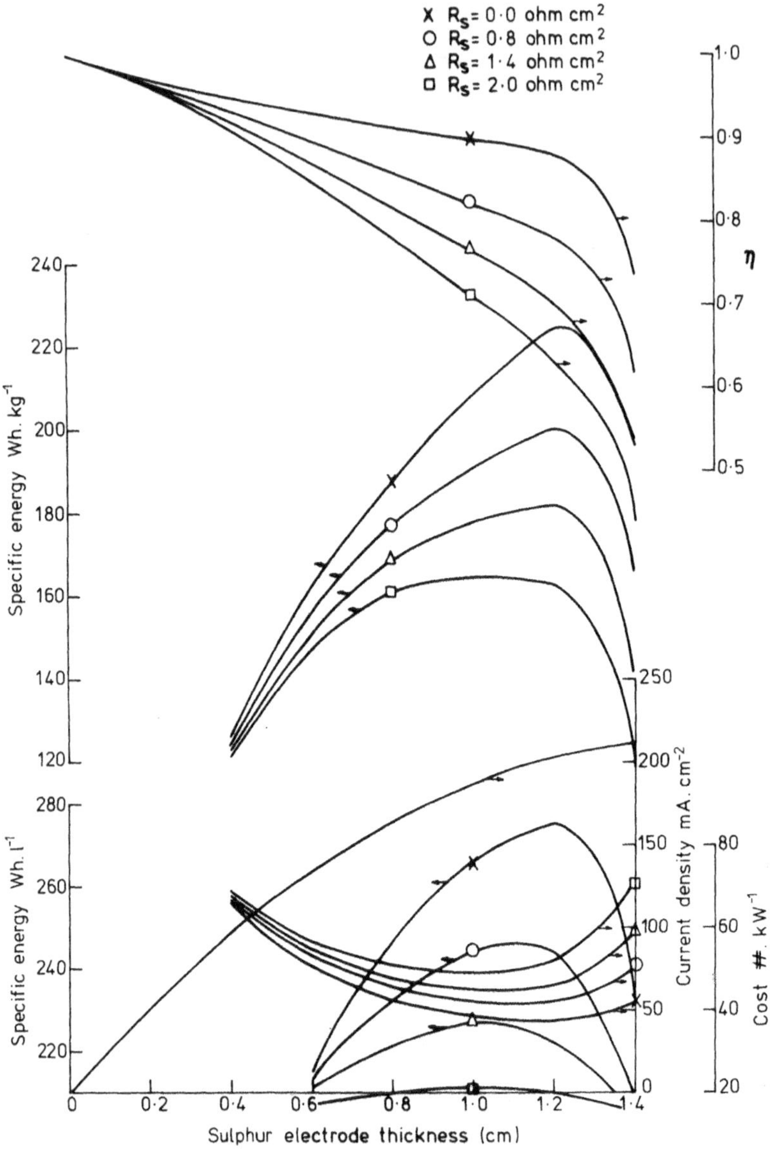

Fig. 9.7 Effect of sulfur electrode resistance (R_s) on specific energy of a central sulfur cell.

both the volumetric and gravimetric specific energies reach a maximum at the same tube diameter. As expected, there is a greater increase in sulfur electrode thickness, from 10 mm for a 2.5 cm diameter tube to 20 mm for a 5 cm diameter tube, since this parameter is largely determined by the cross sectional area of the current collector needed to keep the cell resistance to a reasonable value.

The effect of electrolyte tube length on specific energy for a central sodium cell is shown in Fig. 9.6. Both the volumetric and gravimetric specific energies pass through a maximum at approximately 45 cm. The decline in specific energy for electrolyte tube lengths greater than 45 cm is largely due to the resistance of the cell case and the sodium. If the material used for the cell case is mild steel, its contribution to the cell resistance at $L = 100$ cm is equal to that of the sodium in the wick, and together they account for half the cell resistance.

The corresponding plot for the central sulfur cell is shown in Fig. 9.5. In this case it is the resistance of the aluminium pole together with the sodium electrode resistance which causes the decline in specific energy.

A further use for this type of calculation is to determine the sensitivity of the specific energy to the various cell parameters. Fig. 9.7 shows the effect of varying the sulfur electrode resistance for a central sodium cell.

10 Cell performance

J. L. SUDWORTH

10.1 INTRODUCTION

The important parameters in cell performance are internal resistance, discharge and charge characteristics and the specific energy. Of equal importance, however, is the stability of these parameters over the life of the cell and, of course, the length of the cell life.

The cell resistance is of interest for two reasons: it determines the maximum power available from the cell, and it is the major source of the cell's internal heating. (A discussion of thermal effects will be deferred until Chapter 11 when thermal management of batteries is considered.) The discharge characteristics of a sodium sulfur cell are in fact determined almost entirely by the internal resistance, which is ohmic over most of the discharge range. Thus the capacity of the cell is effectively independent of discharge rate and the electrical energy obtained on discharge can easily be calculated if the cell resistance is known.

The charge characteristics of sodium sulfur cells are determined by the method of charging, the temperature of operation and the need to avoid applying a high voltage and a high current density to the solid electrolyte at the top of charge. It is the charge acceptance of a cell which effectively determines the capacity range over which the cell operates.

Cell failure may be sudden, usually because of failure of the solid electrolyte, or it may be gradual, in which case resistance increase or loss of capacity may be responsible. In all cases the life of a population of cells is amenable to statistical analysis using probability theory, and useful information on life and reliability can be obtained from cell test data. As well as intrinsic reasons for cell failure, external factors must also be considered. These include charge and discharge rate and operating temperature, and although at present little quantitative information is available on the effect of these parameters there is no doubt that they can greatly influence cell life and reliability.

10.2 CELL RESISTANCE

Before the various aspects of cell resistance are discussed it is necessary to define exactly what is meant by this term. When a sodium sulfur cell is charged or discharged, the charge and discharge curves will generally be similar in form to the open circuit voltage curves. Typical charge/discharge curves are shown in Fig. 10.1 (350 °C) (Sudworth, 1981). The cell resistance is the quotient of (1) the

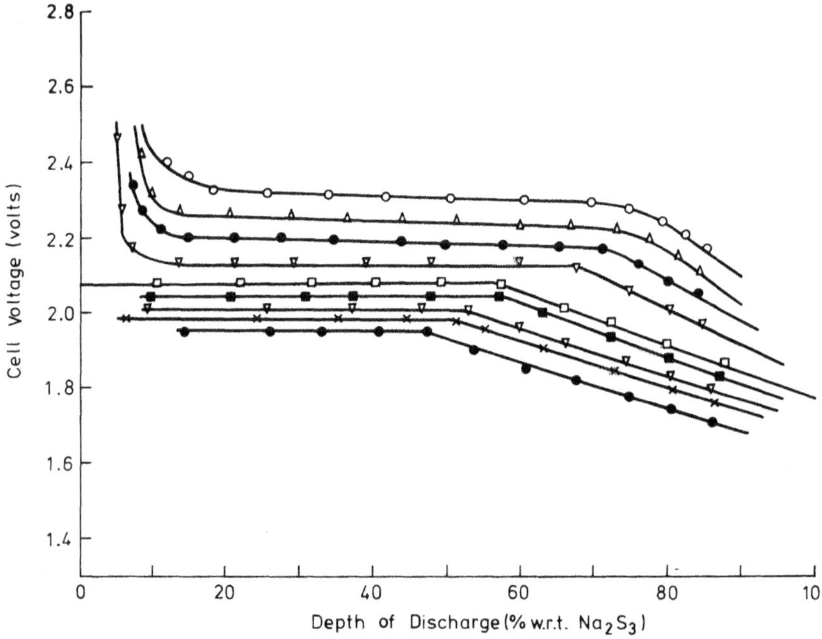

Fig. 10.1 Charge/discharge curves for central sulfur cells at 350 °C. □ open circuit voltage; ■ 2.27 A; ▽ 4.55 A; × 6.82 A; ● 9.1 A; △ 13.65 A; ○ 18.2 A. Reproduced from Sudworth (1981).

equilibrium open circuit voltage (OCV) minus the load voltage, and (2) the current. If accurate coulometry is used to determine the depth of discharge, the equilibrium open circuit voltages determined by Gupta and Tischer (1972) can be used; otherwise the equilibrium open circuit voltage of the cell must be measured. This can take several hours and some workers use an open circuit voltage determined after a fixed time, often seconds or minutes. Such measurements do not give the true cell resistance and should be used for purposes of comparison only.

Inspection of Fig. 10.1 reveals a number of interesting points. On discharge the inflexion in the discharge curve occurs at lower depths of discharge as the current density increases. This must be due to formation of sulfides more sodium rich than is indicated by the overall stoichiometry in part of the sulfur electrode, and is consistent with the potential distribution model of the sulfur electrode discussed in Chapter 5.

A close inspection of Fig. 10.1 shows that, although the discharge curve again becomes almost parallel to the OCV curve in the single phase region, the cell resistance is higher than in the two phase region. This is a direct consequence of the shift of the inflexion in the discharge curves. The manner in which the cell resistance varies with depth of discharge will be characteristic of the sulfur electrode design. For the cells from which the data in Fig. 10.1 were obtained, the cell resistance is constant over a large part of the two phase region, and in the single phase region it is greater by a factor of between 1.3 and 1.6 as shown in Fig. 10.2. Similarly, on charge the inflexion in the curve occurs earlier at higher current densities.

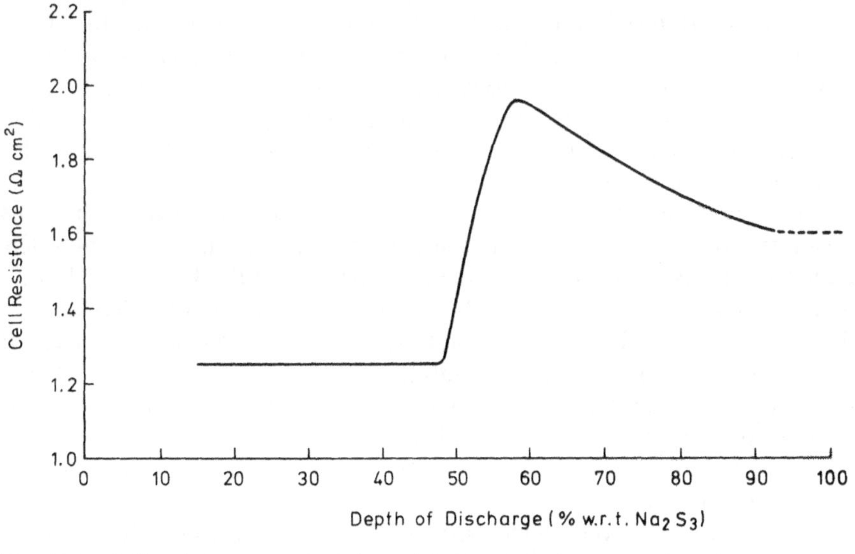

Fig. 10.2 Cell resistance as a function of depth of discharge. Derived from Fig. 10.1; discharge current 9.1 A.

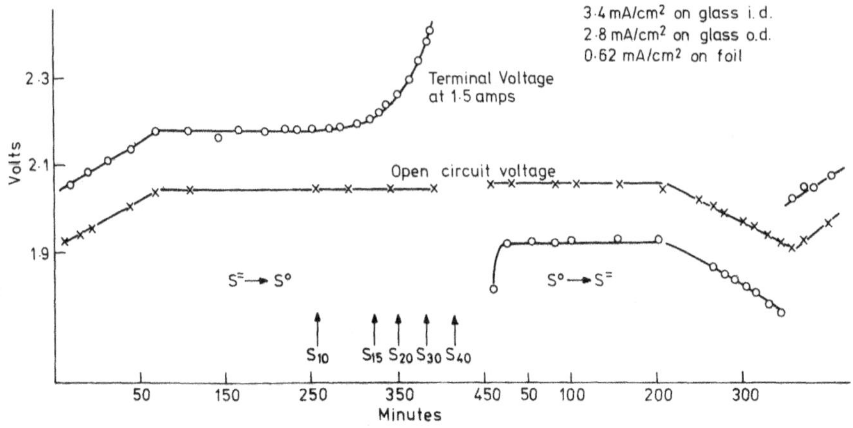

Fig. 10.3 Charge/discharge curve for glass electrolyte cell.

The shapes of the charge/discharge curves for the glass electrolyte cell are very similar to those of the beta alumina electrolyte cells, as can be seen from Fig. 10.3 (Levine, 1983). Because the current density is very low, there is little polarization at the electrodes and the point of inflexion in the discharge curve corresponds to that in the open circuit curve, with the result that the cell resistance decreases continuously during discharge as shown in Fig. 10.4 (Levine, 1983).

Clearly, it is important when reporting cell resistances to state exactly how they were determined and to which part of the charge or discharge they refer. Some workers prefer to report a whole cycle average resistance and there is obvious merit in this; it is not, however, a very convenient value to determine. The discharge resistance in the early stages of discharge, when it is reasonably constant, is easy to determine and is the value which will be used in this discussion of cell resistance.

When reporting cell resistance it is usual to use the units $\Omega\ cm^2$, where the area referred to is the area of the solid electrolyte in contact with the sulfur electrode matrix. This is useful when comparing results from cells having different capacities, but care should be exercised when comparing cells with very different geometries. These units are not appropriate for the glass electrolyte cell and for this the cell resistance in ohms is quoted.

The current/voltage curves for two 88 Ah cells, similar in design to that referred to in Fig. 10.1, are shown in Fig. 10.5 (Sudworth, 1981). These curves are linear and there is no asymmetry between charge and discharge indicating the absence of interfacial polarization at the sodium metal/solid electrolyte interface. Some deviation from linearity is observed for the charge curves at top of charge owing to passivation of the sulfur electrode, and for the discharge curves at the end of discharge owing to precipitation of solid sulfides. These data

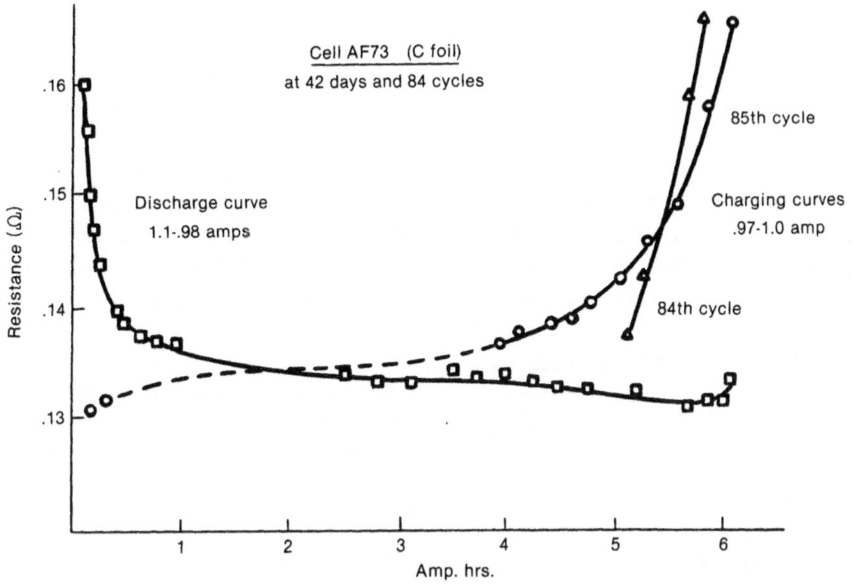

Fig. 10.4 Cell resistance as a function of depth of discharge for glass electrolyte cell.

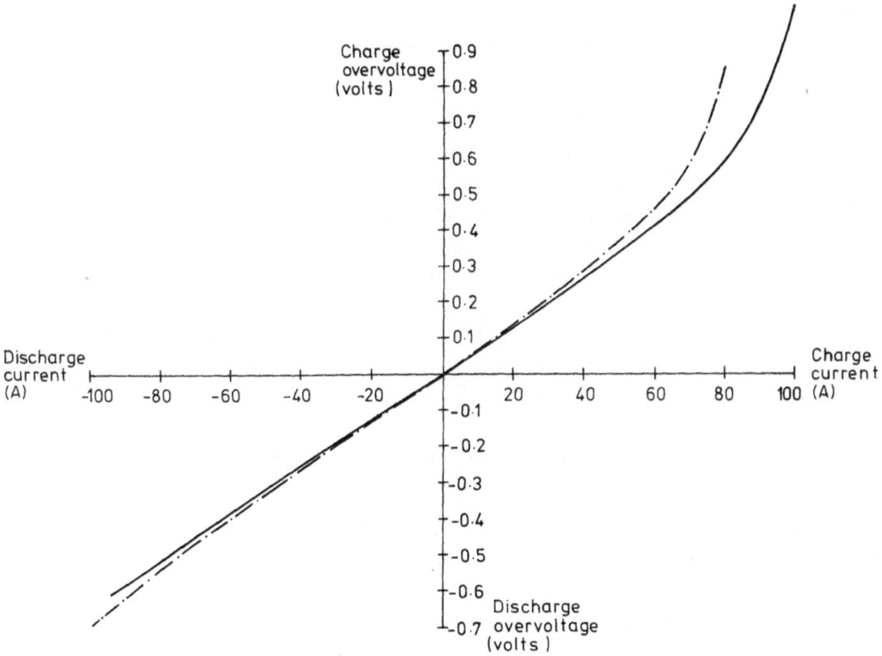

Fig. 10.5 Polarization curves for two 88 Ah central sulfur cells (Sudworth, 1981). Active electrolyte area 220 cm^2, temperature 350 °C.

are consistent with results obtained with smooth carbon electrodes which show a linear current/voltage relationship in the potential range 0.4–0.5 V cathodic and 0.1–0.2 V anodic. For a sulfur electrode resistance of 0.4 Ω cm^2 these values correspond to current densities of 1–1.25 A cm^{-2} on discharge and 0.25–0.5 A cm^{-2} on charge.

The contribution of the two electrodes to the cell resistance can be determined by dismantling the cell and measuring the radial resistivity and average thickness of the solid electrolyte. Data for a number of cells are given in Table 10.1 (Sudworth, 1980a), and it will be seen that the solid electrolyte is a major contributor to the total cell resistance. We saw in Chapter 7 that once interfacial

Table 10.1 Breakdown of cell resistance

		Cell			
		1	*2*	*3*	*4*
Capacity	(Ah)	38	38	38	100
Electrolyte area	(cm^2)	100	100	100	230
Electrolyte length	(cm)	16	16	16	30
Cell resistance	(Ω cm^2)	1.20	1.70	1.50	1.54
Electrolyte resistance	(Ω cm^2)	0.79	1.29	1.09	0.99
Sulfur electrode resistance	(Ω cm^2)	0.41	0.41	0.41	0.55

polarization has been eliminated, polarization of the sodium electrode is negligible, which leaves the sulfur electrode and the positive and negative poles as the source of the electrode resistance. In these particular cells the dimensions of the pole are such that a very low resistance is expected. Thus the electrode resistance is effectively the sulfur electrode resistance.

It is clear from these data that significant reductions in cell resistance will be obtained only by reducing the electrolyte resistance, that is by reducing either the radial resistivity or the wall thickness or a combination of both.

In the glass electrolyte cell also, most of the cell internal resistance is the resistance of the electrolyte (Levine, 1983).

10.2.1 The effect of sulfur electrode dimensions on cell resistance

Very few data are available on the effect of sulfur electrode thickness on resistance. The data in Table 10.1, which refer to cells with a sulfur electrode 10 mm thick, indicate a sulfur electrode resistance of 0.4 Ω cm. The resistance of the sulfur electrode is largly determined by the resistivity of the electrode matrix, and values considerably higher than 0.4 Ω cm have been reported (Birk, 1980).

The effect of cell length on resistance (expressed as Ω cm^2) should be small provided that the resistivity of the positive pole is low. A theoretical curve for a central sulphur cell with a 15 mm diameter carbon pole, derived from the

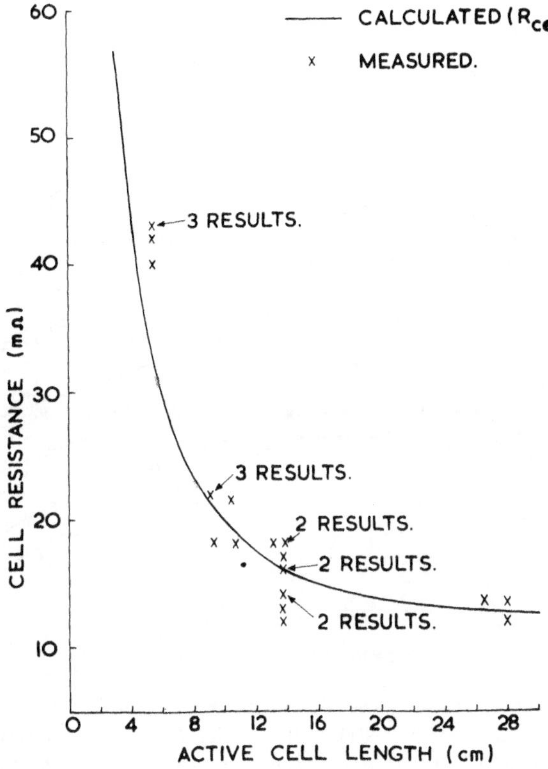

Fig. 10.6 The effect of length on cell resistance. Reproduced from Piercy and Sudworth (1977).

equations developed by Gibson (1974), is shown in Fig. 10.6 (Piercy and Sudworth, 1977). Experimental points obtained from central sulfur cells are included for comparison. Charge/discharge curves for a cell with a 260 mm long sulfur electrode (10 mm thick) are shown in Fig. 10.7 (Sudworth, 1981), and should be compared with the curves for a 130 mm long electrode in Fig. 10.1. The data in Table 10.1 show that there is an increase in sulfur electrode resistance with length. Other features of the longer electrode are noticeable from Fig. 10.5: thus the cell resistance decreases initially then remains constant during most of the two phase region, and at high current densities an increase in resistance is seen near the end of discharge. These variations in cell resistance may be associated with the increased potential gradient down the pole, or with changes in the distribution of voids in the sulfur electrode owing to less effective wicking of the melt. Whatever the reason there is a penalty, albeit small, associated with the use of longer cells.

The resistance of the sodium electrode is not usually a significant factor in cells with beta alumina electrolyte. In glass electrolyte cells, where the bore of

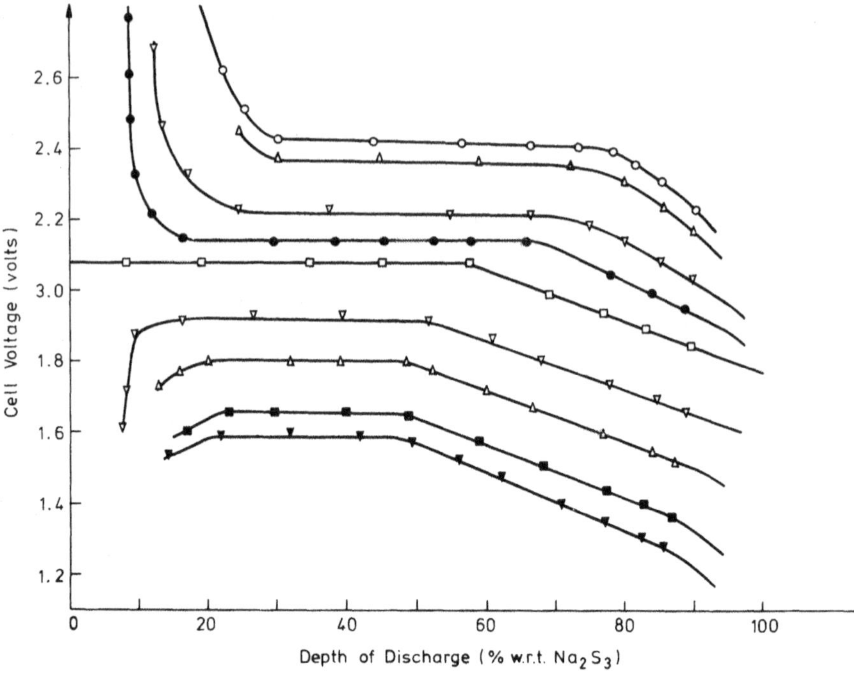

Fig. 10.7 Charge/discharge curves for a cell with a 300 mm long electrolyte tube. Reproduced from Sudworth (1981). Cell capacity 95 Ah, active electrolyte area 214 cm^2, temperature 350°C. □ open circuit voltage; ● 10 A; ▽ 20 A; △ 40 A; ○ 50 A; ■ 60 A; ▼ 75 A.

the hollow fibres is measured in microns, it might be thought that sodium electrode resistance would be a significant factor in limiting the cell length. That this is not the case can easily be demonstrated. Thus a cell designed to deliver 200 A would have 186 000 fibres of internal diameter ~70 microns (Levine, 1983), giving a total cross sectional area of sodium of 7.16 cm^2. To meet the design factor used by Levine (1983), for the permissible voltage drop (0.01 V) the fibre length could be approximately 50 cm.

10.2.2 The temperature dependence of cell resistance

Table 10.2 gives data on the discharge resistance of three central sulfur cells at temperatures between 275 and 400 °C. The same data are shown in graphical form in Fig. 10.8 (Sudworth, 1981), and it can be seen that there is a strong temperature dependence below 350 °C. The contribution of the beta alumina to the cell resistance can be determined by measuring the resistance of a whole tube using reversible electrodes at different temperatures and subtracting this from the cell resistance. The sulfur electrode resistance determined by this method is shown in Fig. 10.9.

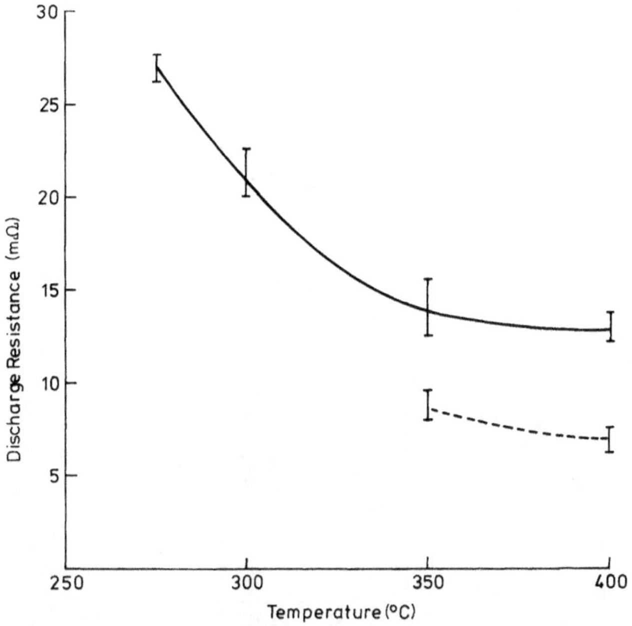

Fig. 10.8 Effect of temperature on cell resistance. Reproduced from Sudworth (1981). —— cell BR16: capacity 37 Ah, active electrolyte area 91 cm^2. - - - cell BR 30: capacity 95 Ah, active electrolyte area 214 cm^2.

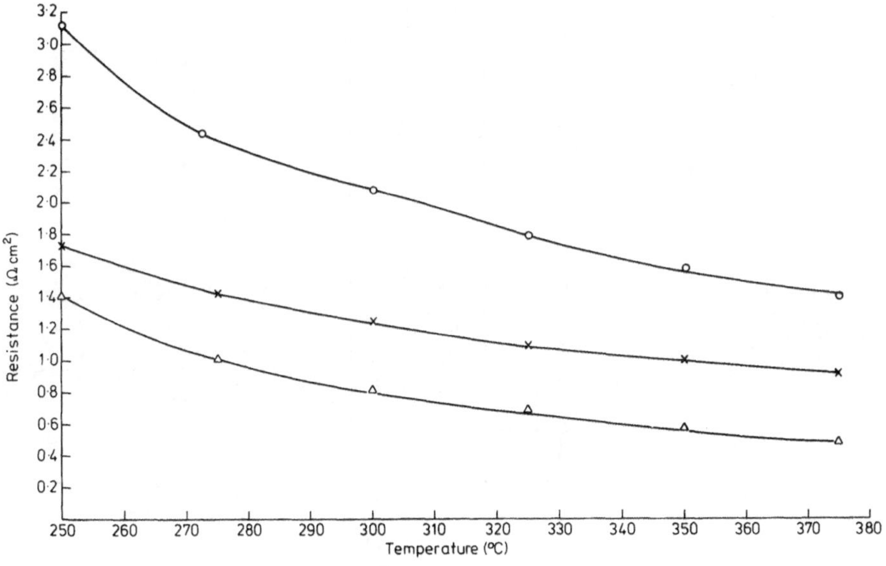

Fig. 10.9 Components of cell resistance obtained by determining resistance of a sodium sodium cell and a sodium sulfur cell as a function of temperature, the same beta alumina tube being used in both cells. △ sulfur electrode resistance; × beta alumina resistance; ○ cell resistance.

Table 10.2 Temperature dependence of cell resistance

	Cell capacity 40 Ah				*Cell capacity 100 Ah*				*Cell capacity 100 Ah*			
Temp. (°C)	*Cycle number*	*Current (A)*	*Resistance (mΩ)*	*Mean Resistance (Ω cm²)*	*Cycle number*	*Current (A)*	*Resistance (mΩ)*	*Mean Resistance (Ω cm²)*	*Cycle number*	*Current (A)*	*Resistance (mΩ)*	*Mean Resistance (Ω cm²)*
275	16	16.15	26.5									
	19	13.65	25.0									
	18	9.10	25.5	2.38								
	17	4.55	26.1									
300	10	16.16	19.6		6	75	8.1					
	9	13.65	19.5		7	60	8.3					
	11	11.38	20.0	1.77	5	40	8.8	1.93				
	5	9.10	18.0		8	30	10.1					
	8	6.83	20.1		9	20	9.5					
	6	4.55	20.1									
	7	2.28	20.2									

350	3	9.10	13.7		4	75	6.5		13	95	6.2	
	2	6.83	13.4	1.28	3	60	7.0	1.48	14	80	6.3	
	4	4.55	14.6		2	40	6.9		12	70	6.2	1.39
	1	2.28	15.1		1	20	7.3		15	60	6.2	
									5	50	6.7	
									11	40	6.2	
									4	40	6.6	
									2	30	7.0	
									3	20	6.9	
									1	10	6.5	
400	13	16.15	12.1						7	50	6.0	
	15	13.65	12.6	1.15					9	40	5.7	
	12	9.10	12.3						6	30	5.9	1.28
	14	4.55	13.6						8	20	5.8	
									10	10	6.7	

10.3 DISCHARGE CHARACTERISTICS

Unlike most secondary batteries, the discharge capacity of a sodium sulfur cell is independent of the discharge rate, as can be seen from Fig. 10.1. Even under short circuit conditions, the whole of the theoretical cell capacity is available as shown in Fig. 10.10 (Tilley, 1980). However, because of the relatively high impedance of the cell, the cell energy is quite dependent upon the discharge rate and is given by the relationship:

$$E(\mathrm{Wh}) = V_0 C - \frac{C^2 R}{n} \tag{10.1}$$

where E is the cell energy, C the cell capacity in ampere hours, V_0 the cell EMF, n the discharge rate in hours and R the average cell resistance in ohms. Thus a 100 Ah cell discharge at the 2 hour rate with an average resistance of 7 mΩ would deliver 165 Wh. The voltage efficiency is given by the relationship:

$$\eta = 1 - \frac{RC}{V_0 n} \tag{10.2}$$

The maximum power available from a cell is given by:

$$P_{\mathrm{max}} = \frac{V_0^2}{4R} \tag{10.3}$$

where R is the cell resistance at a particular depth of discharge. In the two phase region of the discharge this reduces to:

$$P_{\mathrm{max}} = \frac{1.08}{R} \simeq \frac{1}{R} \tag{10.4}$$

These relationships follow from the ohmic behaviour of sodium sulfur cells.

For traction applications it is important to consider the power available at all states of discharge. Fig. 10.11 shows how this and other parameters vary with depth of discharge. The experimentally determined curve of the cell resistance referred to in Fig. 10.2 is shown in Fig. 10.9; the other curves are derived from the data in Figs 10.1 and 10.2. For this cell under these conditions the minimum power occurs at the end of discharge.

Under some conditions there is evidence of a 'knee' in the discharge curve – the lower the temperature and the higher the current density the more pronounced the knee – but often it is completely absent and there is no obvious indication that the end of discharge has been reached. As mentioned in Chapter 5, discharge can continue beyond this point, but this is undesirable for the following reasons. In order to maximize the specific energy, the amount of

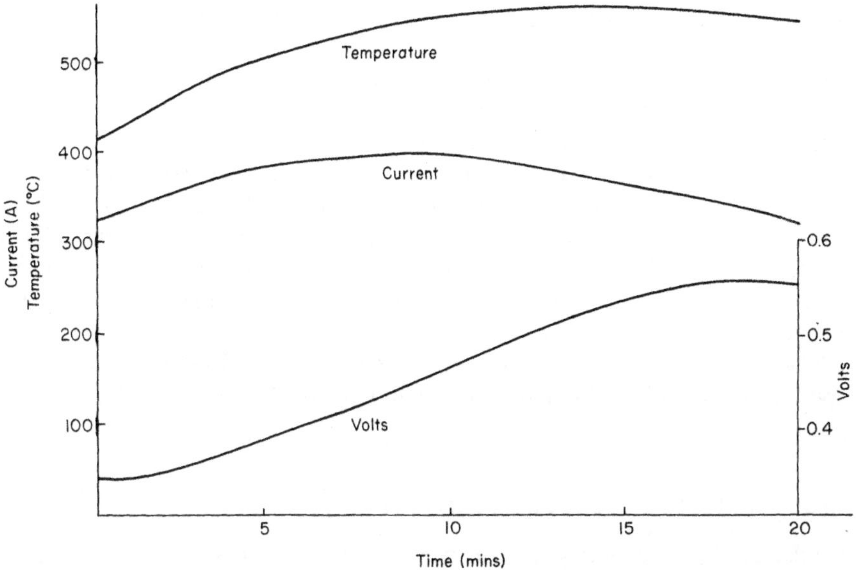

Fig. 10.10 Discharge of a cell under short circuit conditions.

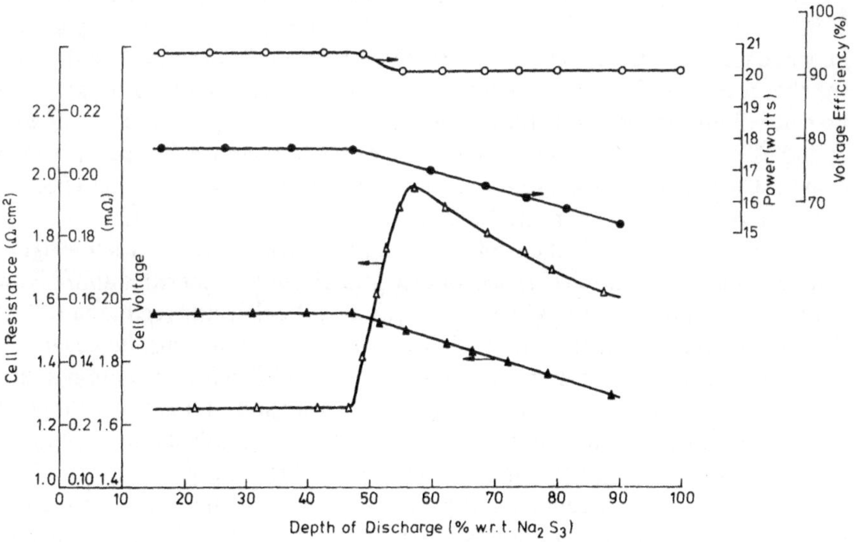

Fig. 10.11 Variation of cell characteristics with depth of discharge (derived from Figs 10.1 and 10.2). Temperature 350°C. ○ voltage efficiency, ● power, △ cell resistance, ▲ cell voltage.

sulfur loaded into the cell initially should be equivalent to the maximum volume of sodium trisulfide which can be accommodated in the electrode compartment; discharge beyond this point will, therefore, lead to rupture of the electrode compartment. In this situation the cell capacity is limited by the positive electrode. If the cell capacity is limited by the negative electrode, i.e. by the amount of sodium in the cell, discharge to the point where sodium depletion occurs will prevent discharge beyond the nominal cell capacity. In negative limited cells it is necessary to terminate the discharge before a significant increase in cell resistance occurs, otherwise current concentration effects at the sodium metal/beta alumina interface (Virkar *et al.*, 1980) will lead to electrolyte failure.

The detection of the end of discharge is very important for battery design and operation and will be discussed further in Chapter 11.

10.4 CHARGE CHARACTERISTICS

Unlike discharge capacity, the charge acceptance of a sodium sulfur cell is dependent upon current density and temperature. The dependence upon current density is least with cells having electrodes containing sodium polysulfide wicks (Demott, 1980), and the data presented here refer to sodium sulfur cells of this type.

The temperature dependence of charge acceptance with current density as a parameter is shown for a 40 Ah central sulfur cell in Fig. 10.12 (Sudworth, 1981). Constant current charging was used. It can be seen that the temperature dependence of charge acceptance is greater at higher charge rates. The shape of the charge curves also changes with temperature: at lower temperatures there is no true voltage plateau and at top of charge the potential rises relatively slowly; at 400 °C a true plateau is observed and the cell voltage rises steeply at top of charge. Clearly, it is possible to get a misleading impression of the effect of parameters such as temperature and current density on charge acceptance; if cells are charged at constant current to a cutoff voltage of, say, 2.5 V a low figure will be obtained for the charge acceptance at high current densities and/or low temperatures. This illustrates in a simple way how the charging regime can affect the charge acceptance; the situation is more complex than this, however.

Constant current charging is a common method of charging sodium sulfur cells, and as was mentioned it necessitates the selection of a cutoff voltage at which charging is terminated. It has the advantage of simplicity, not requiring any sophisticated equipment, and it can result in a high charge acceptance. The converse of constant current charging is constant voltage charging – applying a fixed voltage to the cell and allowing the current to vary. A set of constant voltage charge curves is shown in Fig. 10.13 (Piercy and Sudworth, 1977). The disadvantage of this method is immediately obvious: because of the 300 mV change in the OCV of the cell between the beginning and end of the single phase

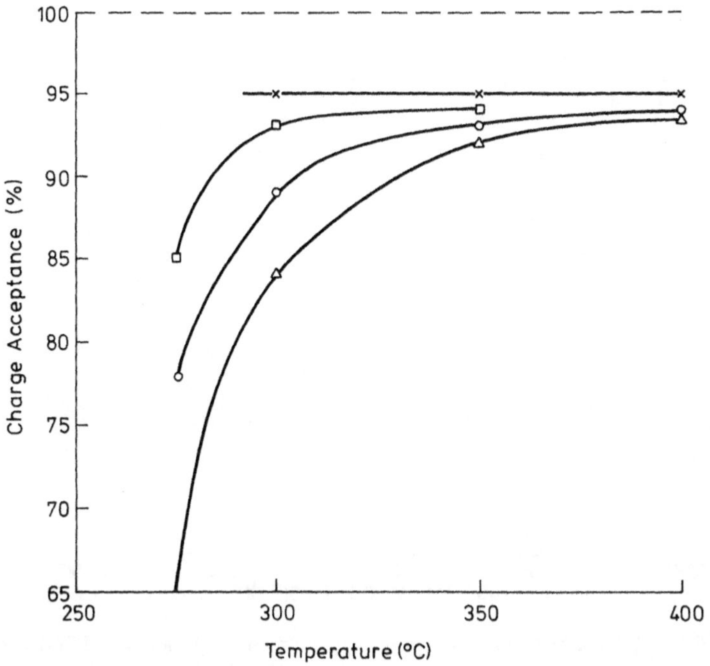

Fig. 10.12 Effect of temperature on charge acceptance. Reproduced from Sudworth (1981). Active electrolyte area 91 cm^2, capacity 37 Ah. × current density 50 mA cm^{-2}; □ current density 100 mA cm^{-2}; ○ current density 150 mA cm^{-2}; △ current density 200 mA cm^{-2}.

region, very high currents will flow at the beginning of charge if a reasonable charge rate is to be obtained in the two phase region.

For single cells the most useful charge regime seems to be a constant voltage charge with a current limit at the beginning of charge, as shown in Fig. 10.14 (Piercy and Sudworth, 1977). Using this method, a charge acceptance of 95% can be obtained in eight hours.

Although it is possible to charge sodium sulfur cells very rapidly (85% charge acceptance in 40 minutes has been obtained: Demott, 1980), this may result in a reduced cell life.

Reference was made to the need to define a cutoff voltage when charging sodium sulfur cells. This arises because the solid electrolyte will fracture if a sufficiently high voltage is applied to it.

In Chapter 2 it was shown that dielectric failure of the electrolyte is inevitable if the cell voltage exceeds 50–70 V (Hooper, 1980). Well before this limit is reached, however, electrolyte failure can be caused by current concentration effects leading to localized degradation of the solid electrolyte (Virkar *et al.*, 1980).

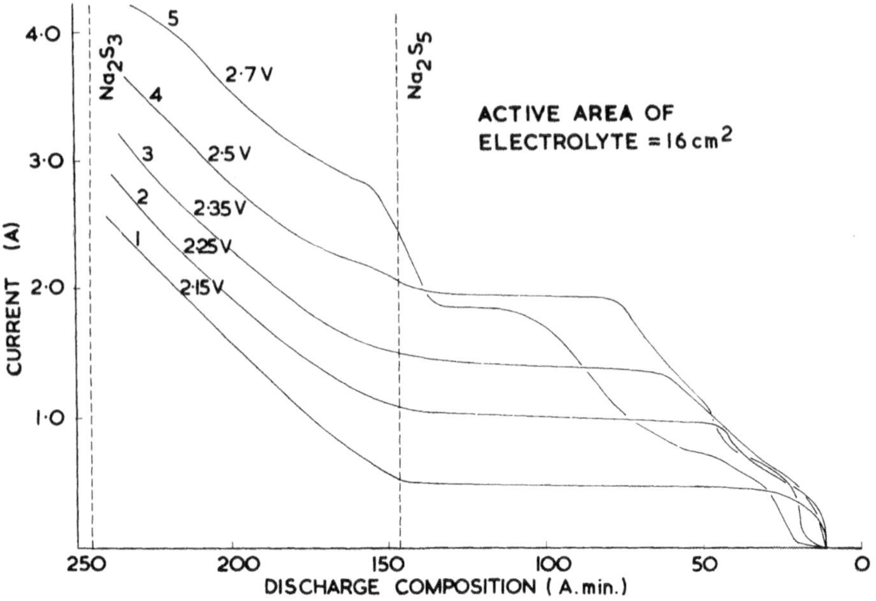

Fig. 10.13 Constant voltage charging. Reproduced from Piercy and Sudworth (1977). Active area of electrolyte 16 cm^2, temperature 350°C.

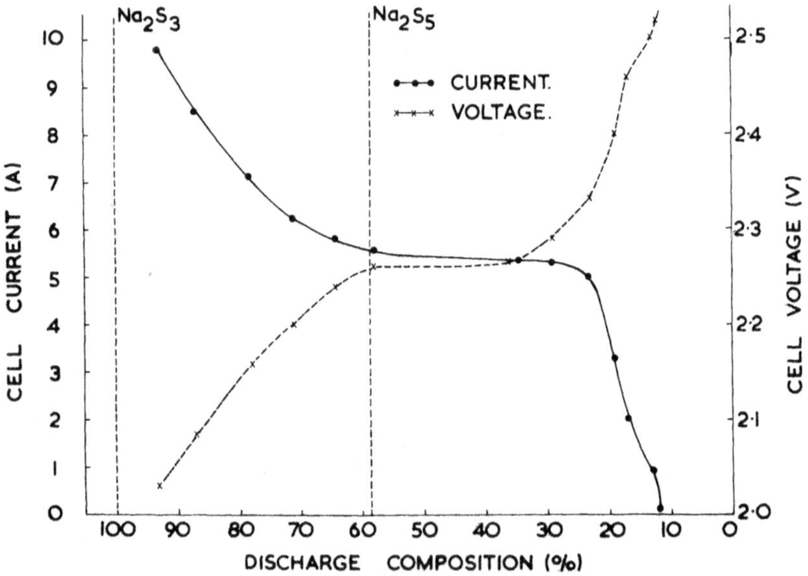

Fig. 10.14 Current limited constant voltage charging. Reproduced from Piercy and Sudworth (1977). Active area of electrolyte 118 cm^2, temperature 350 °C.

The sulfur electrode structure and the charge regime can also affect the breakdown voltage (De Jonghe, 1981). The more homogeneous the interface between the solid electrolyte and the sulfur electrode, the less likelihood there will be of current concentration effects. It is, of course, impossible to eliminate these completely, since transition from an ionically conducting melt to an insulating melt is inevitable when the sulfur electrode becomes fully charged: the rise in cell resistance at top of charge merely reflects the decreasing area of ceramic electrolyte available for ionic transfer. It has been suggested that breakdown of the ceramic might be prevented by interposing an insulating fibrous layer between the solid electrolyte and the bulk of the sulfur electrode (GGE, 1970). This could prevent dielectric breakdown of the ceramic, but it is unlikely to prevent current concentration effects. In any case, an insulating layer adjacent to the solid electrolyte results in unacceptable discharge characteristics, as was shown in Chapter 5, and this method of ameliorating the problem is impractical.

Charging a single cell at constant current and allowing the cell voltage to rise at top of charge will result in cell failure at a certain voltage. It is not necessarily the case, however, that failure is inevitable if this voltage is exceeded during battery charging. When cells are charged in a series chain to a maximum voltage, the first cell to passivate will develop a voltage determined by the voltage limit and the number of cells in the chain:

$$V = nV_c - 2.08(n-1)$$

where V is the passivated cell voltage, V_c is the charging voltage expressed as average volts per cell, and n is the number of cells.

As the passivated cell voltage approaches V, the current will attenuate to zero. Thus, unlike constant current charging, the power dissipated in the passivated cell will reach a maximum at a voltage less than V, and this maximum value will be less than the maximum power dissipated in the case of constant current charging of a single cell. This effect reduces the risk of cell failure when series chains of cells are being charged to a voltage limit on the whole chain, and it is found that series chains comprising nineteen cells can be recharged quite rapidly (less than three hours) to a cut-off voltage of 45 V without cell failure occurring (Sudworth, 1983).

10.5 SPECIFIC ENERGY

The energy of a sodium sulfur cell of a given theoretical capacity is determined primarily by the internal resistance and the discharge rate, but there is an additional factor – the discharge range over which the cell operates. The theoretical capacity, corresponding to the conversion of the sulfur in the cell to sodium trisulfide, will only be obtained if the cell is discharged completely and on recharge all the sodium polysulfide is converted to sulfur. The former can be achieved by coulometric control of discharge, a method which is particularly appropriate for cells in long series chains. If measurement of the load voltage is used to determine the depth of discharge, variations in cell resistance, due for instance to temperature fluctuations, will limit the accuracy to which it can be determined. To ensure overdischarge does not occur it would be necessary to set the discharge limit at something less than the theoretical capacity. Full recharge is difficult, if not impossible, to achieve and a realistic maximum would be 95%. Thus the actual capacity would be between 85% and 95% of the theoretical, depending on the method used to control depth of discharge. For each application and battery configuration, therefore, there will be an appropriate factor by which to multiply the theoretical cell capacity. It is this factored capacity which should be used in Equations 10.1 and 10.2 to obtain the cell energy.

The specific energy of sodium sulfur cells is dependent upon cell design and duty cycle, as was seen in Chapter 9. The weight analyses of two cells, a central sulfur cell and a central sodium cell designed for a traction application, are given in Tables 10.3 and 10.4 (Sudworth and Tilley, 1983). The volumetric and gravimetric specific energies for these cells at various discharge rates are given in Table 10.5. The values are calculated assuming 100% depth of discharge and 95% charge acceptance. Also given in Table 10.5 are maximum specific powers at different stages of discharge.

Table 10.3 Weight breakdown of a central sulfur cell

	Weight		
Component	*g*	*kg kW*$^{-1}$	%
Beta alumina	181.0	2.25	20.18
Reactants (sodium and sulfur)	285.2	3.55	31.81
Carbon felt	34.6	0.43	3.86
Cell case	275.6	3.44	30.75
Current collector	22.8	0.284	2.54
Seal components (alpha alumina, glass seal, flanges, gaskets, disc springs etc.)	97.40	1.21	10.87
Total	896.6	11.16	100

Beta alumina tube 30 cm long, 3 cm diameter, 0.2 cm wall thickness, resistivity 5 Ω cm.
Sulfur electrode thickness 1.2 cm, discharge rate 2 hours.
Cell energy density 179 Wh kg^{-1}.

Table 10.4 Weight breakdown of a central sodium cell

Component	Weight *g*	*kg kW*$^{-1}$	%
Beta alumina	181.0	1.706	17.82
Reactants (sodium and sulfur)	423.7	4.00	41.71
Sodium wick	54.4	0.513	5.36
Carbon felt	38.3	0.361	3.77
Seal components (alpha alumina, glass seal, flanges, gaskets, disc springs etc.)	131.76	1.24	12.97
Cell case	186.7	1.76	18.38
Total	1015.86	9.58	100

Beta alumina tube 30 cm long, diameter, 3.0 cm, 0.2 cm wall thickness, resistivity 5.0 Ω cm.
Sulfur electrode thickness 0.8 cm, discharge rate 2 hours.
Cell energy density 209 Wh kh^{-1}.

Table 10.5 Gravimetric and volumetric specific energies of cells in Tables 10.3 and 10.4

Discharge time (h)	Central sulfur cell *Wh kg*$^{-1}$	*Wh l*$^{-1}$	Central sodium cell *Wh kg*$^{-1}$	*Wh l*$^{-1}$
1	172	276	159	310
2	179	331	209	407
3	218	350	256	440
4	223	359	234	456
5	227	365	239	466
8	231	373	247	480

10.6 CELL LIFE

Before discussing the factors which affect cycle life, it is necessary to define cell failure. Obviously if the solid electrolyte fractures, the sodium and sulfur will react chemically and the cell will lose its capacity and its voltage. After failure the cell resistance may be high or low depending upon the cell design.

This type of cell failure is clear cut and the term *cycle life* or *cell life* is used by many to mean the time to electrolyte failure. Before this stage is reached, however, the cell characteristics may degrade to such an extent as to render the cell performance unacceptable. Thus the cell resistance may increase or the capacity may decline, and limits should be set for these parameters so that cell failure may be defined in terms of a minimum capacity and a maximum resistance.

Effects such as seal leakage, loss of capacity, resistance rise and corrosion tend to result in a gradual decline in cell performance rather than sudden failure (although there are exceptions to this, e.g. a sodium seal leak can result in the cell being short circuited and failing suddenly); they also tend to affect all cells in a similar manner and eventually result in a battery with an unacceptable performance. This is another way of saying that unpredictable failures are caused by electrolyte failures, whereas predictable failures, i.e. cell performance falling below a specified level, are caused by other failure modes. In the following discussion of life and reliability it is implicitly assumed that cell failure is caused by electrolyte failure unless the failure mode is specifically identified.

10.6.1 Life and reliability

The distribution of failures in a population of cells over a period defines life and reliability and is amenable to analysis by statistical methods. One such technique which has proved useful for sodium sulfur cells is a variant of probability analysis known as hazard plotting (Nelson, 1982). The basis of this method is that the distribution of failures can be described mathematically by a limited number of functions: exponential, normal, Weibull etc. The advantage of hazard plotting over probability plotting is that it can be used for multiply censored data. (In life testing cells, the test often has to be terminated before all the cells have failed, or cells have to be removed from test before they fail, resulting in incomplete or censored data.)

The construction of a hazard plot using a simulated population of 50 cells is illustrated in Table 10.6. The cells are first ranked in order of increasing time on test, regardless of whether failure was observed; cells still running are marked with a +, and failures are unmarked. The number in parentheses is simply the reverse order number, known as the K number, and the hazard value for failed cells is the reciprocal of this expressed as a percentage.

The cumulative hazard value for each failure is the sum of the hazard value and the cumulative hazard value of the preceding failure. A hazard paper for a theoretical distribution is chosen. There are hazard papers for the exponential, Weibull, extreme value, normal, and log normal distributions. Different distributions can be tried and the one that fits the data best (a straight line plot) should be used. (Hazard papers are exclusively offered in the catalogue of TEAM (Technical and Engineering Aids for Management) Box 25, Tamworth, NH 03886, USA.)

In this case the exponential distribution gives the best fit for the data and the resulting hazard plot is shown in Fig. 10.15. The exponential cumulative hazard function is:

$$H(x)=x/\theta \tag{10.5}$$

Table 10.6 Simulated exponential data and hazard calculations

No.	Hours	Hazard	Cumulative hazard	No.	Hours	Hazard	Cumulative hazard
1 (50)	3	2.00	2.00	26 (25)	385+		
2 (40)	52	2.04	4.04	27 (24)	385+		
3 (48)	58	2.08	6.12	28 (23)	385+		
4 (47)	71	2.13	8.25	29 (22)	385+		
5 (40)	77	2.17	10.42	30 (21)	385+		
6 (45)	77+			31 (20)	391	5.00	46.49
7 (44)	77+			32 (19)	410	5.26	51.75
8 (43)	77+			33 (18)	562	5.56	57.31
9 (42)	77+			34 (17)	611	5.88	63.19
10 (41)	77+			35 (16)	621	6.25	69.44
11 (40)	101	2.50	12.92	36 (1)	621+		
12 (39)	130	2.56	15.48	37 (14)	621+		
13 (38)	161	2.63	18.11	38 (13)	621+		
14 (37)	180	2.70	20.81	39 (12)	621+		
15 (36)	235	2.78	23.59	40 (11)	621+		
16 (35)	235+			41 (10)	662	10.00	79.44
17 (34)	235+			42 (9)	884	11.11	90.55
18 (33)	235+			43 (8)	1101	12.50	103.05
19 (32)	235+			44 (7)	1110	14.29	117.34
20 (31)	235+			45 (6)	1232	16.67	134.01
21 (30)	301	3.33	26.92	46 (5)	1232+		
22 (29)	306	3.45	30.37	47 (4)	1232+		
23 (28)	319	3.57	33.94	48 (3)	1232+		
24 (27)	334	3.70	37.64	49 (2)	1232+		
25 (26)	385	3.85	41.49	50 (1)	1232+		

Thus the time to failure as a function of the cumulative hazard is:

$$x(H) = \theta H \qquad (10.6)$$

and x plots as a straight line function of H. The slope of the line θ is the mean time to failure, which is also the value of life at $H = 1$. The failure rate $\lambda = 1/\theta$. This type of distribution is typical of random failure as opposed to infant mortality or wearout failure. Random failure of cells, which is usually taken to mean that there are many failure modes operative, is often a consequence of inadequate quality control, and by improving this it is possible to reduce the failure rate considerably and to change the distribution function (Fig. 10.16; Sudworth, 1980a). The data in Fig. 10.15 are plotted on Weibull paper and are described by the equation:

$$H(x) = (x/\alpha)^{\beta} \qquad x \geqslant 0$$

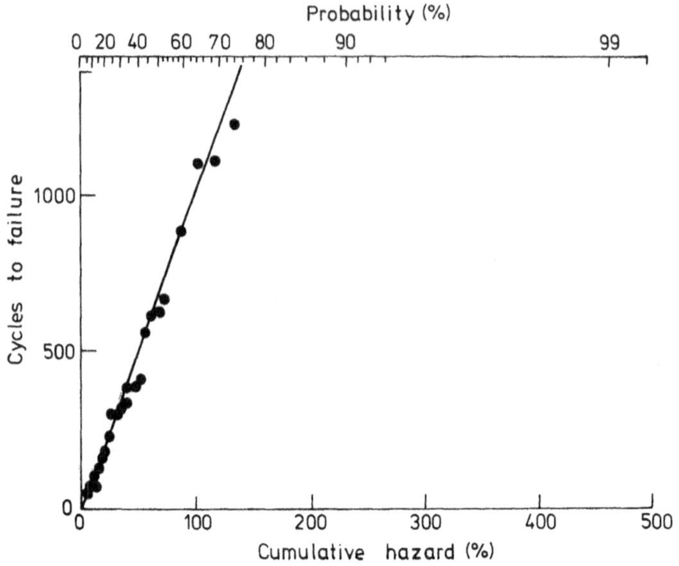

Fig. 10.15 Hazard plot of simulated exponential data (Nelson, 1969). Reproduced from Journal of Quality Control by permission of the author and publisher.

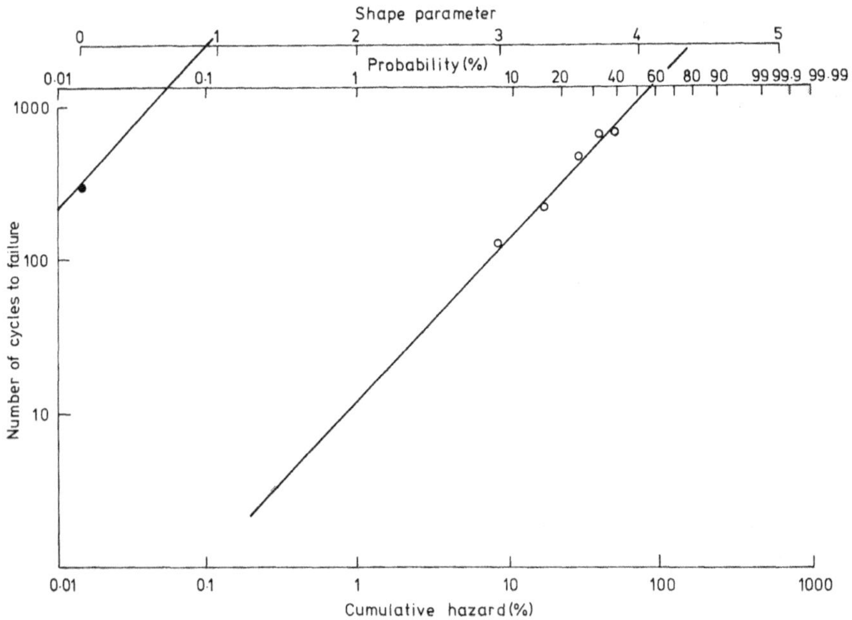

Fig. 10.16 Weibull hazard plot of cycles to failure of BR16 cells. Reproduced with permission from Sudworth (1980a). Temperature 350 °C.

where α is the characteristic time to failure and β is the value of the shape parameter of the distribution. Thus the time to failure x as a function of the cumulative hazard is:

$$x(H) = \alpha H^{1/\beta}$$

Taking logarithms:

$$\log x = \frac{1}{\beta} \log H + \log \alpha$$

Weibull hazard paper is log/log paper and thus x plots as a straight line function. The slope of the line is $1/\beta$ and this fact can be used to estimate the shape parameter from the plot by drawing a parallel line through the reference point marked on all Weibull hazard paper. A value of $\beta < 1$ means a decreasing failure rate and a value of $\beta > 1$ an increasing rate. For $\beta = 1$ the Weibull distribution reduces to the exponential distribution i.e. random failure rate. The value of α is the time at which $H = 1$.

Reduction of the failure rate can have the effect of prolonging the time of the test to the point where a second failure mechanism becomes operative. This shows up on the hazard plot as a change in slope.

In Fig. 10.17 (Sudworth, 1980b) results from a population of 40 cells are plotted on Weibull paper. Two failure modes were observed – electrolyte failure and sodium seal leakage – and the cumulative hazard values have been calculated separately as shown in Table 10.7. It can be seen that the electrolyte failures result in a shape factor of 0.7, which means that the failure rate is decreasing and that the characteristic time to failure extrapolates to 10 000 cycles. The seal failures, however, result in a shape factor of >3, i.e. rapidly increasing failure rate, and the characteristic time to failure is 1000 cycles.

The significance of this is as follows. The electrolyte failures determine the reliability of cells in a battery up to the time at which seal failures begin to occur. These electrolyte failures determine the maintenance requirements of a battery, i.e. the number of cells which have to be replaced during the life of the battery. Improvement in battery life as opposed to reliability will require modification to the cell design: in this case a sodium seal with a longer life is required. This is unlikely to alter the shape of the plot, but rather will increase the characteristic time to failure due to cell seals. If the life of the seal is extended sufficiently, another failure mode might be exposed (e.g. corrosion of the current collector), and this would possibly result in a plot having a different shape factor. The slope of the plot of electrolyte failures would, however, be expected to remain constant.

The preferred failure distribution is a Weibull distribution with a high shape factor which would require elimination of early failures.

Table 10.7 Hazard plot data

k	*Cell life (cycles)*	*Hazard 100/k*	*Cumulative hazard Failure mode* Seal	Ceramic
40	50	2.5		
39	50	2.56		
38	50	2.63		
37	50	2.70		
36	50	2.78		
35	50	2.86		
34	60+	2.94	2.94	
33	102+	3.03		3.03
32	152	3.13		
31	191+	3.23		6.26
30	200	3.33		
29	200	3.45		
28	200	3.57		
27	200	3.70		
26	200	3.85		
25	360+	4.00	6.94	
24	459+	4.17	11.11	
23	519+	4.35		10.61
22	555+	4.55	15.66	
21	562+	4.76	20.42	
20	571	5.00		
19	641	5.26		
18	650	5.56		
17	690+	5.88	26.30	
16	696+	6.25	32.55	
15	697+	6.67	39.22	
14	697+	7.14	46.36	
13	718+	7.69	54.05	
12	766	8.33		
11	789	9.09		
10	792	10.00		
9	795+	11.11	65.16	
8	893+	12.50	77.66	
7	896	14.29		
6	901	16.67		
5	905	20.00		
4	974	25.00		
3	1056	33.33		
2	1070	50.00		
1	1138	100.000		

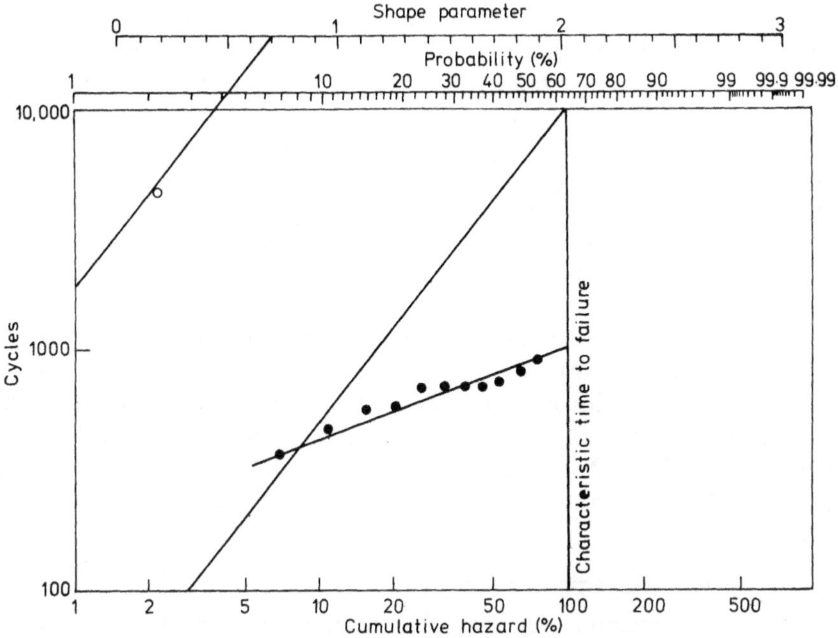

Fig. 10.17 Weibull hazard plot of cell life. Reproduced with permission from Sudworth (1980b). Temperature 350°C. • seal failure; —— ceramic failure.

This change from one failure mechanism to another raises the question of how to express cell life. The hazard plots in Figs 10.14, 10.15 and 10.16 have cycles as the ordinate and the seal failures are also expressed in these terms. For these particular cells, however, it is well established that seal failures depend solely on the time on test, not on the number of electrical cycles (Sudworth, 1980b); cells which have been left on open circuit still fail after approximately one year. Provided that the number of cycles per day is kept constant, it is quite acceptable to express the time to failure in cycles. Obviously, if the cycle regime is changed the point of inflexion will change; it will not, however, affect the characteristic cycle time to electrolyte failure.

It is of course implicit in this argument that electrolyte failure rate is a function of cycles. Certainly electrolyte failure in cells which are not cycling is rare, as might be expected since all the proposed failure mechanisms involve passage of current. It is possible, however, that it is the quantity of electricity passed rather than the number of cycles which is the important factor. For cells having sulfur electrodes of the same capacity per unit area of electrolyte and discharging over the whole capacity range, the two are directly related. It is only when comparing cells with different capacity sulfur electrodes, or cells which are not cycling over the same capacity range, that the distinction is important. There is no general agreement on this point, and both quantity of electricity (expressed as Ah cm^{-2} of electrolyte surface) and cycles are used. From consideration of the mechanism of electrolyte failure, it might be expected that if the charge voltage is strictly controlled to a relatively low value, thus minimizing current concentration effects, the amount of current passed will be the most important factor. If the top of charge voltage is not strictly controlled and current concentration occurs at top of charge, then the length of time or the number of times the cell is exposed to those conditions will be the important factor and the number of cycles will be the appropriate parameter.

The failure modes which result from gradual deterioration of cell performance to an unacceptable level can be looked upon as wearout failures, and it is these which would be expected to determine the ultimate life of the battery. Examples of such failure modes are loss of capacity, resistance rise, seal leakage etc. These have been discussed in earlier chapters, but the more important ones will be briefly reviewed here in the context of cell life and performance. Electrolyte failures, on the other hand, tend to give rise to infant mortality failures and thus determine the reliability of the battery.

10.6.2 Electrolyte failure

The various mechanisms for electrolyte degradation were discussed fully in Chapter 2 and will not be recapitulated here, except to note that the chances of electrolyte failure are greatly increased if current concentration effects are present. In a cell these effects can be due to the sulfur electrode or the sodium electrode, and it is this interaction between electrodes and electrolyte which is most likely to lead to early electrolyte failure.

An extreme case of this effect is what is sometimes referred to as 'sodium starvation'. If a constant current is passed through a cell, and supply of sodium metal to the electrolyte surface is restricted (owing for example to a blockage in the restrictor which is often incorporated for safety reasons) the cell voltage will be driven negative which can cause electrolyte failure (Fally *et al.*, 1973).

A similar effect occurs on charge as described earlier, and both types of failure are easily explained by current concentration effects (Virkar *et al.*, 1980). Factors which are likely to lead to current concentration at the electrolyte are the structure of the sulfur electrode, partial wetting of the electrolyte by sodium metal (Virkar and Miller, 1979), and discontinuities at the glass seal between the alpha alumina collar and the solid electrolyte.

Sulfur electrode structures in which regions of the solid electrolyte are not in contact with carbon felt, or in which there are sharp boundaries between conducting and non-conducting felt at the solid electrolyte surface, are known to cause electrolyte degradation. Gupta and Tennenhouse (1978) reported that a sulfur electrode structure having felt-free areas on the solid electrolyte surface resulted in a pattern of cracks which mirrored the electrode structure. Demott (1980) has reported darkened bands on the surface of beta alumina corresponding to bands of insulating alumina felt in the sulfur electrode. No cracks were observed, but the darkening was regarded as evidence of current concentration effects.

In cells where the glass seal is exposed to electrochemical action, e.g. plate cells, black sodium dendrites are often observed in the solid electrolyte at the edge of the glass seal. The extent of current concentration under these conditions where the geometry is strictly controlled has been calculated (Virkar and Miller, 1979), and factors as high as 50 can occur. In sodium sulfur cells the glass seal is normally kept out of the electrochemically active area by having a region free of carbon felt adjacent to the seal.

In cells having glass electrolyte, impurities in the sodium metal can cause degradation of the glass. If calcium is present as an impurity, it can be discharged at the glass surface and the calcium ions entering the glass can increase its resistivity. If the calcium ion concentration in the glass becomes too great, the membrane will break.

The glass electrolyte is less sensitive to the presence of potassium, and the effect of the potassium level commonly found in liquid sodium, 130 ppm, is relatively small (less than 5% resistance increase). Sodium oxide and hydroxide corrode the glass badly, and as these compounds cannot pass through the electrolyte they are concentrated during discharge of the cell, causing breakage of the glass fibres. It is thus essential to remove oxygen from the sodium and this can be achieved by digesting the sodium at 480 °C with zirconium titanium alloy (Levine, 1983), as described in Chapter 7.

10.6.3 Capacity loss

The problem of capacity loss was briefly touched upon in Chapter 5. It is particularly important that it should be minimal in sodium sulfur cells, since the capacity of a battery of series connected cells is determined by the lowest capacity cell. In fact, loss of capacity by one cell can affect the performance of a battery considerably more than failure of that cell. There are a number of mechanisms for loss of capacity, and attention must be paid to each of them if a cell is to retain its capacity over the whole of its life.

The most obvious way in which loss of capacity can occur is by loss of sulfur from the cell. The vapour pressure of sulfur (BP 444 °C) is quite high at the cell operating temperature and a very good seal is required to prevent loss of sulfur vapour. If the cell temperature rises significantly, venting of sulfur vapour is a distinct possibility in cells with compression seals (Tilley, 1976), and this is one of many reasons for ensuring that the cell operating temperature does not exceed 400 °C.

A second mechanism for loss of sulfur is by reaction with metallic components in the sulfur electrode, i.e. corrosion. This is often more complex than simply a reduction in the quantity of sulfur by an amount proportional to the weight of metal sulfide formed. Thus if stainless steel is used as the current collector, the capacity of the cell attenuates to about 40% of theoretical after no more than 100 cycles. Examination of the cell after test will show the presence of a corrosion product containing iron, sodium and sulfur, but in an amount insufficient to explain the loss of capacity. What appears to happen is that the corrosion product displaces the carbon felt away from the current collector, which may affect the distribution of current within the electrode, and also coats many of the fibres in the carbon felt. Iron sulfides appear to have some solubility in sodium polysulfide melts and can be transported to all parts of the electrode, even being deposited on the beta alumina. Why this should affect the cell capacity is not clear, but the explanation may lie in the effect of the transition metal sulfides on the wettability of the carbon felt by sodium polysulfides. It might be expected that transition metal sulfide would be more readily wetted by sodium polysulfides than by sulfur, and if the corrosion products are concentrated near the current collector, as is the case, the sodium polysulfides in this region would tend to be retained by capillary forces and become isolated from the remainder of the melt. Weiner (1976) has shown that addition of small quantities of transition metal sulfides to the sulfur electrode of a metal-free cell can increase the charge acceptance temporarily. Nickel sulfide also appears to be transported throughout the electrode, whereas chromium sulfide is relatively immobile.

Although the effects of the transition metal sulfides on charge acceptance are not fully understood, it is clear that stainless steel should not be used for the current collector. This is not to say that stainless steel cannot be used in the sulfur electrode; provided that it is not in an electrochemically active region, and therefore is not subject to rapid corrosion, it is an acceptable material. Thus in the glass electrolyte cell stainless steel can be used as the cell case (Levine, 1983). Other transition metal alloys such as nichrome do not cause capacity loss. This appears to be because the corrosion products remain attached to the parent material and are not transported throughout the electrode (Tilley and Wright, 1981).

One method which can be used to differentiate between capacity loss caused by permanent loss of sulfur, and that due to the inability to convert all the sodium polysulfides present to sulfur, is to determine the available capacity. This is done simply by determining the change in open circuit potential of the cell in the single phase region after the passage of a known quantity of electricity. The total capacity of the cell can then be calculated from the data of Gupta and Tischer (1972) and compared with the theoretical capacity.

Loss of sulfur by leakage from the cell or by corrosion reactions with metals cannot explain all cases of capacity loss. Cells which are leak tight and contain no corrodable metal components can still exhibit this phenomenon. Zobel (1978) has shown that small flat plate cells with vitreous carbon current collectors can show a reduction in charge acceptance from 90% to 40% in 20 cycles if charged at a constant current of 100 mA cm^{-2}. No satisfactory explanation for this effect has yet been advanced. It may be that wettability of the carbon felt by sulfur and polysulfides undergoes a subtle change during cycling; certainly the reproducibility of plain carbon felt electrodes with respect to initial charge acceptance is very poor, large variations occurring even with apparently identical felts. It may well be that the surface properties of the carbon felt are such that slight changes can cause it to be preferentially wetted by either polysulfides or sulfur. Hall (1979) has shown that the contact angle of sulfur on carbon fibres depends on the previous history of the fibres, and that it can be changed from wetting ($<90°$) to non-wetting ($>90°$) by cathodic polarization. He ascribes this to the presence of surface impurities which are removed by chemical pretreatment or cathodic polarization. Brennan (1977) has also observed changes in the anodic curves of vitreous carbon electrodes in polysulfide melts after repeated voltammetric sweeps.

One further mechanism for capacity loss is an increase in cell resistance. With constant current charging the charge is terminated when a predetermined voltage – the cutoff voltage – is reached. If the cell resistance increases with cycling, the charge will be terminated even though more charge would be accepted if the cell voltage was allowed to rise above the cutoff voltage. This is an artefact of the test conditions and the resistance rise phenomenon rather than a true loss of capacity. A more subtle form of this effect, which is related to the sulfur electrode, is observed in some cells. This is a sudden increase in the charge resistance when the two phase region is reached. This can be of sufficient magnitude that the cutoff voltage is exceeded, thus terminating the charge. Cells with molybdenum current collectors often exhibit this effect after a number of cycles, as illustrated in Fig. 10.18 (Markin *et al.*, 1979).

Most of these causes of capacity loss are associated with reduced charge acceptance, but in certain circumstances the discharge capacity can be affected. One obvious way this would happen is for the sodium electrode to become depleted of sodium metal before all the sodium in the cell has been discharged. This may be due to progressive clogging of the sodium restrictor by debris from the sodium compartment, or from loss of gas pressure in cells which rely on this

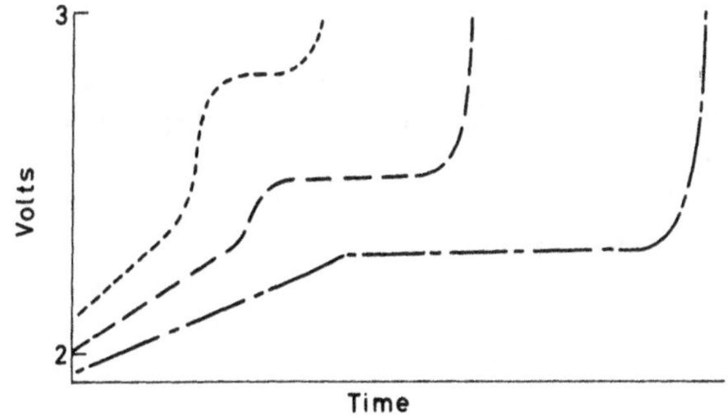

Fig. 10.18 Charge/discharge curves for a cell containing an anodized molybdenum current collector. Reproduced with permission from Markin *et al.* (1979). —-—- cycle 50, 50 mA cm^{-2}; ——— cycle 51, 75 mA cm^{-2}; ---- cycle 52, 100 mA cm^{-2}.

to maintain the sodium level. With good cell design this should not be a problem. Zobel (1978) has reported that cells with a large excess of sodium can polarize on discharge after a number of cycles and this must be due to the sulfur electrode. Such effects are absent from cells having a well designed sulfur electrode.

Historically, capacity loss has been a problem with sodium sulfur cells, but it should not occur in cells if due attention is paid to the points covered. That is, the cell should be leak tight, should have a current collector of which the corrosion rate is minimal or which results in corrosion products that do not migrate into the electrode, and should have a sulfur electrode structure which does not rely on the properties of as-received carbon felt.

10.6.4 Seal leakage

Loss of sulfur through seal leakage has already been mentioned as a cause of capacity loss, but leakage of the sodium seal can have more severe consequences. The cause of seal degradation was discussed in Chapter 8; here we are concerned with the consequences. If the seal becomes permeable to sodium metal it will migrate to the outside of the cell where it will immediately be oxidized to sodium oxide. This will react with the steel components of the seal to form ferrates which have a distinctive green colour. At this stage the cell performance is unaffected, but corrosion of the seal face gets progressively more rapid until one of two things happens. In cells in which the sodium is under pressure, the permeability of the seal can increase to the point where the sodium escapes rapidly enough to short out the cell which, of course, terminates the cell life. Alternatively, the seal becomes permeable enough to allow oxygen ingress and sodium oxide is formed inside the cell. This can attack the beta alumina in the region of the seal, and in some cells which have been removed from test after seal leakage has been observed, the beta alumina wall has been reduced to half its original thickness.

10.6.5 External factors

It has already been mentioned that the life of a cell can be affected by the way it is recharged. Other external factors which might be expected to affect life will be considered here.

The depth of discharge is an important factor in the life of lead acid batteries but there is no evidence of a similar relationship in sodium sulfur batteries. This is not unexpected, since it is the change in morphology of the active material in lead acid batteries which is responsible for the effect on life; in sodium sulfur batteries, both electrodes are liquid and the question of morphological change does not arise.

The operating temperature might be expected to have a considerable effect on cell life, but no systematic study appears to have been carried out; most cell tests are done under isothermal conditions at a temperature of between 300 and 350 °C. In the absence of experimental data it is only possible to speculate on the effect of operating temperature on life. Sufficient is known, however, of the properties of the cell components to enable some general observations to be made.

Increasing the temperature will be expected to adversely affect seal life in a number of ways. In compression seals the metal components under stress will be more susceptible to creep, leading to a reduction in seal stress and the possibility of leakage. In cells with thermocompression bonds, the intermetallic which is formed between aluminium and the transition metal of the seal face (Mitoff, 1980) will grow more rapidly and increase the probability of seal leaks. In all cells there will be a tendency for the glass seal between the alpha alumina collar and the beta alumina tube to densify (Barry *et al.*, 1980), resulting in increased stress and the possibility of cracking. Corrosion of metal components in the sulfur electrode will also be accelerated. For all these reasons, 400 °C is generally regarded as the maximum temperature to which the cell should be exposed for significant periods.

Reducing the temperature will obviously have a beneficial effect on all these factors, but on the basis of some theories of electrolyte breakdown it might be expected to reduce electrolyte life. This effect is unlikely to be important at temperatures above 300 °C.

10.7 SUMMARY

Because of the relatively high resistivity and wall thickness of beta alumina tubes, the resistance of cells containing this electrolyte is rather high. It is this factor which determines the specific power and, to a great extent, the specific energy of the cell. The capacity of the cell is independent of the rate of discharge, and the energy delivered can easily be calculated from the theoretical energy (based on the amount of reactants in the cell) and the internal resistance.

The glass electrolyte cell operates at much lower current densities and the energy delivered by this cell should be less dependent upon the rate of discharge.

Sodium sulfur cells can be recharged very rapidly (less than one hour), but at present little is known of the effect of such high charge rates on cell life. It does seem probable, however, that long charge times (three hours or more) will help to prolong cell life.

The resistance and capacity of sodium sulfur cells have been shown to remain constant for 1000 cycles if interfacial polarization at the sodium/solid electrolyte interface is eliminated (see Chapter 7), and if the sulfur electrode is designed to prevent blocking of the solid electrolyte surface by elemental sulfur during charge.

The life of the cell is certainly in excess of 1000 cycles and the reliability expressed in terms of mean time before failure is measured in thousands of cycles. There is no known fundamental limitation to cell life, and as improvements are made to components such as seals, longer lives should be achieved.

The reliability of the cell is determined by the quality of the solid electrolyte. High quality beta alumina ceramic has been made in the laboratory and pilot plant. The final step is to demonstrate that this quality of material can be produced consistently in large scale production with an economic level of reject material.

The situation for the glass electrolyte cell is somewhat different in that only relatively low capacity cells (12 Ah) have achieved reliable operation for more than 100 cycles. Here again the problem appears to lie with the quality of the solid electrolyte. The problem is not one of scaling up the process, since it is already fabricated by a method which is inherently suitable for large scale production. Once a long life glass electrolyte is produced there should be few problems in putting it into production.

10.8 REFERENCES

Barry, T. I., Schajer, G. S. and Stackpool, F. M. (1980) Deutscher Verlag Fuer Schweisstechnik Conference, Baden-Baden, West Germany.

Birk, J. (1980) DoE/EPRI Beta (Sodium/Sulfur) Battery Workshop IV Proceedings.

Brennan, M. P. J. (1977) *Electrochim. Acta*, **22**, 279.

Bugden, W. G. and Duncan, J. H. (1977) in *Science of Ceramics*, vol. 9 (ed. K. J. de Vries), The Nederlandse Keramische Vereniging.

CGE (1970) British Patent 1 347 990.

De Jonghe, L. C. (1981) EPRI Interim Report EM-2160.

Demott, D. S. (1980) 4th DoE/EPRI Sodium Sulfur Battery Workshop, Derby.

Fally, J., Lasne, C., Lazennec, Y. and Margotin, P. (1973) *J. Electrochem. Soc.*, **120**, 1292.

Gibson, J. G. (1974) *J. Applied Electrochem.*, **4**, 125.

Gupta, N. K. and Tennenhouse, G. J. (1978) Electrochemical Society Extended Abstracts, vol. 78-2, abstract 39.

Gupta, N. K. and Tischer, R. P. (1972) *J. Electrochem. Soc.*, **119**, 1033.

Hall, S. M. (1979) PhD Thesis, University of Southampton.

Hooper, A. (1980) *Trans. J. Brit. Ceramics Soc.*, **79**, 134.

Levine, C. A. (1983) *The Sulphur Electrode* (ed. R. P. Tischer), Academic Press, New York, London.

Markin, T. L., Junkison, A. R. and Bones, R. J. (1979) in *Power Sources*, vol. 7 (ed. J. Thompson), Academic Press, London.

Mitoff, S. P. (1980) EPRI Report EM-1341.

Nelson, W. (1969) Journal of Quality Technology, **1**, 1 January 1969.

Nelson, W. (1982) *Applied Life Data Analysis*, Wiley, New York.

Piercy, R. and Sudworth, J. L. (1977) Symposium on Electrode Materials and Processes for Energy Conversion and Storage, *Proc. Electrochem. Soc.*, **77-6**, 950.

Richman, R. H. and Tennenhouse, G. J. (1975) *J. Am. Ceram. Soc.*, **58**, 63.

Sudworth, J. L. (1980a) 4th DoE/EPRI Sodium Sulfur Battery Workshop, Heidelberg.

Sudworth, J. L. (1980b) Society for Electrochemistry Spring Meeting, Loughborough University.

Sudworth, J. L. (1981) Seminar on Energy Options for Developing Countries, Madras.

Sudworth, J. L. (1983) Sodium Sulfur Battery Workshop V, Washington DC.

Sudworth, J. L. and Tilley, A. R. (1983) in *The Sulphur Electrode* (ed. R. P. Tischer), Academic Press, New York, London.

Tilley, A. R. (1976) Proceedings of Symposium and Workshop on Advanced Battery Research and Design, Electrochemical Society and Argonne National Laboratory Report ANL 76-8, B231.

Tilley, A. R. (1980) 4th DoE/EPRI Sodium Sulfur Battery Workshop, Mannheim.

Tilley, A. R. and Wright, M. L. (1981) 16th IECEC Meeting, Atlanta, vol. 1, 841.

Virkar, A. V. and Miller, G. R. (1979) in *Fast Ion Transport in Solids* (eds P. Vashista, J. W. Mundy and G. K. Shenoy), Elsevier North-Holland.

Virkar, A. V., Viswanathan, L. and Biswas, D. R. (1980) *J. Mater. Sci.*, **15**, 302.

Weiner, S. A. (1976) Ford Sodium/Sulfur Battery Program, National Science Foundation, Washington DC, Contract NSF-C-805.

Zobel, F. G. R. (1978) *J. Power Sources*, **3**, 29.

11 Battery design and performance

J. L. SUDWORTH

11.1 INTRODUCTION

In Chapter 10 the design and performance of cells was described. Most of those data were obtained from cells tested individually in a rigorously controlled environment, but for most applications cells will be connected together to form a battery and all the problems which are conveniently bypassed in single cell tests must then be faced. A practical sodium sulfur battery must maintain itself at the correct operating temperature, be resistant to overcharge and over-discharge and also to the environmental stresses appropriate to the application. It must be capable of easy maintenance and, perhaps most important, it must be safe under all conceivable abuse conditions. Furthermore, all this must be achieved without degrading the specific energy to an unacceptable level and whilst keeping the cost at an economic level.

There are a number of characteristics of the sodium sulfur cell which differentiate it from other secondary cells and which have important implications for battery design. The solid electrolyte, unlike aqueous electrolytes, cannot be regenerated after it has been decomposed and there are no cell reactions possible other than formation and decomposition of sodium sulfides. This means that the current efficiency is 100%, but it also means that charging must stop completely once all the sodium polysulfide has been decomposed to its elements, otherwise the cell voltage will rise until the solid electrolyte is damaged. Similarly, if the cell is allowed to discharge beyond its design capacity, the electrolyte may be damaged by current concentration effects if the cell is negative (i.e. sodium) limited or by mechanical forces if the cell is positive limited. In single cell tests, cell damage can easily be prevented by setting limits to the charge and discharge voltage. In a battery this may be much more difficult.

The operating temperature of the battery presents a number of problems. No current can be drawn from the cell at room temperature, and it must be heated to at least 230 °C before any significant current drain is possible. Thus, heat must be provided to raise the temperature of the battery to its operating level. When a battery is operating, heat may be absorbed or evolved, depending upon the charge/discharge rates, and cooling may be necessary. The method adopted must not, however, result in rapid cooling when the battery is standing on open circuit. If, on the other hand, the temperature rises above the normal operating temperature, the vapour pressure of sulfur becomes significant, reaching one atmosphere at 444 °C, and if the temperature continues to rise sulfur will eventually be released from the cell. Thermal management, therefore, is a vitally important feature of all sodium sulfur batteries.

11.2 CONFIGURATION OF CELLS

There will be very few applications where it will be possible to have all the cells connected in series; it will be necessary therefore to have both series and parallel connections. There are two extreme schemes for intercell connection, and any number of intermediate ones. These are shown diagrammatically in Fig. 11.1

In Fig. 11.1(a), the number n of cells in a series chain equals half the battery voltage and there are no cross connections within the battery. The implications of this arrangement are as follows. If n is large, the failure of one cell will have little effect on the distribution of current amongst the parallel chains, assuming

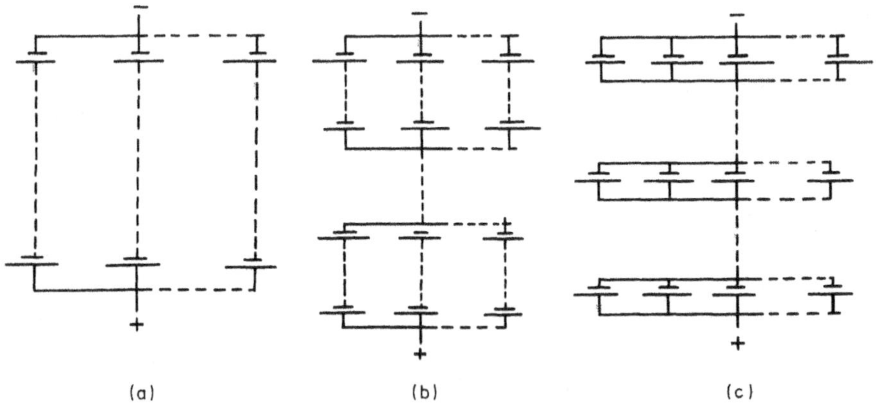

Fig. 11.1 Methods of connecting cells in series/parallel configurations.

the resistance of the failed cell is not very high. If the number of parallel chains is *m*, the larger *m* the smaller the increase in current in the chains in which all the cells are intact. Obviously for a given capacity battery the greater *m* the smaller *n* and vice versa. A disadvantage of this arrangement is that a cell which has a very high resistance after failure will prevent any current flowing through the other cells in that chain and the battery capacity will be reduced by $1/m$. Even if the resistance of the cell is only a few times greater after failure, the Joule heating will produce undesirable local temperature increases in the adjacent cells. This is illustrated in Fig. 11.2. A prerequisite of this battery configuration is that failed cells should have a low resistance. Assuming that this can be guaranteed, there is another problem to overcome. It is essential to prevent any one cell from exceeding certain voltage limits on charge and discharge; the greater *n* the more difficult it is to detect the relatively small changes in individual cell voltages. For example, in a 100 cell chain it is necessary to detect a change of 0.5 V in 200 V if the voltage limit normally adhered to in single cell testing is to be maintained. This is not quite as difficult as it seems, since it is a change in voltage rather than an absolute voltage level which is being detected; nevertheless it is a stringent requirement.

In the discussion so far, it has been implicitly assumed that monitoring of individual cell voltages is not possible. This is almost certainly the case for

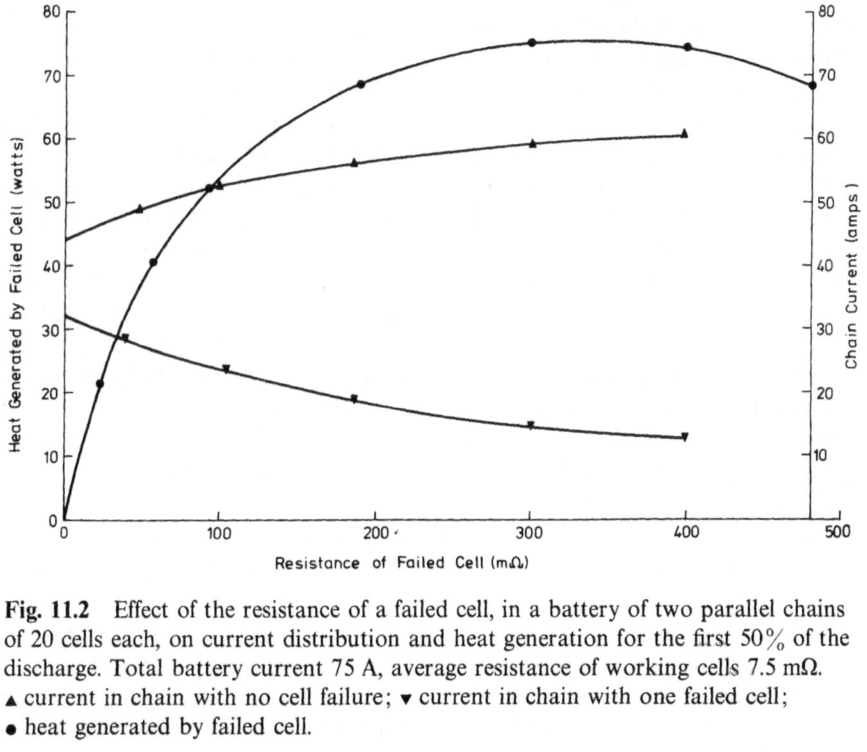

Fig. 11.2 Effect of the resistance of a failed cell, in a battery of two parallel chains of 20 cells each, on current distribution and heat generation for the first 50% of the discharge. Total battery current 75 A, average resistance of working cells 7.5 mΩ. ▲ current in chain with no cell failure; ▼ current in chain with one failed cell; ● heat generated by failed cell.

practical batteries. Apart from the complexity of the monitoring system, the physical difficulties of bringing a wire from every cell in a battery at 350 °C should not be underestimated. There is a real risk of leakage currents leading to an imbalance in cell capacities, and removal and replacement of cells becomes a major problem. It is possible, however, to contemplate monitoring groups of cells, and it has been found possible to cycle twenty cells in series without monitoring individual cell voltages (Sudworth, 1983). The cells in these batteries were well matched for capacity (±2%) and for charge acceptance and resistance, and this facilitated battery control. When constant current charging was used, the end of the charge could be detected by monitoring the rate of change of battery voltage. The cells were of the central sulfur design with a pressure fed sodium electrode, and with this type of cell at low discharge rates there may be little or no indication that the end of discharge has been reached. (This is less of a problem in those central sodium cells in which the Faradaic sodium is contained in a reservoir; here, the electrical connection to the sodium in the reservoir can be a resistance wire, so that when the sodium level falls below the top of the electrolyte tube the cell resistance increases sharply (Mitoff, 1978). The voltage increase arising from this can be used as a signal to terminate discharge.) Control of depth of discharge was achieved by connecting a sodium/sodium coulometer in series with the cells (Hames, 1980) – with a chain length of twenty cells the 5% weight penalty is not excessive. Sodium coulometers depend upon detecting the change in sodium level associated with the passage of current in a sodium/beta alumina/sodium cell.

The incorporation of sodium coulometers in every series chain also overcomes another problem: that of overdischarge due to failure of individual cells. If a battery has the configuration shown in Fig. 11.1(a), failure of one cell, assuming it has a low resistance, will result in the other chains discharging into the chain of cells containing the failed cell when the battery is standing on open circuit. The extent of overdischarge can be calculated and some typical values are shown in Table 11.1 for a battery with eight parallel chains, each having eleven cells connected in series. Because the open circuit voltage remains constant at 1.74 V at sulfur electrode compositions more rich than $Na_2S_{2.7}$, the extent of overdischarge can be considerable and will be directly proportional to the number of failed cells in a series chain and inversely proportional to the number of parallel chains. A coulometer will prevent overdischarge due to this phenomenon. Alternatively the chains may be isolated electrically from each other outside the battery.

When charging a battery having the configuration shown in Fig. 11.1(a), the possibility exists of charging each series chain separately, allowing close control of the charging current. The higher the battery voltage, the more difficult it is to detect the increase in voltage which signals top of charge. Thus, if one cell polarizes before the others in the series chain, it might be difficult to detect this before a dangerously high voltage is applied to that cell, causing it to fail. The maximum number of cells which can be charged in series will depend upon three

Table 11.1 Effect of cell failures on depth of discharge in a battery with eight parallel chains of eleven cells

Number of chains containing failed cells	*One failed cell per chain (% discharge of non-failed cells)*	*Two failed cells per chain (% discharge of non-failed cells)*
1	103	114
2	107	133
3	112	⩾150
4	120	⩾150
5	133	⩾150
6	⩾150	⩾150
7	⩾150	⩾150

things: the sophistication of the charger; the characteristics of the cells, i.e. how well matched they are for capacity; and the charge rate. The last will affect the shape of the charge curve at the top of charge: low charge rates lead to a steep rise in cell resistance, high charge rates to a more gradual rise in cell resistance as top of charge is approached.

Fig. 11.1(c) represents the other extreme to Fig. 11.1(a) for cell configuration. This arrangement can only work if cells fail open circuit: if any current can flow through a failed cell, the other cells in parallel with it will discharge through it, which at best will lead to those cells getting out of balance with the cells in series and to a lower overall efficiency for the battery, and at worst will incapacitate the battery. For this reason, cells connected in this configuration must be individually fused to ensure open circuit after failure. The main requirement for this configuration is that the resistance of the cell immediately after failure should be low enough to result in the fuse current being exceeded. The greater the number of cells in parallel the higher the load voltage of the remaining cells, and thus the higher the cell resistance which can be tolerated after failure. It is important of course that the fuse should not significantly increase the cell resistance (a copper silver alloy (silver braze) is suggested by Mitoff (1980)). This type of cell configuration is usually proposed for large stationary batteries which would have as many as forty cells in parallel. The voltage of each cell in the parallel string is of course the same, but there is a current dispersion reflecting the dispersion in cell resistance. The large number of cells in parallel makes it possible to consider monitoring the voltage of each parallel group of cells and using this to control the charge and discharge of the battery. By this means it is possible to reproduce the conditions obtaining in single cell tests and to eliminate the possibility of overcharge. In central sodium cells with a built-in resistor wire, the current will attenuate to a very low value in any cell which becomes fully discharged, so no individual cell will become overdischarged. In cells in which it is not possible to incorporate a resistance wire, overdischarge can be prevented by terminating the discharge when the cell voltage drops to 1.78 V. This will not allow complete discharge to Na_2S_3: for a cell resistance of 5 mΩ and a discharge current of 20 A, the depth of discharge at a voltage of 1.78 V will be only 85%. It is, of course, possible to set the cutoff voltage at a value corresponding to an OCV of 1.78 V, but any cell which has a lower internal resistance than the mean will discharge beyond the composition Na_2S_3. This situation is ameliorated somewhat by the shape of the open circuit voltage/composition plot, since cells which tend to discharge more quickly owing to their lower resistance will suffer a drop in current as their open circuit voltage falls below the average of all cells. This effect acts to keep cells in step during discharge in the single phase region, despite differences in internal resistances. Mitoff (1978) has analysed this effect and the results of his calculations, which are based on data from actual cells, are shown in Fig. 11.3.

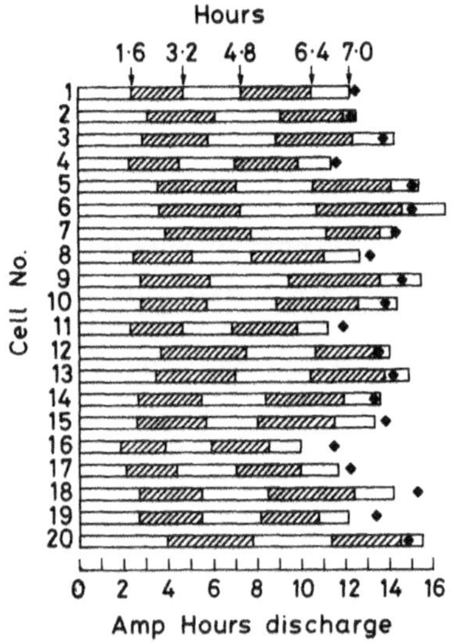

Fig. 11.3 State of charge as a function of hours of discharge for each cell during the first discharge. State of charge after rest period shown by diamonds.

Figure 11.1(b) illustrates the cell configurations which lie between the extremes of Fig. 11.1(a) and (c). In some respects, with batteries of this type one gets the worst of both worlds in that failure of a cell to high resistance effectively cuts out all the cells in that series chain, whilst failure to low resistance allows cells in other chains in parallel to overdischarge.

From the foregoing discussion it is clear that the cell reliability requirements are very dependent upon both the cell configuration in the battery and the mode of failure. If all cells fail low resistance, long series chains are possible; the longer the chain, the higher the permitted cell failure rate. If all cells fail high resistance, a chain length of one cell is the arrangement allowing the highest cell failure rate. The most stringent requirements for cell reliability arise when both low resistance and high resistance failures occur in the same battery. In this case it will almost certainly be necessary to consider maintaining the battery, i.e. replacing failed cells at specified intervals. This reduces the cell reliability requirements and is preferable to building in redundancy in the form of extra cells in the battery, since the battery performance can be returned to its initial value at every service, and there is no weight or volume penalty. If cells are to be replaced during the life of the battery this must be borne in mind at the design stage.

11.3 THERMAL EFFECTS

The maximum electrical work obtainable from a cell is equivalent to the free energy of the cell reaction, which is of course related to the other parameters by:

$$\Delta G = \Delta H - T\Delta S \tag{11.1}$$

Thus $T\Delta S$ is a measure of the energy which is unavailable as electrical work and therefore appears as heat.

The EMF of the cell is related to the free energy by:

$$\Delta G = -nF\ \mathrm{EMF} \tag{11.2}$$

Similarly, the thermoneutral potential E_t, the potential at which the passage of current will result in neither the evolution nor absorption of heat, is given by:

$$\Delta H = -nFE_t \tag{11.3}$$

The heat absorbed or evolved per mole of the cell reactants at the reversible EMF is:

$$Q = T\Delta S \tag{11.4}$$

From Equation 11.1:

$$Q = \Delta H - \Delta G \tag{11.5}$$

which from Equations 11.2 and 11.3 can be written as:

$$Q = nF(\mathrm{EMF} - E_t) \tag{11.6}$$

For a cell discharging or charging for time T at a current I and voltage E, q, the actual heat absorbed or evolved during that time given by:

$$q = IT(E - E_t) \tag{11.7}$$

If the cell reaction is proceeding non-reversibly but at 100% current efficiency, then at any time the rate of heat evolution or absorption P_t is given by:

$$P_t = \frac{dq}{dt} = I(E - E_t) \tag{11.8}$$

In the single phase region, any local deviations from the overall stoichiometry will invalidate this relationship, but as was shown in Chapter 5 such deviations are minimal; thus Equation 11.8 is a good approximation. The key parameter for calculation of heat evolution or absorption from the cell is the thermoneutral potential. This can be calculated from the thermodynamic data and a number of determinations of these have appeared in the literature. Gupta and Tischer (1972) and Cleaver and Davies (1973) determined the temperature coefficient of the EMF of a cell containing sodium polysulfides, a sodium beta alumina electrode and a low surface area carbon electrode. Mitoff (1980) and Knödler (1983) used sodium sulfur cells with carbon felt sulfur electrodes, and Knödler used adiabatic calorimetry as well as EMF measurements. There is reasonable agreement on the thermoneutral potential in the two phase region, but significant differences are apparent between the two earlier determinations and between these and the two more recent determinations in the single phase region. Mittoff's and Knödler's data are in good agreement over the whole composition range, and these data must take precedence over the data of Gupta and Tischer. Fig. 11.4 shows how the thermoneutral potential changes with depth of discharge, and Table 11.2 gives Knödler's values for the thermo-

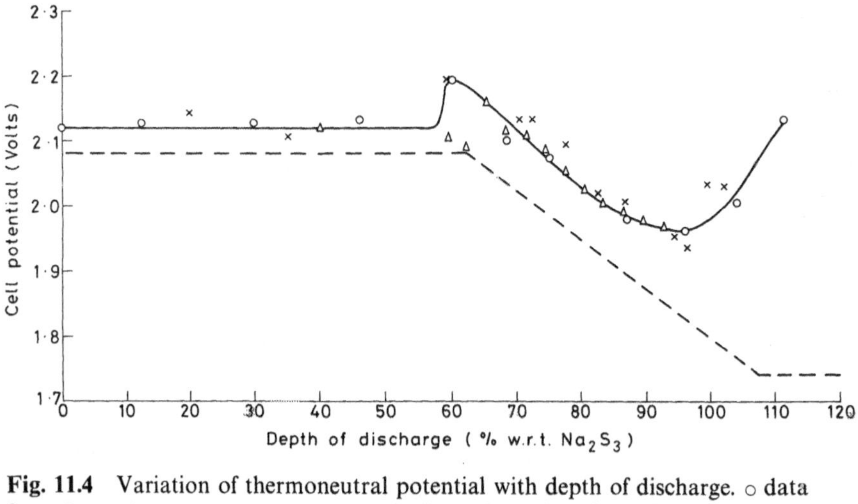

Fig. 11.4 Variation of thermoneutral potential with depth of discharge. ○ data from Knoedler (1984); △ data from Mitoff (1980); × data from Cleaver and Davies (1973). The cell EMF (---) is also shown.

Table 11.2 $-\Delta G$, $-\Delta H$, $-\Delta S$ values calculated from the experimental data

State of discharge	$-\Delta G$ *(kJ mol^{-1})*	$-\Delta H$ *(kJ mol^{-1})*	$-\Delta S$ *(J mol^{-1} K^{-1})*
$E=2.08$ V Two phase region (Na_2S_5+S)	401 ± 3	410 ± 5	15 ± 3
$E\approx2.06$ V Transition region between two phase and one phase region	397 ± 5	430 ± 8	50 ± 8
$E=1.95$ V One phase region (Na_2S_4)	375 ± 5	398 ± 5	35 ± 6
$E\approx1.73$ V Deep discharge region (Na_2S_x with $x<3$)	334 ± 10	420 ± 15	130 ± 20

dynamic functions ΔG, ΔH and ΔS for the cell reaction. From these data the mean value of the thermoneutral potential over the whole discharge range is 2.14 V.

The aim of thermal management is to keep the battery temperature within a range which will permit charge and discharge of the battery at rates appropriate to the application. A lower temperature limit of 300 °C and an upper limit of 400 °C are considered appropriate for traction applications. A secondary aim is to minimize the rate of heat loss when the battery is standing idle. This is desirable for two reasons: first, a battery which cooled down rapidly would have a limited number of applications; second, heat loss is equivalent to self-discharge of the battery and adversely affects overall efficiency.

Clearly, there are conflicting requirements for thermal management: to minimize heat loss, highly efficient thermal insulation is required, but for high rate discharge the heat evolved from the cells must be removed to prevent the temperature from rising unduly. The reconciliation of these two conflicting requirements is a major factor in battery design.

Before discussing the design of batteries, it is worth while considering how the design and characteristics of the cells affect thermal management. The cell resistance is obviously important and should be as low as possible. This is more important for high rate batteries such as traction batteries where I^2R heating is considerably greater than $T\Delta S$ heating, and thus reduction in heat generation is almost proportional to the reduction in cell resistance. Thus for sodium sulfur traction batteries there are two compelling reasons for minimizing cell resistance: reduction of heat load and increase in power. A second, less obvious parameter is the energy: weight ratio of the cell: the higher this is, the faster the rate of temperature rise for a given heat generation rate, simply because of the lower mass of the cell. Thus for stationary applications where weight is less important there may be advantages in keeping the energy: weight ratio relatively low.

The design and dimensions of the cell determine the extent to which the temperature of the cell will be uniform. Central sulfur cells with annular sodium storage (Gibson and Tilley, 1978) in which sodium metal is adjacent to the cell wall are less likely to develop temperature gradients than central sodium cells. Obviously, short cells will have less of a temperature difference along the length of the cell than long cells.

11.4 BATTERY DESIGN

11.4.1 Introduction

A sodium sulfur battery can be described as a group or groups of electrically interconnected cells in a thermally insulated container or containers, provided with means for heating and cooling the cells. Many different designs have been proposed, some of which have been constructed and tested. These designs can be divided into two categories – integral designs and modular designs. In integral battery designs all, or a substantial fraction, of the cells are packed closely together in a thermally insulated container; in the modular design the battery is split up into a number of subunits or modules, each comprising a number of cells in a thermally insulated container. In an integral battery the ratio of inert material, i.e. container, thermal insulation etc. to cells is minimized, which leads to high energy: weight and energy: volume ratios. Modular batteries have the disadvantage that the ratio of inert material to cells is greater but, to offset this, thermal management is easier, safety problems are minimized and maintenance (i.e. replacement of failed cells) is simplified.

11.4.2 Thermal management

(a) *Thermal insulation*

In both integral and modular batteries, the thermal insulation and the cooling system can take a number of forms. Mineral fibre insulation is low in cost but has relatively poor insulating properties, leading to thick insulation and corresponding low energy: volume ratio for the battery. Compressed powders are more effective, particularly if they are evacuated. The most efficient form of insulation is multifoil insulation, which consists of multiple layers of metal foil separated by thin layers of fibrous material and contained in an evacuated metal container. Reiss (1980) has listed the following criteria for thermal insulation:

(1) Low thermal conductivity at 300 °C;

(2) stable long term performance, e.g. if evacuation is necessary a residual gas pressure of >1 mbar must be tolerated;
(3) cost of insulation should be less than 10% of the battery cost; thus insulation which requires cleanroom conditions for its preparation is not permissible;
(4) for traction applications the weight must be low, i.e. the material should have a low density and the side walls containing the insulation should be light and require no support.

Multifoil insulation satisfies only the first of these criteria: the heat flow characteristics of a 20 mm thick sheet are shown in Fig. 11.5. At pressures less than 10^{-1} mbar, this type of insulation has the lowest thermal conductivity achievable ($\lambda = 10^{-4}$ W m^{-1} K^{-1}). This insulation has been in use for several years, mainly in aerospace applications. Unfortunately, it does not satisfy

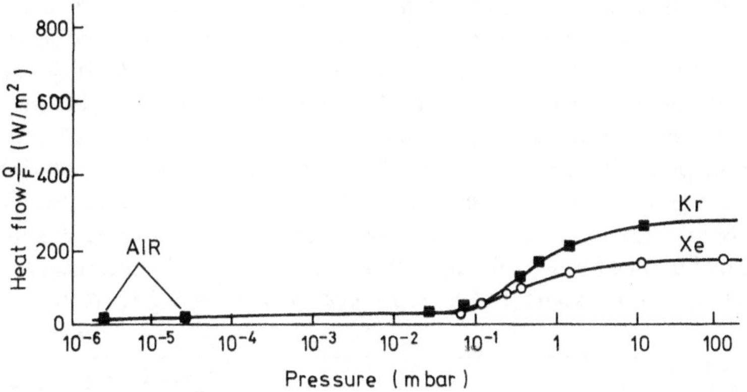

Fig. 11.5 Heat flow characteristics at 320 °C of a 20 mm thick sheet of multifoil insulation containing 42 foils.

Reiss's other criteria: as can be seen from Fig. 11.5, a low residual gas pressure (less than 10^{-2} mbar) must be maintained during the lifetime of the battery, and the insulation must be prepared under cleanroom conditions which leads to a high cost. If the cost could be reduced, the other disadvantages could be ameliorated enough to make this insulation acceptable. Thus the limitations on the area of the panels arising from the deformation of the walls due to atmospheric pressure can be eased by ribbing the metal or by making the insulation inside the panels load bearing. The vacuum can be maintained by the use of getters: Gross and Birnbreier (1979) have proposed that palladium hydride be used as a reversible getter, so that by activating a resistance heater the hydride could be decomposed, thus increasing the pressure in the multifoil insulation and increasing the heat flow rate through it. An order of magnitude increase can easily be achieved, which would go some way towards combining the low heat loss required for open circuit standing, and the higher heat loss required when the battery is discharging. One disadvantage of this proposal is that the temperature of the outer skin of the battery box would rise considerably unless an effective cooling system was installed. Gross and Birnbreier also suggested a simplified form of multifoil insulation in which the inner wall of the container was dispensed with by evacuating the whole container. This means that vacuum tight electrical lead-throughs are required and that the cells must not give off any gases which could lead to pressure increases.

Compressed powders can satisfy Reiss's criteria if they are evacuated, provided that they cut down the infrared radiation transfer in the same way that metal foils do in the multifoil insulation, and are strong enough to withstand atmospheric pressure on the side walls without deformation occurring. Heat transport due to infrared radiation can be reduced by the addition of opacifiers – materials which coat the particles and make them reflective to infrared radiation. For desirable heat flow properties a mean particle diameter of less than 1 micron is desirable, but pressures of up to 10 mbar are acceptable.

Evacuated opacified fibrous insulation (Ziegenbein, 1983) is another promising alternative to multifoil insulation. Easily handled fibre boards can be made by compressing the fibres and heating the compact to just below the softening point of the material. Board fabricated from borosilicate paper need only be maintained at a 5 N m^{-2} vacuum level (Fig. 11.6) compared with the 10^{-2} N m^{-2} level required for multifoil insulation. The lower level should be easily maintained for several years, whereas that required for multifoil would be much more troublesome. Evacuated fibrous materials appear to be far more

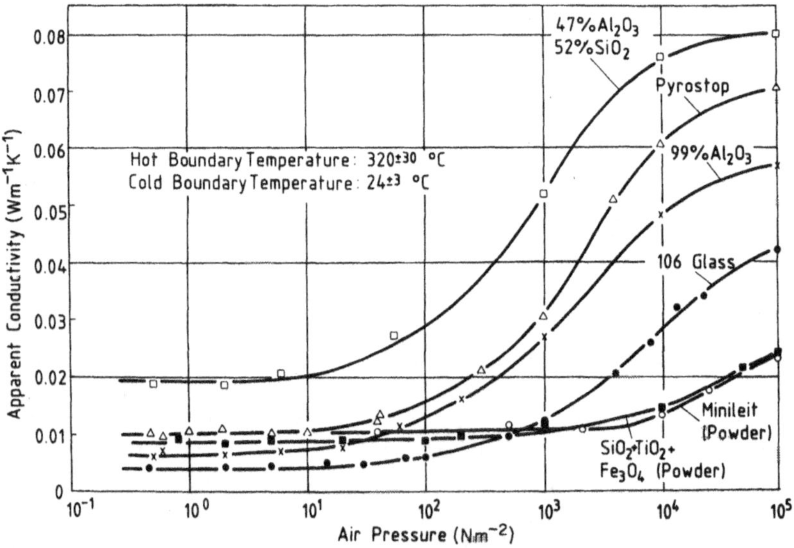

Fig. 11.6 Thermal conductivity – air pressure relationships, fibres and powders.

suitable than evacuated powders for load bearing high temperature battery insulation because of their lower solid conductivity, low residual gas desorption rates and easier handling. The extinction coefficient can be substantially improved by dispersing opacifiers in the fibrous structure.

(b) *Battery cooling and heating*

Both evacuated multifoil and evacuated compacted powder insulations will allow long periods of open circuit stand without a serious fall in temperature and will result in a battery skin temperature close to ambient. Two problems arise, however, when battery cooling is introduced. First, introduction of the coolant, whether gas or liquid, provides a path for heat loss when the battery is standing idle and thus increases the rate of cool-down. Second, temperature gradients within the battery will be severe unless the inlet temperature and outlet temperature of the cooling medium are quite close. Cooling methods fall into two main categories, gas cooling and liquid cooling, and these can be considered separately.

Gas cooling has the advantage of simplicity, particularly if air can be used. A disadvantage, however, arises from the low specific heat of gases. The simplest method of gas cooling is to pass cold air over the surface of the cells. Depending upon the discharge rate of the battery, the air may be circulated by natural convection or forced flow. Because of the low specific heat of air it rapidly heats up to the cell temperature, and considerable temperature gradients can occur. These effects can be minimized if the direction of flow is along the axis of the cells, since heat transfer along a cell is much easier than heat transfer from cell to cell. Also, the shorter the cell, the less the temperature difference which can be sustained along its length. Mitoff (1980) has described a design for a load levelling battery in which air circulates vertically by natural convection, the flow rate being controlled by flaps activated by bimetallic strips. Fischer (1980) has described a similar system for a traction battery in which the cells are packed into a double walled container insulated by evacuated compacted powder, the cooling air being blown into the box at the base of the cells and forced to flow from the bottom to the top of the cells. With this system the temperature difference down the cells, which are only 25 cm long, is only 10–20 °C (Reschke, 1983). One objection to this method of cooling is that of safety; if a cell in the battery should breach, the reaction of sodium and sulfur with air will increase the severity of the incident. For this method to be used, therefore, it is necessary virtually to eliminate the possibility of cell breaching and to stop circulation of air immediately cell rupture occurs (Hartmann, 1983). Minck (1983) has described a single pass air cooling system which minimizes temperature gradients. In an integral battery of close packed cells the busbars are hollow and segmented and cooling air is blown through as shown in Fig. 11.7. This contraflow system results in a relatively even temperature profile down the busbar and the cells lose heat to this cool surface by radiation from their ends.

For large stationary batteries it is possible to consider recirculating an inert

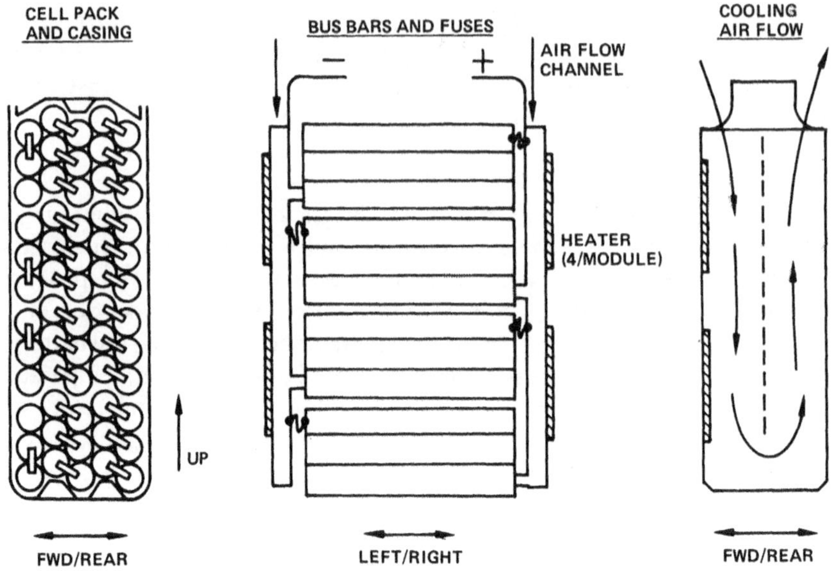

Fig. 11.7 Schematic diagram of Ford car battery.

gas such as nitrogen. Good thermal control can be obtained if the temperature difference between inlet and outlet gas is less than 50 °C, which can be achieved by passing the gas through a heat exchanger. This method of cooling has the advantage that the cells are completely enclosed in an inert atmosphere, so that the effect of a cell breaching is much less than in an air cooled system. The disadvantage of this approach is that, because of the relatively small temperature difference between the inlet gas temperature and the cell temperature, very large volume flow rates are required.

Liquid cooling has advantages over gas cooling: the higher specific heat of liquids means that much lower volume flow rates can be used, and liquid/gas heat exchangers are smaller and more efficient than gas/gas heat exchangers. The choice of liquids, however, is severely limited by the high operating temperature. Mitoff (1980) has described a liquid cooling system for a stationary battery in which liquid is circulated through copper pipes which are electron beam welded to each cell (for this system, groups of cells have to be connected in parallel). The system was never used because of the difficulty of finding a liquid having sufficiently high flammability limits and long term stability at 300–400 °C. Pompon (1980) has proposed the use of the latent heat of vaporization of a liquid to cool cells in a battery. The paraffins diphenyl and terphenyl ($17 < C < 23$) have boiling points in the range 300–400 °C, and a terphenyl mixture $C_{18}H_{14}$ melting at 80 °C and boiling at 357 °C was selected as most appropriate. To minimize the volume of liquid, each cell had a wick on the outside of the case so that the whole cell did not need to be immersed in the liquid. The same objections of flammability and flashpoint apply to this cooling system.

So far the discussion has been restricted to integral batteries. For modular systems a different approach is possible. Sudworth (1983) reported the construction and testing of a battery comprising two modules of forty cells. These cells were arranged in two parallel rows inside a thermally insulated box; thus each cell with the exception of the four end cells was in the same thermal environment. The battery was designed to operate at the two hour discharge rate, and for each cell the rate of heat generation calculated from Equation 11.8 was 15 W for the first 60% of discharge, increasing to 40 W by the end of the discharge. The module was enclosed in 20 mm thick Microtherm® insulation, which resulted in a total heat loss of 800 W at a temperature gradient of 300 °C. This module was then placed alongside an identical module. The two modules, separated by a 10 mm gap, were enclosed in a thermally insulated box (10 mm thick Kaoboard®). The resulting modular battery could be discharged at the two hour rate if the skins of the modules were maintained close to ambient temperature. This was achieved by blowing air past them. The temperature gradient along the skin of the module in the direction of the air flow is dependent upon the flow rate, but in principle it can be negligible provided a high enough flow rate is used. Even if a temperature gradient is present along the module skin, there will be a smaller gradient along the row of cells.

When the battery was standing idle the fans were switched off, with the result that the module skin temperature rose until it reached an equilibrium value determined by the natural convective air flow over the skin and the rate of exchange of this air with ambient air from outside the battery. In this battery, when the air vents were left open the temperature of the module skins rose to 180 °C. Thus ΔT across the module insulation was almost halved, with a commensurate reduction in heat flow from the cells to the module skin. With this battery the ratio of heat removal at the two hour rate to that on open circuit stand was 2:1, but higher ratios should be achievable by using better insulation inside the outer battery box and closing the air ducts. To maintain the cell temperature constant when the battery was standing on open circuit, resistance heaters located between the two rows of cells could be energized by the battery. With this battery, which comprised seventy-six cells having a capacity of 88 Ah, the power required to keep the temperature constant was 630 W, which meant that the fully charged battery could maintain its operating temperature for 20 hours before it began to cool down. Further details of this battery are given in the next section on battery performance.

At lower discharge rates the cell temperature could be maintained constant by intermittent use of the fan and, when the rate was reduced so that the heat produced by the cells was less than 630 W, by use of the resistance heaters at the appropriate power level. On charge it was necessary to increase the heater power to 740 W at the eight hour charge rate – a direct result of the $T\Delta S$ contribution.

Liquid cooling can also be used; either cooling pipes can be attached to the module skin, or a double skin with integral cooling channels can serve as a common skin for two adjacent modules. Liquid cooling results in a much more even skin temperature, and because the liquid coolant can easily be pumped through a radiator the coolant temperature can be accurately controlled. Thus the rate of heat removal from the cells, which is determined by ΔT across the insulation, can be varied by varying the coolant temperature. The advantage of cooling the module rather than the cells is that it is not necessary to have a liquid which is stable at the maximum cell temperature, since the maximum temperature at the module skin can be fixed at any desired value by use of the radiator system. The maximum coolant temperature will be determined by the stability of the liquid and this temperature will determine the rate of heat loss from the battery when it is standing idle. As with air cooling, the thermal insulation inside the skin of the outer battery box should be as effective as possible to minimize heat loss. One advantage of liquid cooling is that the warm air from the radiator can be used directly for space heating.

A disadvantage of this thermal management system is that it is not possible to remove heat more rapidly than the ΔT across the insulation will allow. Thus if a cell short circuits and its temperature rises rapidly, increasing the supply of cooling fluid to the module skin will have little effect. It is essential, therefore, that the modules should be safe even when cells are short circuited or when cells fail. In Chapter 12 it will be shown that the modules described here are safe under these conditions.

So far in this discussion of thermal management, thermal insulation and battery cooling have received the most emphasis; battery heating, however, is equally important. It is necessary to have heaters in a battery to ensure that the battery can be raised to its operating temperature and maintained at this temperature during open circuit stand and charging at low rates. Resistance heaters are normally used, but gas or liquids can be used as heat transfer media for heating the battery. Whichever method is used, it is vitally important that the rating of the heaters should be such that it is not possible for them to raise the battery temperature to a dangerously high level. It is prudent to design the heating system so that a temperature of 400 °C cannot be exceeded. Under no circumstances should the heaters be capable of raising the battery temperature above 600 °C. Above this temperature, fully charged cells are likely to burst owing to the vapour pressure of sulfur, leading to catastrophic failure of the battery. With electric resistance heaters this can be avoided by selecting the appropriate heating elements, and as an additional precaution the circuit can be fused.

Whichever thermal management system is adopted there will be some degree of thermal cycling, i.e. regular fluctuations in temperature as the battery charges and discharges. Provided that this does not result in phase changes in the cell reactants or in high electrical polarization of the cells, it is not necessarily undesirable, and in fact using the thermal mass of the battery as a heat storage system will reduce the load on the battery cooling system and increase the overall efficiency of operation. Thus cells should be designed to allow a reasonable amount of thermal cycling, which means paying particular attention to seal design.

A more severe thermal cycle which batteries may have to survive is the freeze/thaw cycle which theoretically occurs when the battery temperature falls below 284 °C or 120 °C depending upon the state of charge of the battery. In fact, as was mentioned in Chapter 5, sodium sulfide melts are very susceptible to supercooling and a discharged cell can be cooled down to 200 °C without any evidence of freezing of the sulfide melt. It is known that individual cells can be subjected to repeated freeze/thaw cycles without affecting their performance (Hattori *et al.* (1977) but little information is available on batteries. Minck (1980) reported that when a group of twenty-six central sodium cells was accidentally cooled down to room temperature, twenty-five cells operated properly after reheating. Thus a freeze/thaw cycle may lead to extra maintenance but would not be catastrophic. In integral batteries with close packed cells it may be necessary to cool the battery to room temperature if cells need to be replaced (Minck, 1983), and cells intended for this type of battery must be able to withstand repeated freeze/thaw cycles.

11.4.3 Other design considerations

From the foregoing it can be seen that thermal management is a major factor in determining the design and construction of sodium sulfur batteries. There are, however, a number of other matters which must be given careful consideration at the battery design stage.

(a) *Cell spacing*

For maximum volumetric and gravimetric specific energy, cells should be packed as closely together as possible, but if direct air or gas cooling of cells is adopted, sufficient space must be allowed between individual cells for the gas flow required. Another factor which, if found to be necessary, will affect cell spacing is incorporation of barriers between individual cells for the prevention of failure propagation in the event of a cell rupturing. Mitoff (1980) has described a design for a load levelling battery in which individual cells are located in a hole, whose diameter is slightly greater than that of the cell, in a block of refractory material. A gas is forced through the annulus formed between the cell and the refractory material when cooling is required, and the cell spacing is determined by the thickness of the refractory safety barrier and the dimensions of the annular gap.

For traction batteries where weight and volume are more critical, it is necessary to space the cells at the most one or two millimetres apart and any safety barrier must be thin enough to fit in this gap. When cells are closely packed it is essential to have good electrical insulation between each cell. Micaceous resin sheets of the type used to support heating elements in domestic appliances are quite effective and glass fibre sheaths have also been used (Minck, 1983).

(b) *Electrical insulation*

The importance of good electrical insulation between adjacent cells and between the cells and the battery box cannot be overemphasized. Any insulating material used should retain sufficient strength at the battery operating temperature to resist mechanical forces which are likely to arise during operation. Particular attention should be paid to the insulation of any leads incorporated for monitoring temperatures and voltages. The number of such leads should be kept to an absolute minimum to reduce the risk of leakage current as much as possible. A precaution which will reduce the risk of short circuits is to ensure that the maximum potential difference between any two adjacent cells is 2 V when the battery is open circuit.

(c) *Maintenance*

An important consideration in battery design is ease of maintenance. This is less important if the battery is to be maintained by replacing modules rather than individual cells, as the failed cell can be removed from the module in a convenient location, but since individual cells will have to be removed from the battery or from a module (unless the module is a permanently sealed unit) it is sensible to make this job as simple as possible. The intercell connection system used will greatly affect ease of maintenance. Bolted connections have not been found to be satisfactory in sodium sulfur batteries other than as a temporary measure, and welded joints are preferred. If aluminium cell terminals are used, the type of connection illustrated in Fig. 11.8 is possible. Removal of a cell is simply a matter of cropping the two welds and lifting out the cell. A new cell (previously heated to the battery operating temperature) can then be inserted, and the terminals cropped to the same length as the adjacent terminals and TIG welded *in situ*. This operation takes only one or two minutes. It is of course necessary to ensure that the replacement cell is at the same state of discharge as the other cells in series with it, and to minimize the risk of the operation it is preferable to have the battery and the replacement cell fully discharged.

Before maintenance can be carried out it is necessary of course to determine which cells in the battery have failed. In batteries connected as shown in Fig. 11.1(a), the current in the chain containing the failed cell will fall and this can be used as an indicator for a failed cell.Fig. 11.9 shows this for a battery comprising two series chains of six and seven cells which simulates the case of one cell failing to zero resistance (Fischer, 1980). It can be seen that as the discharge proceeds the current in the chain with seven cells falls and the current in the other chain rises. This is due to the decline in open circuit voltage of the sodium sulfur cell in the single phase region. This effect results in each chain delivering almost its full capacity, in this case only 10% less than a battery with equal numbers of cells in each chain (Fischer, 1980). On charge the opposite

Fig. 11.8 Intercell connection using aluminium terminals.

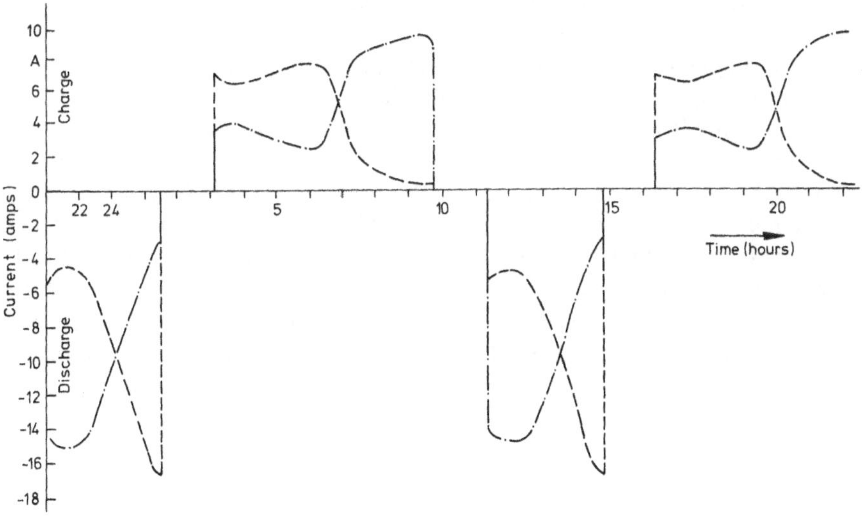

Fig. 11.9 Current distribution in a battery with two parallel chains, one of which contains a failed cell. ——— branch 1, 7 cells; —·—· branch 2, 6 cells.

occurs: the chain with the fewest cells takes the highest current in the first part of the charge and the lowest current in the later stages of charge.

In batteries connected as in Fig. 11.1(c), each cell would be fused and a failed cell would result in a drop in load voltage of the bank in which it was located. In batteries connected as in Fig. 11.1(b), failure of an individual cell would be difficult to detect and location of the failed cell would be difficult unless current shunts were incorporated in the battery proper. Whichever battery configuration is used, the exact location of the failed cell will probably have to be determined by removing the battery or module cover and measuring the voltage of the relevant cells. The safest time to do this is when the battery is fully discharged.

(d) *Instrumentation and control*

It is implicit in the foregoing that some system of instrumentation and control is required in a sodium sulfur battery. The minimum requirement is a thermocouple to monitor battery temperature, and an indication of battery voltage and current. By monitoring these three parameters and adjusting the heating and cooling manually, it is possible to charge and discharge the battery. The battery can be left unattended on open circuit by the use of mains powered resistance heaters of a fixed power equivalent to the heat loss from the battery. The advantage of such a simple system is high reliability. It is of course possible to devise a control system which will perform these functions automatically, e.g. by the use of thermostats, proportional controllers etc., but two things should be borne in mind. First, the more complex is the system, the lower is the reliability. Second, the greater the number of monitoring leads, the greater the chance of leakage currents and even short circuits, and the more difficult maintenance becomes. A control system for a battery, therefore, will be a compromise between simplicity for reliability and complexity for greater control, and ideally should not require connections to individual cells.

In some applications, resistance to mechanical shock will be an important requirement and must be considered at the cell design stage and in battery design. The orientation of the cell in the battery and the direction in which mechanical forces are likely to be applied can together have a large effect on shock resistance. Thus a battery installed in a road vehicle can expect to be subjected to vertical forces transmitted through the suspension system. If the cells in the batteries are mounted vertically, the bending moment on the ceramic electrolyte tube, which is normally supported rigidly at one end, will be far less than would be the case if the cells are mounted horizontally. Conversely, lateral forces which might result from an accident would be more likely to lead to fracture of the ceramic electrolyte in vertically mounted cells.

11.5 BATTERY PERFORMANCE

Because the characteristics of a battery are so dependent upon the battery configuration, it is difficult to generalize on battery performance. Few details of the performance of batteries have been published and therefore, for the most part, the characteristics of batteries built and tested in the British Rail laboratories will be used to illustrate the important features of battery performance.

The first large battery constructed at British Rail (Sudworth and Hames, 1978) was an integral design comprising 176 cells (Piercy and Sudworth, 1977). These 36 Ah cells were connected together to give eight parallel strings of twenty-two cells connected in series (Fig. 11.10), giving a battery capacity of 288 Ah. The eight chains were connected at the eleventh cell to give sixteen individual chains, and the voltage of groups of three and four cells was monitored.

It was realized that cooling would be necessary at the higher rates of discharge, so the thermal insulation around the battery walls was limited to 2.5 cm of mineral wool, the battery temperature being maintained by resistance heaters on the inside surface of the battery box. It was intended to control the battery temperature by varying the output of these heaters. This led to a

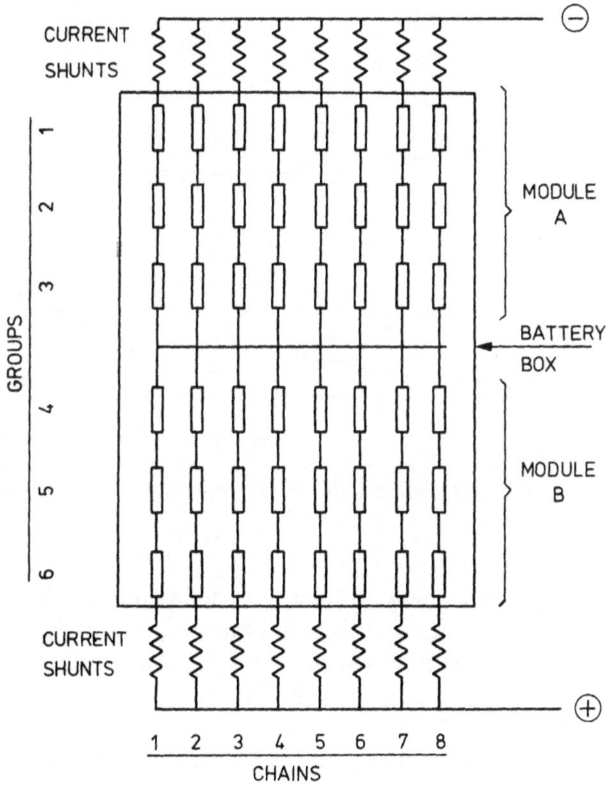

Fig. 11.10 Configuration of 11 kWh battery.

relatively high skin temperature and, as will be described later, temperature gradients within the battery.

The battery was heated to 350 °C at a rate of 30 K h^{-1}. As a precaution against internal short circuits the final intercell connections were only made after the battery reached its operating temperature. During activation one cell failed, but its resistance remained low and it was decided to leave it in the battery.

The first discharge, at the six hour rate (48 A) resulted in a capacity of 11 kWh. The temperature variation is shown in Fig. 11.11, and it is seen that the range of temperatures fell from 40 K at the beginning of discharge to approximately 20 K at the end of discharge, mainly as a result of the cooler cells heating up.

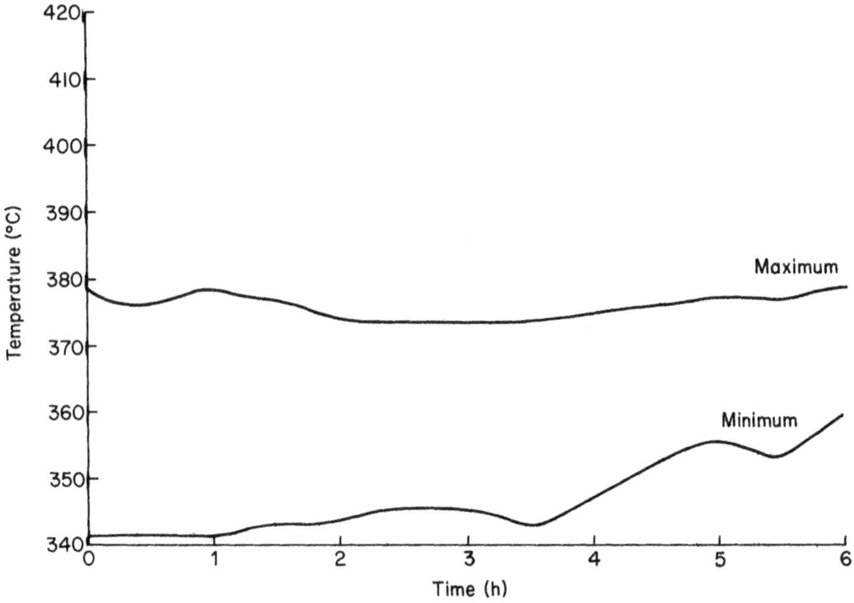

Fig. 11.11 Temperature variation during discharge of an 11 kWh battery.

The distribution of current between the chains in module B is shown in Fig. 11.12 and it can be seen that there is a wide dispersion of current. This was due to the wide variation in resistance of the particular cells used, and in one chain to the presence of a failed cell. During the discharge, however, the distribution of current changed considerably, with the result that the seven complete chains reached almost the same state of discharge. After the external load had been disconnected, internal currents continued to flow for 18 hours until all the complete chains had attained the same depth of discharge

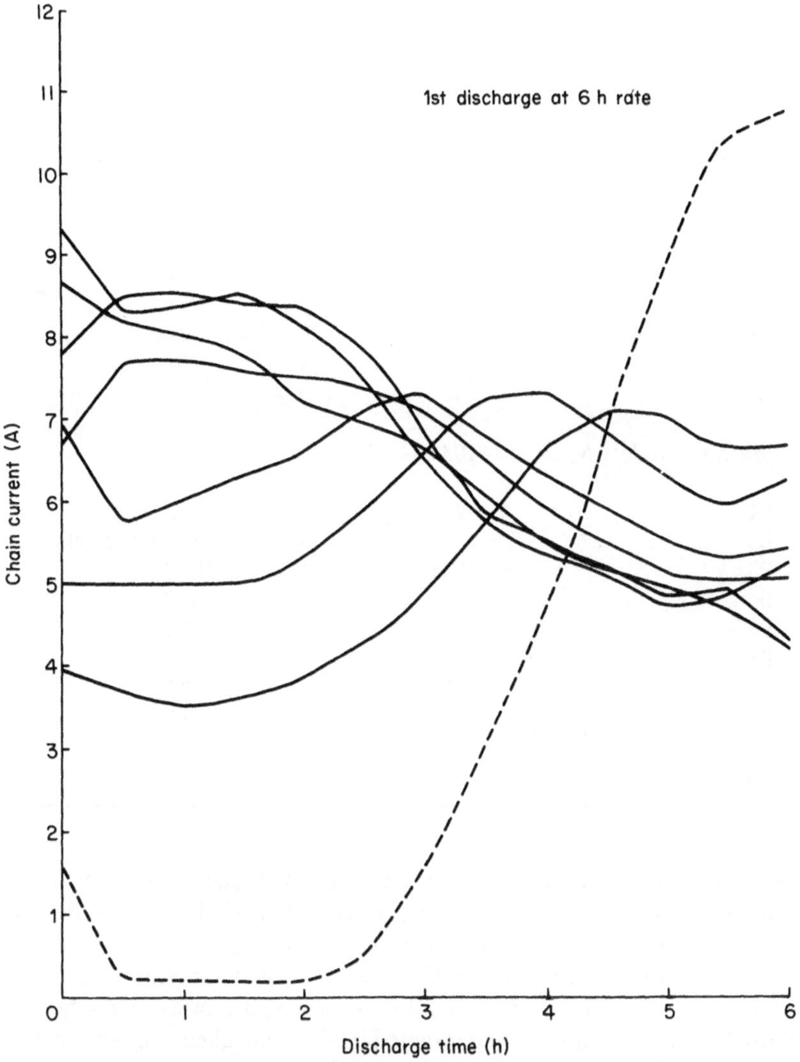

Fig. 11.12 Distribution of current between chains in 11 kWh battery.

(Table 11.3; Hartley, 1980). No equalizing currents flow after termination of charge, because all cells are in the constant voltage region of the OCV curve.

The battery was subjected to eighty charge/discharge cycles at discharge rates varying between two hours and six hours. During this time a number of problems manifested themselves.

For the first ten cycles no more cell failures occurred, but on the eleventh cycle four cells failed and were replaced. The temperature of the battery was controlled thermostatically, the control thermocouples being strapped to

Table 11.3 State of discharge of cells at end of discharge in 11 kWh battery

Chain Number	*% Discharge at end of discharge*	*% Discharge after 18 h equilibration*
A1	105.6	99.7
A2	110.4	99.6
A3	95.3	99.4
A4	118.4	99.6
A5	104.0	99.6
A6	104.4	99.6
A7	94.6	99.6
A8	96.1	99.6
Mean of Module A	103.6	99.6
B1	97.9	102.2
B2	61.7	77.4 failed cell
B3	100.6	102.2
B4	100.9	102.2
B5	100.2	102.2
B6	105.1	102.2
B7	103.3	102.2
B8	110.4	102.2
Mean of Module B	97.5	102.2

individual cells. The temperature fluctuations induced by removing the thermally insulated lid made it essential to switch off the heaters during the maintenance period. Conversely, when the maintenance was complete and the heaters were switched on again, the non-uniform temperature distribution resulted in parts of the battery cooling down whilst other parts heated up rapidly. These problems were aggravated by the time taken to change cells; this was inordinately long, especially when the failed cell had a voltage lead and/or a thermocouple attached. The sheer number of instrumentation leads (over fifty voltage connections and over a dozen thermocouples) led to problems with leakage currents.

The construction and operation of this battery highlighted the problems of operating a high temperature battery, and also showed that a considerable improvement in both cell and battery design was required if a reasonable specific energy was to be achieved. The specific energy of the cell was not high (90 Wh kg^{-1} at the two hour rate), but that of the battery (60 Wh kg^{-1} at the six hour rate, first discharge) was significantly lower. This was further degraded by the low charge acceptance (70%) to approximately 40 Wh kg^{-1}.

Clearly, a cell with a much higher specific energy and charge acceptance was required, together with a radical redesign of the battery. A high performance cell with a nominal capacity of 100 Ah (designated BR30) was developed and its characteristics have been described in Chapter 10. It was decided to adopt the modular battery design described in Section 11.4.2.

To evaluate the performance of the cells in these conditions, a module was constructed having a single chain of nineteen sodium sulfur cells and one sodium/sodium cell to act as a coulometer (Sudworth, 1983). As a safety precaution, each cell was contained in a box fabricated of a composite material made by sandwiching graphite foil between two resin bonded mica sheets. This composite material, of which details will be given in Chapter 12, was an effective barrier to propagation of cell failure in the event of a cell rupturing, and it also served as an electrical insulator between cells. Electrical resistance heaters made by winding resistance wire on to a mica sheet, were inserted between the 50 mm thick Kaowool insulation and the cells. The intercell connections were made by TIG welding together the aluminium cell terminals. The finished module, with the cover removed, is shown in Fig. 11.8. The battery was activated by heating to 350 °C and then the temperature was reduced to 330 °C. For this first module each cell voltage was individually monitored and six thermocouples were inserted between the cells.

The average cell resistance decreased over the first twenty-five cycles from an initial value of 16 mΩ to 9 mΩ and then fell to a value of 8 mΩ over the next 225 cycles (Fig. 11.13). There was very little scatter in individual cell resistances

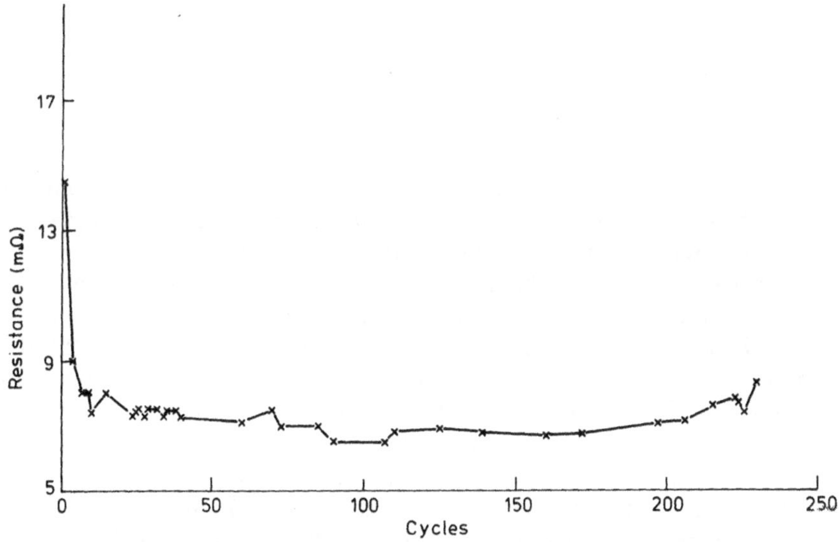

Fig. 11.13 Variation of mean cell resistance with cycles for nineteen cell module.

and capacities, and it was possible to cycle the battery module over 77% of its capacity at charge and discharge currents of 30 A and 44 A respectively.

Although the performance of the battery was satisfactory, cell reliability was not. The cell failure rate was high and expressed as mean time to failure (MTTF) was only fifteen cycles. This problem was highlighted by the construction and testing of a second battery module, which was similar to the first module except that 20 mm thick Microtherm® was used as thermal insulation and individual cell voltages were not monitored. The resulting module was more compact (Fig. 11.14), but the performance was similar to that of the first module. Again cell reliability was poor.

Fig. 11.14 Second twenty cell module.

The charge and discharge curves for this battery for a range of currents at 330 °C are shown in Fig. 11.15. The effect of different charge regimes is shown in Fig. 11.16. It can be seen that some form of step charging could result in higher charge acceptance at the expense of an increase in charge time.

The cause of the cell failures was identified as sodium depletion during discharge: the flow rate of sodium to the annular gap between the beta alumina tube and the cell wall was less than the sodium ion flux through the beta alumina. In a single cell test this would merely have resulted in the discharge current falling if the cell was being discharged through a resistive load, or in the cell being switched out of circuit if the cell was being cycled at a constant current between voltage limits. In a long series chain, however, neither of these possibilities exists and the cell is eventually driven negative which, in the author's experience usually results in failure of the beta alumina. The reduced sodium flow rate was found to be due to blockage of the flow restrictor,

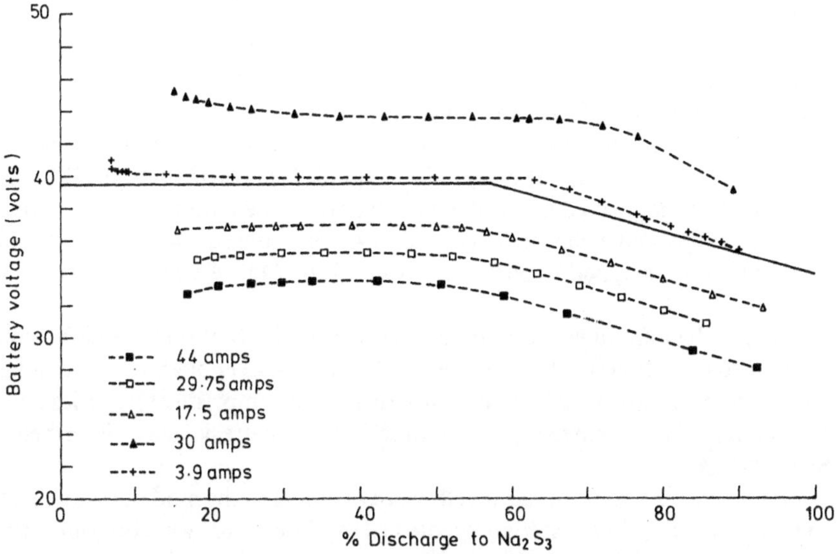

Fig. 11.15 Charge/discharge curves for second twenty cell module.

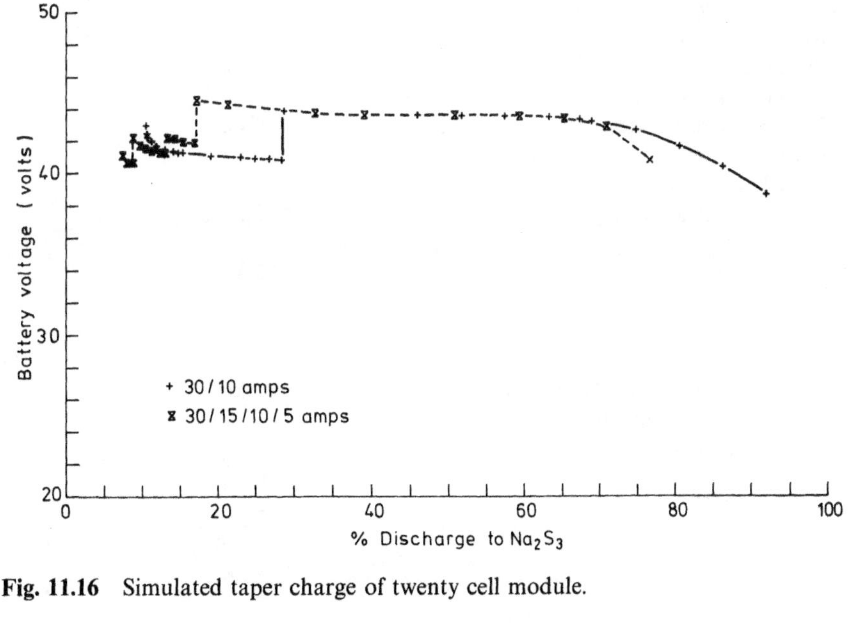

Fig. 11.16 Simulated taper charge of twenty cell module.

incorporated as a safety feature, by sheets of flexible graphite located on either side of the steel plate to protect it from corrosion. The problem was overcome by spacing the discs from the restrictor plate by pieces of wire mesh and increasing the size of the orifice from 0.5 mm to 1 mm.

The effect on cell reliability of modifying the sodium restrictor was to increase the MTTF to 150 cycles. At cycles 134 (module 1) and 107 (module 2) the two modules were connected in parallel and cycled as one battery with a discharge current of 88 A and a charge current of 50 A. One consequence of this was that the failure rate dropped in both modules and the MTTF was increased to 300 cycles.

The conclusion drawn from these experiments was that these modules could be used to construct a vehicle battery. A battery large enough to power a light van could be constructed from two modules, each comprising two series chains of twenty cells connected in parallel. This battery was described in Section 11.4.2.

The seventy-six sodium sulfur cells were identical to the modified design used to replace the cells in the second twenty cell module. There were no connections between the series chains within the battery, and with the exception of one cell in each chain, no voltage leads on indiyidual cells. When the battery was standing on open circuit, the temperature was maintained around 330 °C by the resistance heaters.

Table 11.4 Characteristics of vehicle battery

Breakdown of weights	
Double module	
Aluminium box	4.6 kg
Insulation	8.2 kg
40 baffle boxes at 133 g	5.3 kg
2 inner Filamic boxes	2.0 kg
Heater	0.6 kg
2 Na/Na clocks at 900 g	1.8 kg
38 Na/S cells at 1060 g	40.3 kg
	63.0 kg
Outer box	
Outer box lined with Kaoboard	16 kg
Outer box lid lined with Kaobaord	11 kg
Air ducting (aluminium sheet)	3 kg
Fans	1.6 kg
Miscellaneous	0.4 kg
	32.0 kg

Table 11.4—contd.

Total weight of crate	
Two double modules	126 kg
One outer box	32 kg
	158 kg
Gravimetric energy density	
Cell capacity	88 Ah
Cell resistance	7 mΩ
Available energy at 5 h rate	165 Wh per cell
Using 90%	148.5 Wh per cell
For 76 cells	11.3 kWh
Gravimetric energy density	89.7 Wh kg^{-1} for module
Gravimetric energy density	71.5 Wh kg^{-1} for crate
Volumetric energy density	
Volume of module	
Length	1.04 m
Width	0.16 m
Height	0.60 m
Volume	0.09984 m^3
Volume of crate	
Length	1.26 m
Width	0.38 m
Height	0.67 m
Volume	0.3208 m^3
Volumetric energy density	56.7 kWh m^{-3} for module
Volumetric energy density	35.2 kWh m^{-3} for crate

Details of the weight and volume of the battery and its component parts are given in Table 11.4 together with gravimetric and volumetric specific energies of the cell, the module and the battery.

The battery temperature was controlled from five thermocouples connected in parallel to give an average battery temperature. An additional seventeen thermocouples were installed to monitor cell temperatures in various parts of the battery and the temperature of the outer surface of the modules.

The total voltage and one cell voltage were monitored for each of the four chains. The voltages of the sodium/sodium cell coulometers were also monitored. The battery was discharged at currents up to 98 A and charged at currents up to 48 A. After fifty-one cycles laboratory testing was terminated and the battery was installed in an electric vehicle.

During the fifty-one cycles completed in the laboratory test, three cells had to be replaced. One cell polarized prematurely on discharge after twenty cycles. When the cell was cooled and examined, it was found that the 1 mm hole in the steel sodium restrictor plate was partially blocked by a piece of debris. The number of cell failures during the first fifty cycles was lower than would have been predicted from the failure rate in the two modules tested previously. It is only possible to speculate on the reasons for this, but two differences do stand out. First, the number of parallel chains was doubled from two to four and when the two previous modules were connected in parallel the failure rate dropped. The second difference was the charge rate. Owing to limitations in the charging equipment available (50 A maximum) the charge current was equivalent to 12.5 A per cell, compared with 25 A per cell in the two string battery. Which, if any, of these two factors was responsible for reducing the failure rate cannot be determined, but one would expect that increasing the number of parallel chains would promote current dispersion as top of charge is approached and reduce the likelihood that individual cells will polarize.

After forty-five cycles the electrical connections of the two modules were changed so that one module could be connected in series with the other module during discharge to make an 80 V battery. Unfortunately, owing to a thermocouple sheath shorting to the battery case, an internal short circuit occurred which resulted in the failure of one cell; when this was being replaced it was noticed that another cell in the same module showed signs of seal leakage, and this was also replaced.

The average cell resistance in the battery was similar to that in the two single modules. Details of the performance of the battery in the electric vehicle will be given in Chapter 13.

11.6 SUMMARY

The design of a sodium sulfur battery is determined to a large extent by the application for which it is intended. For load levelling where large battery capacities are required, groups of as many as forty cells are connected in parallel and these groups are then connected in series. Each cell is fused and the voltage of the group can be closely monitored to ensure that it stays within safe limits. Variation in the resistances of individual cells will result in current dispersion, but this will be ameliorated on discharge by the change in open circuit voltage in the single phase region of cell operation. Thermal management of such batteries can be relatively simple, but cell reliability and longevity must be high. The cost of the cell must be low if such batteries are to be competitive with other methods of load levelling.

Vehicle batteries will be considerably smaller than load levelling batteries, and if conventional operating voltages are to be used the number of cells in parallel will be low. In these batteries groups of cells are usually connected in series and these series strings are then connected in parallel with other strings.

The length of the strings varies but is usually in the region of fifteen to twenty cells. In this configuration it is essential that cells do not develop a high resistance after failure, otherwise the battery could become incapacitated after failure of a few cells. This is a problem which at the time of writing still awaits a solution.

Thermal management is very important for vehicle batteries and more work is required in this area.

The status of battery development at the time of writing is described in Chapter 13, and it will be seen there how the different groups developing this battery have approached the problem of battery construction.

11.7 REFERENCES

Cleaver, B. and Davies, A. J. (1973) *Electrochim. Acta*, **11**, 733.
Fischer, W. (1980) 4th DoE/EPRI Sodium Sulfur Battery Workshop, Heidelberg.
Gibson, J. G. and Tilley, A. R. (1978) US Patent 4 194 448.
Gross, F. and Birnbreier, H. (1979) UK Patent Application 2 020 087.
Gupta, H. and Tischer, R. P. (1972) *J. Electrochem. Soc.*, **119**, 1033.
Hames, M. D. (1980) 4th DoE/EPRI Sodium Sulfur Battery Workshop, Derby.
Hartley, D. G. (1980) 4th DoE/EPRI Sodium Sulfur Battery Workshop, Derby.
Hartmann, B. (1983) 5th DoE/EPRI Sodium Sulfur Battery Workshop, Washington DC.
Hattori, S., Yamaura, M., Kimura, S. and Iwabuchi, S. (1977) Proc. SAE Congr. 1977 paper 770281.
Knödler, R. (1983) *J. Applied Electrochem.*, **14**, 39.
Minck, R. W. (1980) Private communication.
Minck, R. W. (1983) 5th DoE/EPRI Sodium Sulfur Battery Workshop, Washington DC.
Mitoff, S. P. (1978) *Power Sources*, vol. 7 (ed. J. Thompson), Academic Press, New York and London, p. 733.
Mitoff, S. P. (1980) EPRI Report EM-1341.
Piercy, R. P. and Sudworth, J. L. (1977) Symposium on Electrode Materials and Processes for Energy Conversion and Storage, *Proc. Electrochem. Soc.*, **77–6**, 950.
Pompon, J. P. (1980) 4th DoE/EPRI Sodium Sulfur Battery Workshop, Marcoussis.
Reiss, H. (1980) 4th DoE/EPRI Sodium Sulfur Battery Workshop, Heidelberg.
Reschke, R. (1983) 5th DoE/EPRI Sodium Sulfur Battery Workshop, Washington DC.
Sudworth, J. L. (1983) 5th DoE/EPRI Sodium Sulfur Battery Workshop, Washington DC.
Sudworth, J. L. and Hames, M. D. (1978) Fall Meeting of the Electrochemical Society, Pittsburgh, Extended Abstract 94.
Ziegenbein (1983) *High Temp. – High Pressures*, **15**.

12 Safety

A. R. TILLEY

12.1 INTRODUCTION

The use of sodium sulfur batteries for traction and load levelling applications has presented several new and demanding safety problems and great emphasis has had to be placed on the development of safe batteries.

The safety of sodium sulfur batteries has often been queried by developers of other battery systems and by the popular scientific press. These doubts about safety are easy to understand, as the consequences of sodium and sulfur burning in air and the explosive reaction of sodium with water are well known. First thoughts concerning the safety of, say, a 50 kWh sodium sulfur battery powered vehicle which would be carrying approximately 40 kg of liquid sodium and 60 kg of liquid sulfur at approximately 350 °C are bound to be somewhat sceptical. It is because these doubts have often been expressed about the safety of this battery that developers have endeavoured to demonstrate that sodium sulfur batteries can be safe even under the most extreme conditions.

The fact that the large amounts of sodium and sulfur in a battery are subdivided into many smaller sealed cell packages is the key to sodium sulfur battery safety, and the safety of the single cell is therefore paramount. The situation which is potentially the most hazardous is one in which direct reaction between sodium and sulfur occurs, and thus failure of seals and welds etc. is of minor importance compared with fracture of the beta alumina electrolyte.

The chemical reaction of sodium and sulfur liberates heat a rate dependent upon the kinetics of the reaction and/or the rate of mixing. The enthalpy of formation of Na_2S_3 is 432 kJ mol^{-1} (Mills, 1974), and if adiabatic conditions apply temperature rises of greater than 2000 °C are possible, with correspondingly high internal pressures being generated. To prevent cell rupture, it is essential to limit the rise in cell temperature after beta alumina failure. If, despite this, cell rupture occurs, the battery should be designed so that a single cell failure will not propagate throughout the battery.

The criteria required for safe operation of cells and batteries have to be defined. To some extent these criteria depend upon the battery application, but the following are considered to be essential by most people working in this field:

Internal criteria

(1) Cell rupture due to beta alumina failure, short circuit effects etc. must have a very low probability;
(2) if cell rupture does occur it should not propagate to adjacent cells;
(3) any secondary containment (module, containers etc.) should also withstand any possible cell failure.

External criteria

(1) For vehicle applications, battery containers and the vehicle structure around the battery should resist impact damage of defined limits. These limits should be based on vehicle accident statistics and be such that in practice the possibility of battery damage from collision accidents will be acceptably low;
(2) in the event of a fire external to the battery, it should not significantly increase the hazards of that fire;
(3) total or partial immersion of the battery in water should not result in a hazardous situation.

For batteries to meet these criteria, certain cell and battery design principles are essential. How these principles were evolved, tested and incorporated into practical cells and batteries will now be discussed.

12.2 GENERAL PRINCIPLES OF CELL SAFETY

The earliest reported safety work appeared in 1975, eight years after research on the sodium sulfur system had started (Hames and Tilley, 1975). Initial work in most laboratories was undertaken with small numbers of low capacity cells, and case rupture occurred only in a few instances on beta alumina failure. As the cell development work progressed, however, larger capacity and higher specific energy cells were built and tested in larger numbers, but it was only when battery development started that the full implications of cell safety were realized. The experiences at the British Rail laboratories in the early 1970s were probably typical.

At British Rail between 1972 and 1974 two small batteries were destroyed by cell failure propagation, even though in single cell electrical testing no cell breaching had occurred. Because of these battery failures, detailed cell safety work involving the testing of cells with deliberately fractured beta alumina was commenced and this work led to the establishment of the three safety principles essential for the production of reliably safe cells (Hames and Tilley, 1975). A safe cell was considered to be a cell which, under all but the most improbable conditions, did not allow its reactive contents to escape in the event of cell failure or an accident. To achieve a safe cell, Hames and Tilley attempted to limit the extent of the reaction which occurred when sodium had direct access to sulfur owing to beta alumina failure. The first safety principle is related to the amount of sodium and sulfur available for reaction:

The quantity of sodium and/or sulfur immediately available for reaction after beta alumina failure must be minimized.

The reactants that have immediate access to each other cause a very rapid temperature rise which is governed by the amount of reactants, the overall heat capacity of the system and the rate of heat dissipation. Of these factors, the thermal capacity and the rate of heat dissipation are of limited use in controlling the temperature of this initial reaction. If 1 kg of Na_2S_3 is produced in a failure reaction, then under adiabatic conditions approximately 3.8 kg of stainless steel or other material of similar heat capacity would be needed to limit the temperature rise to 200 °C (350–550 °C). Some heat dissipation to surrounding cells in a battery is possible, but high rates of heat generation and dissipation to other cells is undesirable owing to the possibility of causing cell failure propagation. To keep the initial temperature rise to a minimum in a cell of high specific energy, the amounts of sodium and sulfur that can immediately react have to be kept to the absolute minimum.

The second principles concerns the remaining bulk of the reactants:

The bulk of the cell reactants must be effectively separated and the flow of the bulk sodium and/or sulfur to the reaction sites must be restricted.

If the rate of the secondary reaction can be limited, then the heat generated can be dissipated over a reasonable time. The minimum flow rate at which sodium can be supplied to the beta alumina surface is set by the maximum discharge rate required under normal operating conditions. The battery insulation and heat balance control would normally be designed to dissipate the heat produced by all the cells in a battery at this power level. It is permissible for the flow rate of the reactants to exceed that required for electrochemical reaction at the maximum discharge rate, as this will still allow reasonable heat dissipation assuming that only a few cells fail simultaneously.

By incorporating these two principles, Hames and Tilley found that the initial failure temperatures could be dramatically reduced and that several cells could be failed safely. Externally measured temperatures were reduced from over 1000 °C to below 600 °C. Breaching of cells, however, still occurred occasionally and it was found that the breaching was due not to simple melting of the outer stainless steel case but to rapid corrosion of the stainless steel by sodium polysulfides at high temperature. A third safety principle was therefore required to ensure safe failure of cells:

The outer cell case must be protected against corrosion by sodium polysulfides at elevated temperatures.

To determine how essential each safety principle was, groups of cells containing only two of the required features were safety tested. It was found that the safety of cells which did not incorporate all the safety principles could not be guaranteed. This early work at British Rail utilized 50 Ah central sodium cells, and one of the advantages of the UK developers' decision to change to the central sulfur cell in 1975 was the easy incorporation of safety features into this new design (Dell *et al.*, 1976). The safety features are essential: thus the effect of any safety feature on cell performance, reliability and cost has to be minimized, and the ease of incorporation of the safety features into various cell designs is extremely important. Before discussing this the various safety test methods that have been used to study cell failures will be briefly outlined.

12.3 CELL SAFETY TEST METHODS

A safety test may be required for several reasons, such as to determine whether a particular cell design is safe for routine laboratory testing, to study the effect of various safety features, or to investigate whether cell safety is affected by any minor changes in cell design. To ensure that the cell safety is really being tested, and to reduce the number of experiments required, the safety test must simulate the most severe failure conditions.

Three safety test methods have been used by nearly all the developers, and these are:

(a) *The precrack method*

This was the first method to be used to test cell safety (Hames and Tilley, 1975). Cells were assembled with fractured beta alumina tubes and then heated until the failure reaction occurred.

(b) *The mechanical fracture method*

This test was first developed by UKAEA Harwell (Montgomery, 1975) for safety tests on central sulfur cells. A screw device was used to apply lateral force via two steel pins to the inside of the beta alumina tube. This replaced the central current collector in the cell. The device which could be remotely operated to fracture the beta alumina tube when the cell had reached operating temperature is shown in Fig. 12.1.

(c) *The electrical breakdown method*

This technique is the most convenient method of failing cells. This type of failure

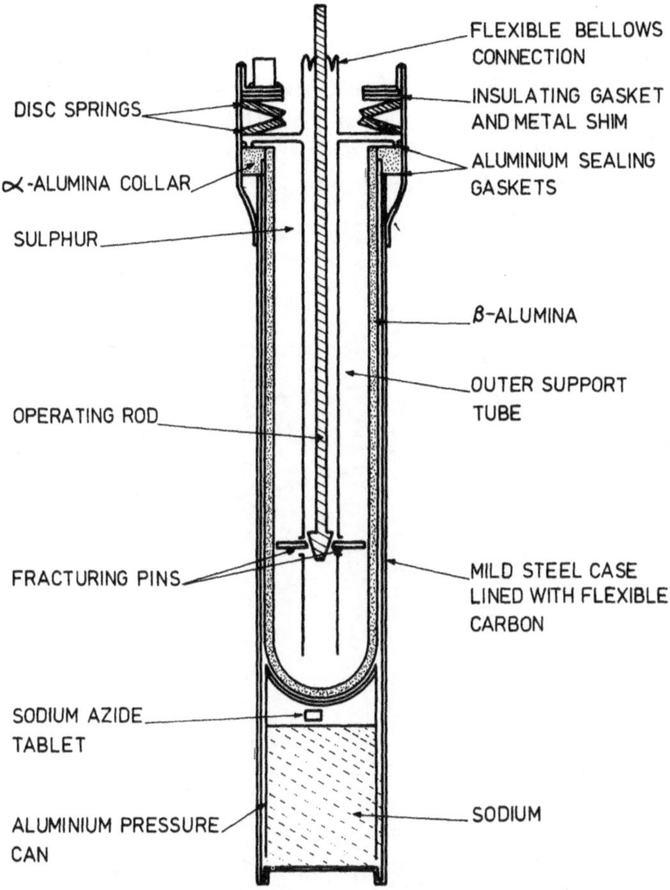

Fig. 12.1 A central sulfur cell incorporating a beta alumina fracturing device (Tilley, 1980). Reproduced with permission of the Electric Power Research Institute.

is performed by discharging the cell and then recharging at a high rate. At the top of charge the voltage is allowed to rise until failure occurs, and then the current is allowed to flow for a further period to ensure that the reaction is not inhibited by solid sodium polysulfides healing the cracks in the ceramic. This failure method is now widely used by all the developers. It has several advantages:

(1) No modifications to standard cells are required;
(2) no remote mechanical actuation is required (the method is ideally suited to multiple cell failure tests as required in battery safety experiments);
(3) cells can be electrically cycled prior to failure to ensure they are performing satisfactorily;

(4) this mode of cell failure is likely in an operational battery.

A comparison of the various test methods (Tilley, 1980) shows that they give approximately the same failure characteristics, as shown in Fig. 12.2. Although the electrical breakdown test is universally used for routine safety tests, the other two methods are still used when particular areas of a cell, such as vulnerable areas near seals and welds, require testing.

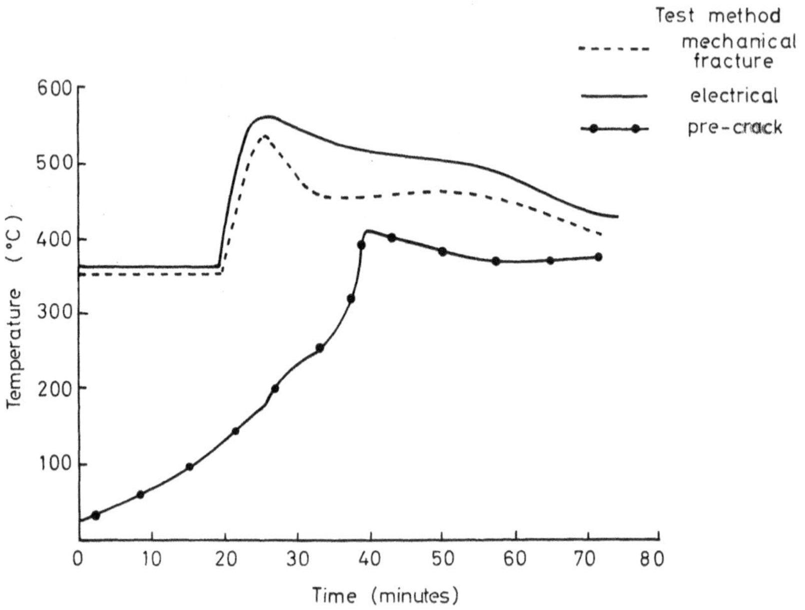

Fig. 12.2 A comparison of various safety test methods (Tilley, 1980). Reproduced with permission of the Electric Power Research Institute.

12.4 INCORPORATION OF SAFETY FEATURES INTO PRACTICAL CELL DESIGNS

The first safety principle concerns the minimization of the sodium or sulfur that is available for immediate reaction after beta alumina failure. The tubular central sulfur design can easily incorporate this principle as shown in Fig. 12.3. By reducing the annular gap between the outer cell case and the beta alumina tube to approximately 1 mm, only a fraction of the sodium required for full discharge of the cell is available to react immediately on beta alumina failure. The minimum size of this annular gap, however, is determined by the straightness and ovality of the beta alumina tube and outer cell case, and in higher capacity, longer cells the amount of sodium contained in this annulus can be quite significant. The space can be filled with inert material to reduce the sodium volume and restrict the sodium flow (Tilley *et al.*, 1978). The effect of the

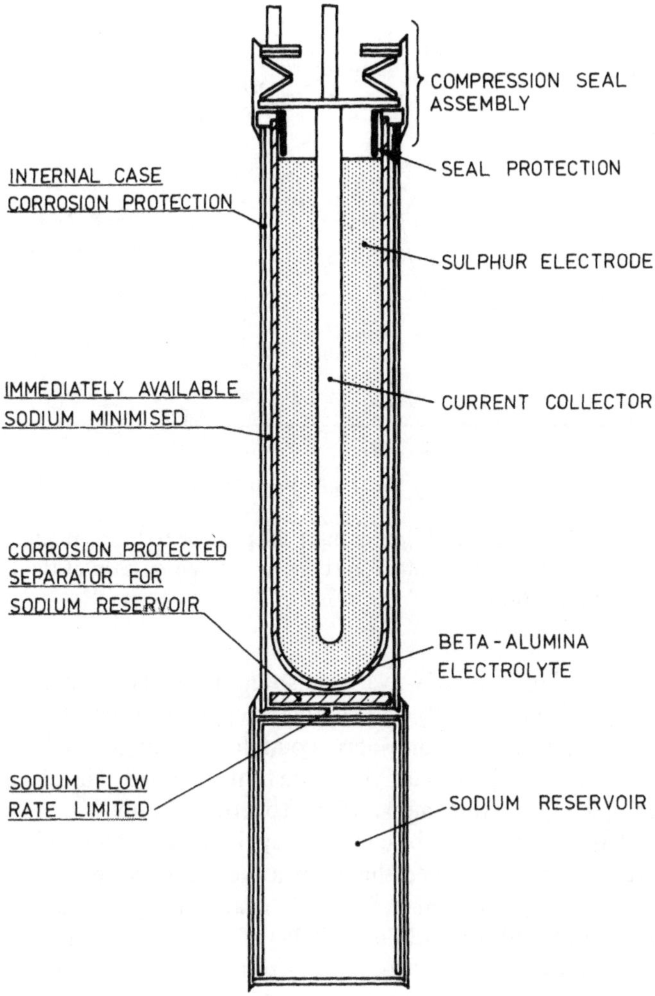

Fig. 12.3 Incorporation of the various safety features into a central sulfur cell design (Tilley, 1980). Reproduced with permission of the Electric Power Research Institute.

addition of annulus fillers on the failure characteristics is shown in Fig. 12.4. The maximum temperature on cell failure is reduced by approximately 100 °C and the rate of temperature rise is considerably reduced. Fillers which have been used include carbon powder, carbon spheres, silicon carbide powder, beta alumina powder and zircon sand (Tilley *et al.*, 1978; Stackpool, 1981). The fillers can provide a measure of sodium wicking which can be advantageous in some cell designs.

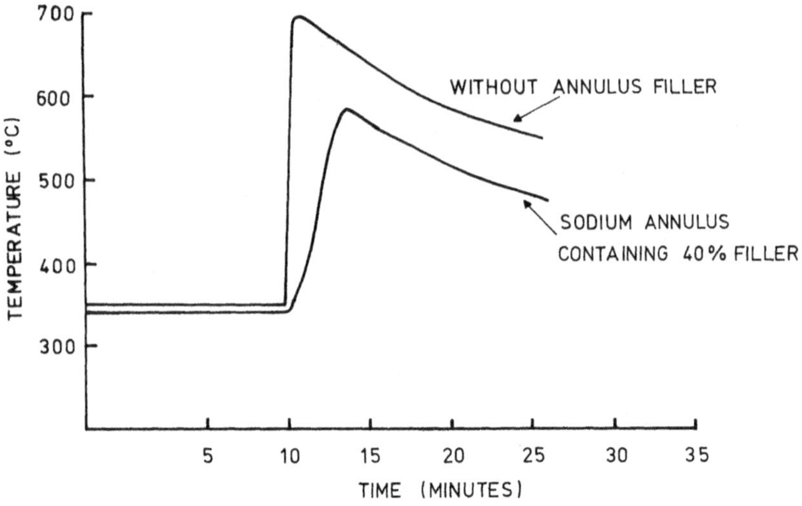

Fig. 12.4 The effect of the addition of inert filler to the sodium annulus on cell failure characteristics (Tilley, 1980). Reproduced with permission of the Electric Power Research Institute.

The amount of immediately available sodium can also be reduced by using capillary wicks (Lomax, 1979). The capillary region between a foil wick and the electrolyte tube provides the only access path for sodium to reach the surface of the electrolyte and so the sodium flow rate is also limited. The disadvantage of relying completely on thin metal wicks to provide only a small amount of sodium is that they can easily be corroded by the sodium polysulfides produced in a failure reaction, and when this occurs the cell may become unsafe if the sodium annular gap is not small. In an alternative design of central sulfur cell shown in Fig. 12.5 (Gibson and Tilley, 1978) the bulk of the sodium is contained between two concentric steel tubes. Corrosion protection of the inner tube is essential if the main bulk of sodium is to be prevented from readily reacting on beta alumina failure.

Incorporation of the first safety principle into central sodium cells is more difficult. A sodium displacement tube is used by most developers of this cell design (Dorrscheidt, 1980; Mitoff, 1980; McClanahan, 1980). A typical arrangement is shown in Fig. 12.6. As with the central sulfur cell design, the amount of sodium available for immediate reaction is determined by the beta alumina and safety tube tolerances. It is more difficult to further reduce the annular sodium volume by the addition of inert fillers; although such systems have been patented (Montgomery and Blake, 1978; Montgomery, 1978a) they are complex and far less practical than simple in-fill systems used in central sulfur cells. The effect of using different materials for tubular inserts and of varying the sodium annular gap has been investigated (Dorrscheidt, 1980). The

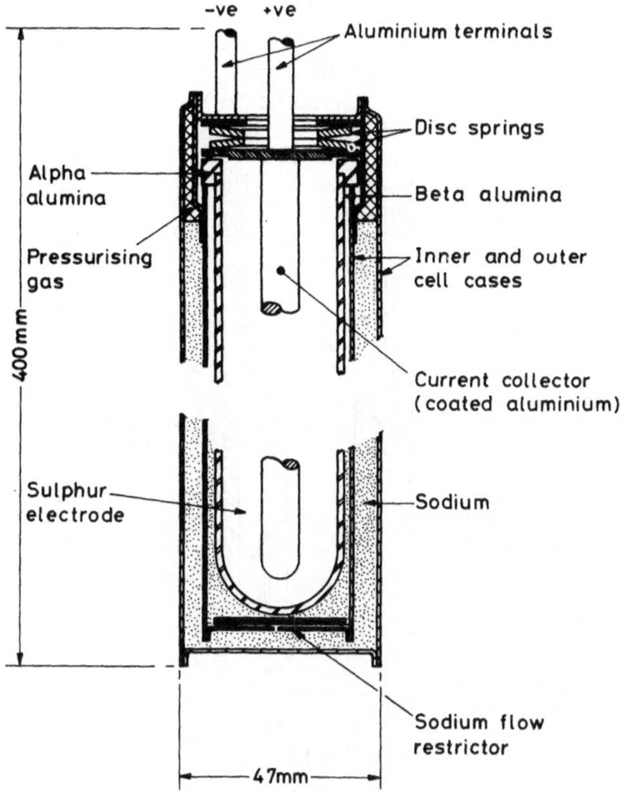

Fig. 12.5 A central sulfur tube cell design incorporating annular sodium storage (Gibson and Tilley, 1979). Reproduced with permission of British Rail Research.

effects are shown in Fig. 12.7. For tubular inserts to be effective safety features in central sodium cells, it is essential that they are protected from corrosion by sodium polysulfides that are formed during the failure reaction. Stainless steel safety tube inserts have been shown to be particularly vulnerable to corrosion or melting during mechanical crushing safety tests (Mitoff, 1980). To ensure a metal tubular insert is going to be effective, it needs a protective layer of inert material such as carbon or alumina. The second safety principle – separation of the bulk reactants and restriction of reactant flow – is also achieved by the tubular insert in a central sodium cell. Restriction of sodium flow is provided by using a small outlet hole or tube in the end of the insert. In the General Electric (GE) 168 Ah central sodium cell, the sodium insert tube has a 0.4 mm diameter exit hole as shown in Fig. 12.6. Again, it is essential that this flow limiting method is not compromised by corrosion by polysulfides at high temperatures. GE have seen severe corrosion of flow restrictors in central sodium cells (Mitoff, 1980).

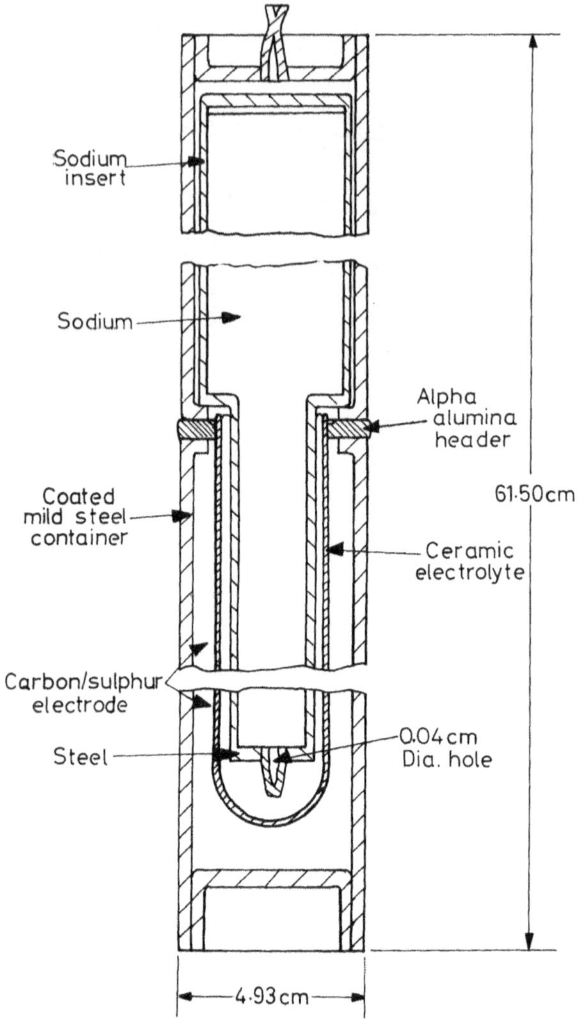

Fig. 12.6 A central sodium cell design incorporating a sodium displacement tube (Mitoff, 1980). Reproduced with permission of the Electric Power Research Institute.

As shown in Fig. 12.3, the central sulfur design lends itself to incorporation of the second safety principle: the bulk of the sodium is contained separately from the sulfur electrode compartment and a flow restrictor is easily interposed between the two reactant containment areas. In larger capacity cells, a more refined separator and flow restrictor is needed and a welded-in restrictor/separator plate has been found to be necessary. The British Rail BR30 cell arrangement is shown in Fig. 12.8. The rather complicated arrangement is essential if separation and restriction to sodium flow are to be maintained

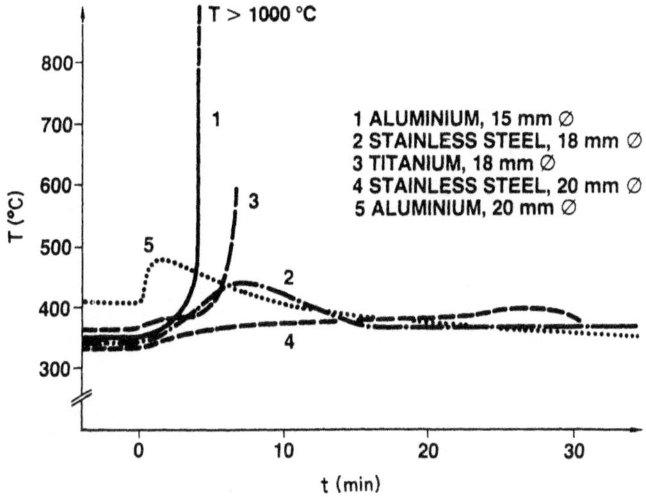

Fig. 12.7 The safety characteristics of central sodium cells (beta alumina tube diameter 25 mm) using various sodium displacement tubes (Dorrscheidt, 1980). Reproduced with permission of Brown Boveri and Cie AG.

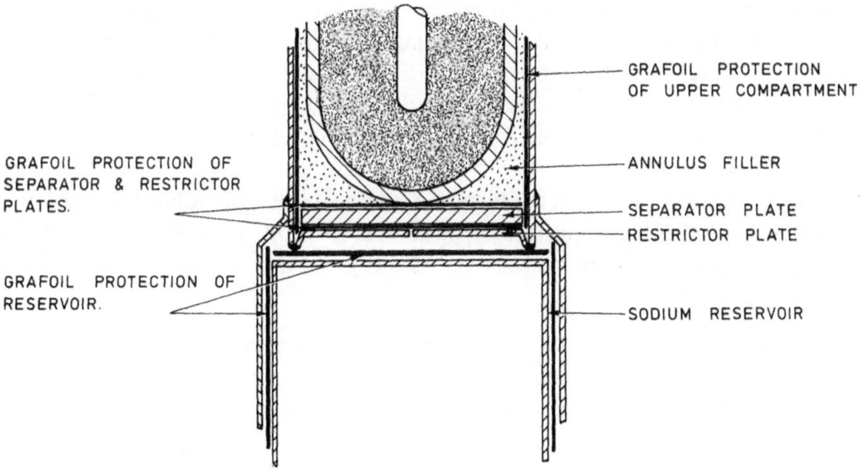

Fig. 12.8 Constructional details of a sodium reservoir separator and flow restrictor (Seeds *et al.*, 1980). Reproduced with permission of Academic Press Inc. (London) Ltd.

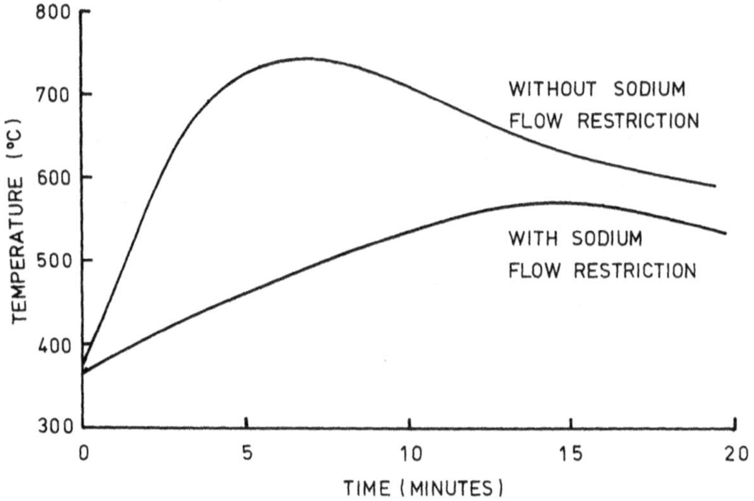

Fig. 12.9 The effect of sodium flow restriction on cell failure characteristics (Tilley, 1980). Reproduced with permission of the Electric Power Research Institute.

during cell failure. Post mortem radiographs of failed cells which used separators only spot welded into place showed that movement of the separator occurred on cell failure. This allowed rapid flow of sodium to the sulfur and resulted in several cells breaching. The effect of varying the sodium flow rate on cell failure characteristics is shown in Fig. 12.9.

It is not only sodium flow that has been shown to be a critical factor in cell safety. The use of low carbon felt densities for sulfur electrode construction, which allows very free movement of sulfur, has been shown to be a contributing factor to cell rupture on beta alumina failure (Sudworth *et al.*, 1977; Chatterji, 1976). The use of sulfur electrodes with felt-free channels is not, therefore, recommended.

The third safety principle, protection of the cell case against high temperature corrosion of sodium polysulfides, is the most important. The cell case protection requirements are different for the two cell designs. For the central sulfur cell the requirements are:

(1) Long term resistance to sodium/sodium oxide attack;
(2) thermal shock resistance;
(3) short term protection against sodium polysulfide corrosion at high temperatures (>450 °C).

The slightly different requirements for the central sodium cell design are:

(1) Long term protection against sodium polysulfide corrosion at cell operating temperature;

(2) if the outer cell case is the current collector the protective layer must be electrically conducting;
(3) thermal shock resistance;
(4) short term protection against sodium polysulfide corrosion at high temperatures (>450 °C).

Several materials which are fairly resistant to sodium polysulfides at cell operating temperatures, such as chromized mild steel, Inconels, nichrome, etc., are rapidly corroded at the temperatures which occur on cell failure. Joint work by Harwell and BR (Montgomery and Tilley, 1975) identified several promising coatings and materials. The technique used to screen the various materials and coatings is shown in Fig. 12.10. Molten sodium was allowed to mix with sulfur in a container which was fabricated from, or coated with, the material under test. Several materials passed this screening test. These were:

(1) Flame and plasma sprayed Al_2O_3 and TiO_2 on mild and stainless steel;
(2) glass coatings on mild and stainless steel;
(3) ceramic cements;
(4) flexible graphite paper linings;
(5) coatings of sodium silicate and carbon powder on stainless steel.

These preliminary tests also showed that carburized mild steel, molybdenum coatings, nichrome, chromized mild steel and stainless steel were totally ineffective.

The successful materials were tested by British Rail in cell safety tests. The

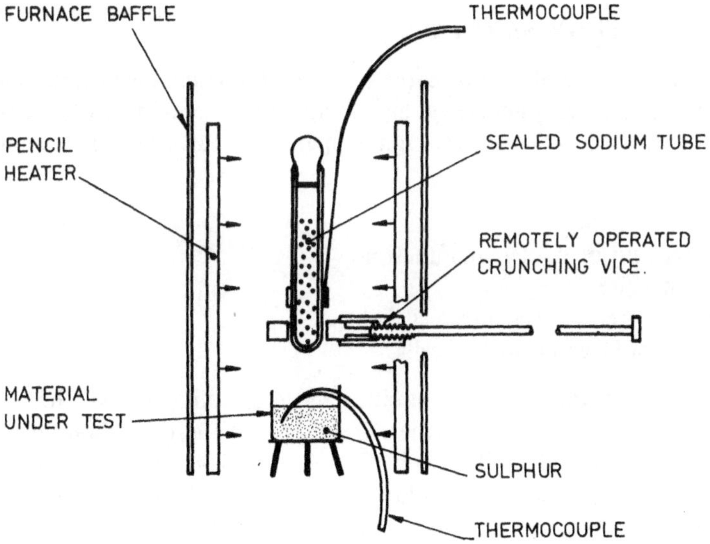

Fig. 12.10 Apparatus for mixing liquid sodium and liquid sulfur in open conditions (A. Seeds, 1980). Reproduced with permission of British Rail Research.

most successful corrosion protection was provided by the flexible graphite paper and several different brands of this material have been successfully used (Grafoil, Sigraflex, Flexicarb). Most of the other protective coatings were susceptible to thermal shock, attacked by sodium when used in central sulfur cells or extremely difficult to apply to the internal wall of the cell case. The flexible graphite paper is easily incorporated into cells and as it is flexible it offers some protection in cell impact tests where other coatings break and spall off. Flexible graphite paper has also been shown to be extremely effective as an isopressed layer on aluminium current collectors in central sulfur cells (Tilley and Wright, 1981). In central sodium cell designs graphite paper could be used to provide the dual function of a corrosion resistant current collector surface and safety corrosion protection.

Graphite paper is the only satisfactory material for safety protection in the cell, and its only disadvantage is its high cost. This is expected to be reduced, however, if the material is used on a large scale, particularly if lower purity materials are produced.

The incorporation of all the safety features into cell designs results in cells which are safe even in the worst case cell safety test conditions. At British Rail, for instance, several thousand central sulfur BR16 cells have been tested in safety tests and by electrical cycling, and these cells have given a breach rate of less than 0.1% of failed cells. Central sodium cell designs, especially cells of high specific energy in which the sodium is totally contained within the beta alumina tube, are not yet as safe as the central sulfur designs.

12.5 BATTERY SAFETY

When cells are assembled into large batteries, the safety hazards are increased. Cell safety is markedly affected at high temperatures, and in a battery conditions which could lead to excessively high temperatures can occur. These conditions are:

(1) Short circuits;
(2) heater and/or cooler malfunction;
(3) external fire;
(4) high cell resistances after cell failure;
(5) failure of large groups of cells.

Cell case breaching on beta alumina failure is unlikely if all the cell safety features are incorporated into the cell design, but there is a finite probability, albeit small, that cells will breach. Two failure propagation mechanisms are possible in a battery when cell failure is accompanied by rupture of the cell case (Hames *et al.*, 1978):

(1) Rupture of the cell cases of adjacent cells, due to corrosion by hot sodium polysulfides escaping from the failed cell;

(2) failure of adjacent cells due to thermal shock of the ceramic electrolyte caused by ejection of hot sodium polysulfides.

The two battery safety problems of excessively high temperatures and failure propagation will be dealt with separately.

12.5.1 High temperature safety

The effect of excessively high temperatures on cells can be disastrous. Fig. 12.11 shows a simplified plot of case bursting pressure and sulfur vapour pressure versus temperature. Many cells which use glass or aluminium diffusion bond seals could fail catastrophically at temperatures well below the crossover point of approximately 700 °C, as the seals weaken at lower temperatures. Some form

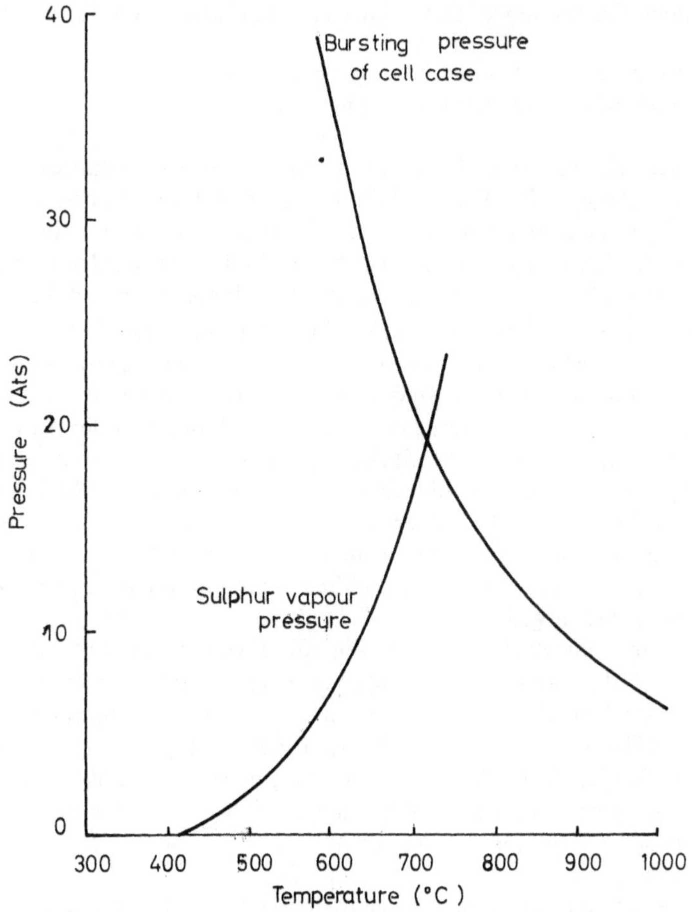

Fig. 12.11 Cell case bursting pressure and sulfur vapour pressure versus temperature.

of controlled pressure release can be incorporated into the cell design either by a relief valve or bursting disc or by seals leakage. Some compression sealed cells, for instance, will leak in a fairly controlled manner above 450 °C (Tilley, 1980), but this pressure release is quite slow and would not help in rapid cell heating. Electrical breakdown safety tests conducted at temperatures higher than 350 °C have also shown that the risk of cells rupturing after failure increases with increasing temperature. With the BR30 cell design, cell safety cannot be guaranteed above 420 °C (Seeds *et al.*, 1980). It is these high temperature safety problems that have probably been the cause of the battery failures experienced by some groups in the early stages of battery development. Although sulfur pressure release prevents violent rupture of the cell, the large amounts of released sulfur have to be dealt with.

A possible solution to the problem of high pressure at elevated temperatures is to reduce the amount of free sulfur. This can be achieved in two ways:

(1) By discharging cells into the single phase region;
(2) by combining sulfur with other chemicals.

Discharged cells can be safely heated to a much higher temperature (~800 °C) than fully charged cells (Tilley, 1980). It may, therefore, be possible to render cells safe by discharging them in a controlled manner, thus removing free sulfur from the cell. If this is to be successful, the cell discharge should not result in cell temperatures above 800 °C. The controlled discharge could be activated externally, and could possibly utilize high current resistors that would be required in vehicle battery control systems, such as regenerative braking 'dumping' resistors. This technique has the advantage that the battery is eventually rendered completely safe, as virtually all the free sodium and sulfur is eliminated. Any other safety devices involving thermal fuses and in-cell thermally activated isolation devices (Bones and Harbar, 1978; Meinhold *et al.*, 1979) would not be effective in an accident which gradually gets more severe, such as an external fire becoming uncontrollable. In this type of situation, eventual bursting of cells and release of free sulfur and sodium is inevitable if the cells are not discharged.

The second approach of combining sulfur with other materials is being developed. Sulfur and sodium polysulfides are readily sorbed into certain zeolites and small laboratory test cells have been tested using such electrodes (Coetzer, 1980; Coetzer and Thackeray, 1980). High performance cells using this type of sulfur electrode construction may never be available and certainly have not yet been demonstrated, but the possible safety advantages do make this type of construction appear advantageous on safety grounds for some applications (Coetzer and Thackeray, 1980).

The possible situations that can cause a high temperature excursion have to be considered separately and an overall cell and battery safety strategy then determined.

(a) *Short circuits*

These could occur as a result of impact damage, control malfunction etc. Cell tests at very high currents show that cells can withstand repeated cycling at currents far higher than that necessary in practice. Fig. 12.12 shows a high rate discharge of a 100 Ah central sulfur cell. The maximum current was approximately 400 A (1.75 A cm^{-2} electrolyte area). The most direct solution to the electrical short circuit problem is the inclusion within the cell of a current overload protection device (Bones and Harbar, 1978). This of course, would prevent the use of an external safety discharging system as described previously.

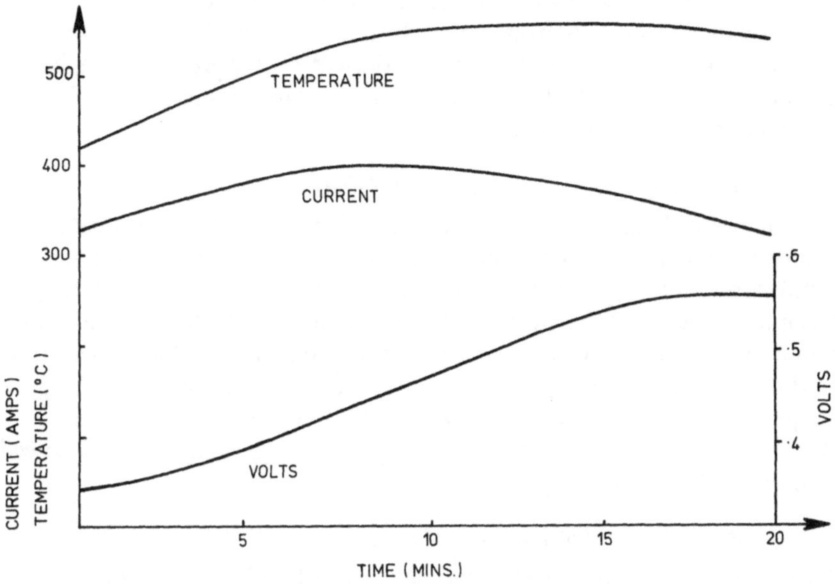

Fig. 12.12 Central sulfur cell (100 Ah) high rate discharge characteristics (Tilley, 1980). Reproduced with permission of the Electric Power Research Institute.

Most developers appreciate the need for external current fuses or trips to cater for gross overload conditions (Zalosh, 1980). Tests at British Rail in battery modules have shown that the 6 mm aluminium rod intercell connectors will only allow currents of 350 A for short periods before the aluminium melts (Seeds, 1980).

(b) *Heater and/or cooling malfunction*

To allow for heater controllers failing in use, the battery heaters should be rated so that no gross overheating can occur with the heaters permanently switched on and the cooling permanently off. Interlocks between cooling and electrical control systems will prevent discharge or charge of a battery when it is outside specified temperature limits. In an accident the normal battery cooling systems may be damaged and cannot be relied upon to keep a battery at a safe temperature. Automatic systems for battery cooling which operate at a predetermined temperature are possible, such as the thermal shorting of battery insulation by bimetallic strip systems, automatic ingress of air or release of hydrogen into evacuated insulation systems (Gross and Birnbreier, 1979).

(c) *External fire*

Because of the thermal insulation around a battery, an external fire would not immediately affect the cells. However, if a forced air cooling system continued to operate it could make matters worse. One possible solution to this situation would be to activate a controlled discharge system externally to the battery, or even use individual cell self-discharge devices. If the external fire became uncontrollable, it would be preferable that no free sodium or sulfur were available to increase the danger of the situation.

(d) *High cell resistance after cell failure*

In a long series chain of cells, it is desirable that a cell fails to a low resistance, otherwise that chain will contribute very little power to the battery. Such a cell could also result in a hazardous situation by heating up the surrounding cells through its own Joule heating effect (see Chapter 11). If the cell failure mode cannot be guaranteed to be low resistance, it would be desirable to incorporate in the cell a device which would short circuit the cell if a predetermined temperature were exceeded. This device could be activated by the initial cell failure or by subsequent heating on electrical cycling owing to the failed cell's own internal resistance.

(e) *Failure of large groups of cells*

This type of failure is covered largely in Section 12.5.2. To cater for large scale battery failure, compartmentalization of a battery is desirable. By dividing a battery into sections or modules, where each module has its own cooling and insulation, propagation of failure due to high temperature effects will be contained within modules.

The hazards associated with excessively high battery temperatures at first sight appear extremely daunting, and it is perhaps more realistic to compare them with the hazards of internal combustion engines (ICEs) or alternative battery power sources. The main objective is to ensure that a sodium sulfur battery is as safe or safer than other power sources. In an external fire, the sodium sulfur battery could be considered safer than other systems since it is designed for high temperature operation, unlike a lead acid system or an ICE fuel system. The sodium sulfur battery would be thermally insulated from a fire and if a controlled discharge were possible the battery could be rendered relatively safe. The short circuit hazard and malfunction of heating/cooling systems can be easily overcome by careful design and incorporation of failsafe devices. The problem of high cell resistance after failure may be solved by using certain cell design features which ensure that cells fail low resistance or by using auxiliary devices which react to cell failure and short the cell out.

The final problem of battery heating due to multiple cell failure can be easily overcome by using a modular construction to make up battery crates and by having effective means of preventing cell failure propagation within the modules. The problems of failure propagation will be discussed in the next section.

12.5.2 Cell failure propagation safety

To prevent reaction products from ruptured cells corroding through adjacent cell cases or failing adjacent cells by thermal shock, an external protective coating has to be applied to the cells or, alternatively, cells have to be separated by material impermeable to hot cell reaction products. External coatings have been shown to be effective in battery propagation tests (Hames *et al.*, 1978). Any external cell coating, however, must not interfere with thermal management of the battery, nor must it prevent failing cells from dissipating heat otherwise the resulting high internal temperatures could result in the cell rupturing. The thickness of the external coating will be a compromise between protection against thermal shock and generation of high internal cell temperatures. Several of the protective coatings developed jointly by British Rail and Harwell (Montgomery and Tilley, 1975; Tilley *et al.*, 1978) were evaluated in small battery safety tests using deliberately breached cells. With the BR16 cell design, it was found that a duplex layer of flame sprayed alpha alumina and stainless steel was satisfactory. A large battery safety test was conducted at British Rail to evaluate all the internal and external safety features (Hames *et al.*, 1978). The battery contained forty-eight BR16 cells close packed in an 8 × 6 array. Six cells in the battery (a group of four and two isolated cells) were assembled without safety features. The six unprotected cells were deliberately failed consecutively over a period of 30 minutes whilst the battery was at 350 °C. Five of these cells ruptured, as expected, with a maximum recorded failed cell temperature of 625 °C. The remaining forty-three cells did not breach and the external duplex cell coating was entirely effective. The maximum recorded temperature on a cell adjacent to one of the ruptured cells was only 532 °C, and although failure of some of these adjacent cells occurred, possibly due to thermal shock, it presented no safety problems as these cells contained the full internal safety features. Throughout the test, no toxic fumes were detected and the outer battery box temperature never rose above 55 °C. Other, cheaper protective coatings made from slurries of carbon or talc with sodium silicate were also shown to be effective on BR16 cells (Wright, 1979). British Rail workers, however, found that these protective coatings were totally ineffective with the larger capacity, higher energy density BR30 cells (Seeds *et al.*, 1980). A multiple layer coating was developed that was effective for BR30 cells, but this degraded the safety of the individual cells and limited the maximum temperature at which cells could be failed safely. BR solved the safety protection problem with the higher energy density BR30 cells by using graphite paper which had been shown by the earlier corrosion tests (Montgomery and Tilley, 1975) and by subsequent cell safety tests to be an extremely effective barrier to polysulfide corrosion. The graphite paper was not applied as a coating on cells, but sandwiched between mica sheets to form a composite material which was interposed between cells as baffles (Tilley, 1982). For the double row battery modules tested at British Rail (see Chapter 11) the composite sheets were formed into boxes. The boxes provided cell location and electrical insulation between cells, and became an integral part of the battery module construction.

Propagation tests have been carried out at General Electric (Mitoff, 1980). Special cells were used to generate a jet of reaction products which could be directed at dummy cells. Tests were conducted in an oven purged with nitrogen. At a cell spacing of approximately 16 mm, using foundry sand between cells, severe attack of the dummy cell was found. Graphite sheet separators were also shown to be ineffective, but it was thought that they would be more effective if backed with a solid material (as in the BR mica/graphite composite). A castable refractory material did prevent propagation, but this material would undoubtedly interfere with thermal management unless individual cell cooling was used, and also higher individual cell failure temperatures would be expected if cells were surrounded with this material.

Other battery safety methods published in patents describe the use of porous inert filler materials between cells. Mixtures of alumina, magnesia, silicon carbide or packed pulverized fuel ash (Montgomery, 1978b) and carbonaceous foams (Clegg and Lauder, 1978) have been suggested. The work at British Rail has shown that scaling up from 40 Ah to 100 Ah cells requires much more effective safety barriers between cells. Some developers are proposing extremely large capacity single and multitubular cells with capacities over 400 Ah (Mitoff, 1980; Williams, 1980). Not only has the single cell safety of large capacity cells to be demonstrated, but from the present results on cell failure propagation in batteries it is quite probable that any advantage in cost and energy density of such large capacity cells will be offset by the need for extremely effective battery safety features.

12.6 COST OF CELL AND BATTERY SAFETY FEATURES

It is not possible to give costs of the various safety features for all the different cell designs in use. One feature is, however, common to all cell designs: this is the close tolerance between the beta alumina tube and the outer cell case (or safety tube insert for a central sodium cell design). The amount of sodium contained in this annular gap markedly affects the safety of a cell, and it is essential to reduce this sodium volume to a minimum. In some cell designs it is possible to use inert fillers to reduce this sodium volume (Seeds *et al.*, 1980; Tilley, 1980). The straightness tolerance requirements on the beta alumina tube and the steel tubing can increase cell costs considerably; as the cell length is increased the reject rate of the various assemblies increases.

A typical cost of safety features for the British Rail BR30 design has been given (Tilley, 1980). In this costing, which was for laboratory test quantities (~1000), the highest cost item was the flexible graphite sheet corrosion protection. This material is now much cheaper, and as it becomes more widely used in large quantities for other sealing applications (especially automotive applications) a further reduction in price is expected. At the latest bulk price (250 m^2) for 0.2 mm thick sheet, it represents approximately 6% of the cell cost for a British Rail BR30 100 Ah cell.

Flow restrictors and reactant separators can be made from mild steel sheet and protected by a layer of flexible graphite (Seeds *et al.*, 1980), and these features do not add significantly to the cell components or assembly costs.

Cell and battery safety features are essential, and safety cannot be compromised because of the cost of the required safety features. The approach by some of the developers of making these necessary safety features perform other functions appears to be the most satisfactory approach. Typical examples of this are the safety tube insert that is used as a simple sodium filling method (Mitoff, 1980), the battery safety cell propagation protection being used as an integral part of battery module construction (Tilley, 1982) and the corrosion resistant current collector developed for the central sodium cell that is also an adequate cell case protection on cell failure (Hartmann, 1980).

12.7 SAFETY OF THE DOW CELL

The Dow cell design, which has been described in Chapter 4, requires safety features similar to those of sodium sulfur cells which use beta alumina electrolyte. Owing to the construction of the cell, which uses many thousands of small hollow glass fibres approximately 70 μm OD, the amount of sodium and sulfur immediately available for reaction on fibre fracture is very small. Bulk mixing of the reactants can occur on failure of the tube sheet assembly, and to prevent any rapid reactant mixing if tube sheet fracture occurs, sodium flow restriction or retainment has to be incorporated in the sodium reservoir (Levine, 1978, 1980). The coated aluminium foil current collector is an additional hazard with this cell design, because if the temperature of the cell exceeds 660 °C the aluminium melts and a violent aluminium/sulfur reaction can take place which increases the temperature and pressure within the cell considerably (Levine and Anand, 1977).

The present Dow cells incorporate all the main safety features of other sodium sulfur cell designs, i.e. minimization of the immediately available sodium and sulfur on fibre fracture, separation of the main bulk of reactants and sodium flow restriction. Problems can occur at high temperatures owing to the aluminium foil/sulfur reaction above 660 °C and the low softening point of the tube sheet glass. The workers at Dow have recognized these problems and have considered using controlled energy release devices sensitive to high pressures or high temperatures, and their 40 Ah cells have incorporated pressure sensitive relief devices (Levine, 1978).

The safety test conditions outlined for Dow cells are basically the same as specified for sodium sulfur cells which use beta alumina (Zalosh, 1980), and although no battery safety testing data are available for Dow cells, the battery safety principles outlined in Section 12.5 apply.

12.8 SUMMARY

At first sight, the idea of a battery containing large quantities of sodium and sulfur at high temperatures is alarming. However, by dividing the reactants into smaller packages, i.e. cells, and by containing these cells in battery modules and in an insulated outer box, construction of a safe battery becomes possible. The key to overall battery safety is to develop a safe cell and to ensure that cell failure propagation cannot occur. This has largely been achieved, and satisfactory solutions to the cell and battery safety problems have been demonstrated. The penalties of incorporating safety features into cells and batteries are not particularly severe, and by testing the safety of cells and batteries under 'worst case' conditions, the developers are ensuring that sodium sulfur installations will be extremely safe.

12.9 REFERENCES

Bones, R. J. and Harbar, J. R. (1978) UK Patent 1 529 854.
Chatterji, D. (1976) EPRI Report EM-266, Project 128-3.
Clegg, G. A. and Lauder, D. R. (1978) UK Patent 1 534 408.
Coetzer, J. (1980) UK Patent 2 029 999.
Coetzer, J. and Thackeray, M. M. (1980) UK Patent 2 030 351.
Dell, R. M., Sudworth, J. L. and Jones, I. W. (1976) 11th IECEC Conference, Nevada, paper 769085.
Dorrscheidt, W. (1980) 4th DoE/EPRI Sodium Sulfur Battery Workshop, Mannheim.
Gibson, J. G. and Tilley, A. R. (1978) US Patent 4 104 448.
Gross, F. and Birnbreier, H. (1979) UK Patent Application GB 2 020 087.
Hames, M. D., Hartley, D. C. and Hudson, N. M. (1978) Proceedings of the 11th International Power Sources Symposium, Brighton, *Power Sources*, vol. 7 (ed. J. Thompson), Academic Press.
Hames, M. D. and Tilley, A. R. (1975) 24th ISE meeting, Marcoussis, Paris.
Hartmann, B. (1980) 4th DoE/EPRI Sodium Sulfur Battery Workshop, Mannheim.
Levine, C. A. (1978) Dow/DoE Sodium Sulfur Review, Walnut Greek, California, USA.
Levine, C. A. (1980) 4th DoE/EPRI Sodium Sulfur Battery Workshop, Mannheim.
Levine, C. A. and Anand, J. (1977) 152nd Electrochemical Society Meeting, Atlanta, vol. 77-4, p. 292.
Lomax, G. R. (1979) UK Patent 1 548 263.

McClanahan, M. L. (1980) Sodium/Sulfur Battery Review, Ford presentation to DoE, Newport Beach, Ca.
Meinhold, H., Haar, W. and Fischer, W. (1979) UK Patent Application GB 2 020088.
Mills, K. C. (1974) *Thermodynamic Data for Inorganic Sulphides, Selenides and Tellurides*, Butterworth, London.
Mitoff, S. P. (1980) EPRI Report EM-1341, Project 128-4, 5.
Montgomery, A. G. (1975) Unpublished work.
Montgomery, A. G. (1978a) UK Patent 1 530 274.
Montgomery, A. G. (1978b) UK Patent 1 526 588.
Montgomery, A. G. and Blake, H. V. (1978) UK Patent 1 530 273.
Montgomery, A. G. and Tilley, A. R. (1975) Unpublished work.
Seeds, A. D. (1980) 4th DoE/EPRI Sodium Sulfur Battery Workshop, Derby.
Seeds, A. D., Wright, M. L. and Hames, M. D. (1980) Proceedings of 12th International Power Sources Symposium, Brighton; *Power Sources*, vol. 8 (ed. J. Thompson), Academic Press.
Stackpool, F. M. (1981) UK Patent Application GB 2 089 559.
Sudworth, J. L., Tilley, A. R. and Bird, J. M. (1977) 152nd Meeting of the Electrochemical Society, Atlanta.
Tilley, A. R. (1980) 4th DoE/EPRI Sodium Sulfur Battery Workshop, Mannheim.
Tilley, A. R. (1982) UK Patent Application GB 2 086 645.
Tilley, A. R., Hames, M. D., Sudworth, J. L. and Bird, J. M. (1978) UK Patent 1 511 152.
Tilley, A. R. and Wright, M. L. (1981) UK Patent Application GB 2 068 632.
Williams, W. J. (1980) Sodium/Sulfur Battery Review, Ford presentation to DoE, Newport Beach, Ca., USA.
Wright, M. L. (1979) UK Patent Application GB 2 046 502.
Zalosh, R. (1980) 4th DoE/EPRI Sodium Sulfur Battery Workshop, Mannheim.

13 Design and performance of prototype batteries

J. L. SUDWORTH

13.1 INTRODUCTION

The characteristics of the sodium sulfur battery preclude its use in applications where small batteries are required for intermittent discharge. This is because the energy required to maintain the operating temperature would exceed the electrical energy delivered during discharge. There are, however, several applications where the high specific energy and the potentially low cost of this battery make it a strong contender – electric vehicles, load levelling, solar energy storage and space vehicle power supply. Several prototype batteries have been constructed and tested to demonstrate the feasibility of using the sodium sulfur battery for certain applications, and these will be described briefly in this chapter.

13.2 LOAD LEVELLING BATTERIES

Load levelling batteries are of interest to electric utilities for reducing the cost of electricity generation. The size of these batteries is in the range 5–100 MWh and thus, to keep the number of cells to a manageable number, the cell capacity should be as high as possible. Long life and low cost are the main requirements for these cells; specific energy is less important.

Minck (1983) has reported the construction and operation of a 100kWh load levelling battery. This represented a full current, low voltage module of a full size battery. Each parallel chain contained sixty-four cells and two of these chains were connected in series to form a 25 kWh submodule (Fig. 13.1). Four of these, in separate thermally controlled compartments, formed the complete battery. The voltages, currents and temperatures of the 512 cells were individually monitored.

Fig. 13.1 Schematic diagram of Ford car battery.

At the beginning of its life in February 1980 the battery yielded 102 kWh (7500 Ah at 13.6 V) at the five hour discharge rate but a number of problems arose during subsequent operation. The most serious problem encountered concerned the behaviour of the cells after failure. Each cell was fitted with a fuse which was designed to blow when the cell failed thus preventing the rest of the cells in the parallel array from discharging through the failed cell. Unfortunately, some of the cells failed with such a high resistance that the fuse current was not exceeded and the continued leakage through the faulty cell forced its submodule out of charge balance with the other three submodules connected in series with it. This limited the capacity available from the battery and several methods of charge control were investigated in an effort to overcome the problem.

During the first month of its life the battery operated well but a number of equipment malfunctions occurred and additional cell failures were induced in one of the modules, leading eventually to its disconnection from the battery.

The remaining 384 cell battery had, at the time of writing, attained 675 cycles in thirty-two months of operation, with failure of about 20% of the cells owing to electrolyte failure.

13.3 ELECTRIC VEHICLE BATTERIES

The first sodium sulfur battery to power an electric vehicle was constructed by

Jones and his co-workers at the Electricity Council Research Centre (ECRC), Capenhurst, England, in October 1972 (Churchman, 1975). The cells, which were the central sodium type and had a capacity of 30 Ah and a specific energy of 160 Wh kg^{-1}, suffered from a number of defects. Thus the stainless steel case was corroded by the polysulfide melt and as a result the cells lost capacity rather rapidly; the seal was of the compression type using flexible graphite to seal the cell case to the alpha alumina collar. As sodium intercalates with graphite, this seal had a limited lifetime. Despite these design defects the initial performance of the cells was quite adequate and 960 were connected together to form a battery. Each cell was enclosed in a sleeve of woven glass fibre and twenty-four cells were packed closely together in an aluminium crate. Aluminium straps were spot welded to the stainless steel case via nickel strip and adjacent terminals were TIG welded together to form the intercell connectors. Forty of these crates were connected in series and parallel to give a 100 V, 50 kWh battery, and enclosed in a double walled container with 10 mm of Microtherm® insulation between the two walls. The specific energy of the battery was 60 Wh kg^{-1}. The air within the battery box was circulated by means of a fan to reduce temperature gradients within the battery and, if necessary, cooling air could be introduced. When standing on open circuit the battery temperature dropped less than 5 °C per hour. The battery was installed in an 18 cwt (900 kg) Bedford van (Fig. 13.2) and the performance of the battery is shown in Table 13.1. This vehicle, which had a range of between 60 and 150 miles, was driven on the public roads without incident and demonstrated unequivocally that the sodium sulfur battery was capable of powering a road vehicle. The life of the battery, however, was only 20 cycles because of the limitations of the cell design.

A second battery was constructed in April 1973. In this battery the same cells were packed more closely to give thirty-six cells in a module. Only twenty-four of these modules were needed, giving a 25% reduction in weight and a 20% reduction in volume. This second battery, which had a specific energy of

Table 13.1 Performance of traction battery NaS1

Open circuit voltage	100 volts		
Capacity	50 kWh		
	V	*A*	*kW*
Average performance	84	186	15.5
During maximum acceleration of vehicle	72	330	24
Peak power	50	581	29
External volume including thermal insulation	1.52 m^3		
Weight of cells	317 kg		
Weight of containment auxiliaries	483 kg		
Total weight	800 kg		
Capacity/unit weight	63 Wh kg^{-1}		
Capacity/unit volume	33 kWh m^{-3}		

Fig. 13.2 ECRC van.

83 Wh kg^{-1}, 42 kWh m^{-3}, was installed in the same vehicle and again operated for 20 cycles.

In October 1977 the Yuasa Battery Company issued a press release in which they described a demonstration of a sodium sulfur battery powered light van. The cells used in this battery were again of the central sodium type with a capacity of 27 Ah. The seal design was considerably more advanced than that used in the Electricity Council cells: a glass/metal seal was made to the alpha alumina collar providing a completely hermetic seal. Unlike the integral design of the ECRC battery the Yuasa battery was a modular design. Thus twenty-eight cells were connected together in a thermally insulated container with provision for heating and cooling. This 1.3 kWh module had a specific energy of 91 Wh kg^{-1}. A 20 kWh battery was constructed and installed in a vehicle (Fig. 13.3) for road tests, but no information is available on range. The use of a metal/glass seal resulted in a much longer life than the ECRC battery and a life of 220 cycles was achieved with one of the twenty-eight cell modules.

Workers at the Shanghai Institute of Ceramics constructed a 6 kW battery

Fig. 13.3 Yuasa van.

which was installed in a road vehicle but no details are available of the performance of this vehicle. The cells used in the construction of this battery were unusual in that they were multitubular cells with six beta alumina tubes in each cell. In these cells, which had a capacity of 120–130 Ah, the sodium was contained inside the beta alumina tubes.

In July 1980, the 11 kWh battery built at the British Rail laboratories in Derby and described in Chapter 11 was transferred to an electric vehicle. This was a Smith's NCB van which had previously been part of the electric vehicle fleet at the Atomic Energy Research Establishment, Harwell (Fig. 13.4). The vehicle was originally equipped with two 208 Ah, 36 V lead acid batteries with a total weight of 750 kg. The DC traction motor was rated at 9.1 kW at the one hour rate and the controller was the contactor type with parallel/series changeover. The unladen weight without batteries was 990 kg and the payload was 900 kg. The range with the original lead acid batteries was quoted as 40–50 miles (64–80km) unladen and the maximum speed was 25 mph (40 km h^{-1}).

The sodium sulfur battery was fixed to the deck of the vehicle and the two modules were connected to the contactor controller, thus retaining the parallel/series changeover. All the thermocouples in the battery were connected in parallel to give an average battery temperature which was displayed in the cab. Two voltmeters and an ammeter were installed in the cab to allow the

Fig. 13.4 British Rail van and car.

driver to monitor the voltage of the two modules, and the total current. Power for the auxiliaries (lights, horn, contactor coils and battery fans) was provided by a 24 V lead acid battery.

Although the only suitable site for testing the vehicle was a car park, it was possible to maintain the maximum speed of the vehicle on a figure-of-eight circuit. The maximum power drawn by the vehicle was in excess of 8 kW and the battery temperature rose during continuous operation. Normally the temperature at the start of a run was 330 °C, and if the fans were switched on when the temperature reached 340 °C the battery temperature did not exceed 360 °C during the whole of the run. Because the vehicle suspension system was designed for a full payload plus 750 kg of battery, the ride was very hard and bumpy and to overcome this the van was loaded with between 400 kg and 600 kg of sand. The range of the vehicle in these tests is given in Table 13.2.

As the vehicle could only be run in the evenings when the car park was empty, the battery was charged during the day in the parallel configuration at a constant current of 50 A. At this charge rate the battery actually absorbed heat for part of the charge cycle and the battery heater power was increased to 740 W from the 630 W necessary to maintain the temperature at 330 °C when the battery was standing on open circuit. For simplicity, the battery heaters were wired through an autotransformer. This system was very reliable and there was little variation in the battery temperature even when the vehicle was parked outside in subzero temperatures. The vehicle was still being operated as late as November 1980 and to maintain the cab temperature at a tolerable level the cooling air from the battery was ducted into the cab.

Table 13.2 Performance of battery vehicle Beta 1

Total battery miles	1172
Total cell failures	19
No. of cycles in vehicle	59
Cell failure rate (in vehicle)	0.22% per day
Miles per failure	61.7
Summary of cell failures	
Broken friction weld	2
Seals leaks	10
Other causes	7
	19
(Broken weld between cells	1)

Load	*% Battery capacity used*	*Miles travelled*	*Range using 85% battery capacity*	*Average range*
0	60	25	35.4	
0	43	19	37.6	
				ca 36
400 kg	55	21	32.5	
400 kg	54	20	31.5	
400 kg	45	17	32.1	
				ca 32
600 kg	70	23	27.9	
600 kg	57.5	20	29.6	
600 kg	71.5	24	28.5	
600 kg	65	24	31.4	
600 kg	67	25	31.8	
				ca 30

After 61 cycles in the van the battery was transferred to an electric car, a PGE3. This had a thyristor chopper controller and a much better suspension system. The car performed satisfactorily although it was not possible to attain its maximum speed of 40 mph (64 km h^{-1}) in the restricted area available. Unfortunately, the thyristor controller developed a fault so the battery was transferred back to the van.

The total mileage exceeded 1000 miles (1600 km) and the total number of cycles including laboratory testing reached 100 before testing was stopped.

Fig. 13.5 BBC van.

In 1983 Reschke of Brown Boveri reported on the performance of a sodium sulfur battery which was used to power a VW van with a total weight of 2400 kg (Fig. 13.5). The cells used to construct the battery were 40 Ah central sodium cells in which all the sodium was contained within the electrolyte tube. The battery consisted of two modules each consisting of 224 cells in a vacuum insulated container. The total heat loss from each module was 240 W and the time taken to cool from 350 to 300 °C was determined to be seven hours. During discharge the battery was cooled by forcing air along the vertically oriented cells. This resulted in a vertical temperature difference of only 15 K along the 22 cm long cell and a horizontal temperature difference of only 10 K.

The power of each module at two thirds open circuit voltage was 12 kW giving a total of 24 kW for the battery. The battery voltage was 116 V on open circuit and the total capacity approximately 280 Ah. The range of the vehicle was 115 km at 45 km h^{-1} and the average energy consumption at this speed was 185 Wh km^{-1} which means that the battery energy was 21 kWh. The time required to accelerate to 40 km h^{-1} was 17 s and the top speed of the vehicle was 62 km h^{-1}. No data are available for battery reliability but it is known that the capacity loss of cells used in this battery is 5% per 100 cycles although no increase in cell resistance is observed.

The vehicle batteries described so far, with the exception of the British Rail battery which for a short period was installed in an electric car, have been used to power vans. The reason for this is quite simply that, although high gravimetric specific energies have been achieved, the volumetric specific energy of all these batteries was rather low (usually around 40 kWh m^{-3}). In a car, space is at a premium and much higher volumetric specific energies are required. Minck (1983) has described the first sodium sulfur battery specifically designed to power an electric car. The design goals of this battery are shown in Table 13.3 and from this it can be deduced that the specific energy required is 127 kWh m^{-3}. This stringent requirement arose from the need to fit the battery within an existing experimental vehicle and it was achieved by packing the 40 Ah central sodium cells tightly together in the horizontal configuration. Fig. 13.6 shows one of the eight modules which went to make up the battery. The sixty cells in each module were connected up as four fifteen-cell strings in parallel and the eight modules were connected in series to give a battery voltage of 250 V. With such a high battery voltage the possibility of internal short circuits was increased and, therefore, each cell was wrapped in a woven glass fibre sleeve.

To save space, the busbar system and the air cooling ducts were combined as shown in Fig. 13.1; the problem of temperature gradients in the direction of air flow was tackled by dividing the air duct into two compartments so that the hot air exited adjacent to the cold air input. The thermal conductivity of the aluminium busbars flattened the lateral temperature gradient and the cells lost

Table 13.3 Design goals electric vehicle engineering Model-1

Nominal voltage	240 VDC
Rated capacity	166 Ah at C/3
Delivered energy	36 kWh at C/3
Electrical efficiency	89% at C/3 discharge, C/6 charge
Peak power (30 sec. pulse)	54 kW at 0–50% DOD 44 kW at 80% DOD
Continuous power	16 kW to 80% DOD
Weight	884 lb (402 kg)
Volume	10 cu ft (283 l)
Thermal loss	300 W at 350 °C
Life	200 cycles; 1 year

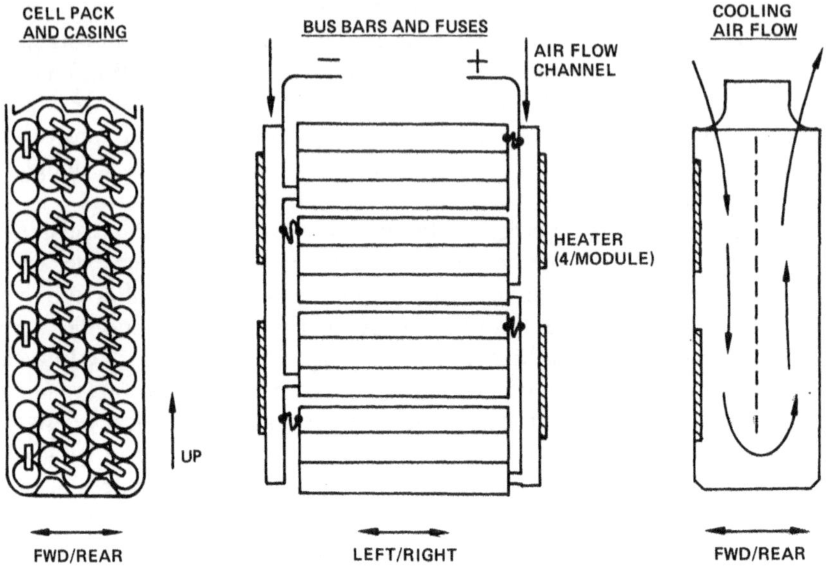

Fig. 13.6 Ford load levelling battery module.

heat to the cool surface by radiation from their ends. The electrical heaters required to heat up the battery were also attached to the busbars.

The battery was tested on the bench in a thermal enclosure constructed of mineral fibre insulation. When the battery was heated up to its operating temperature, fifteen cell failures occurred which was somewhat unexpected based upon experience with single cell freeze/thaw tests. Four of the failures occurred in one string and as these cells failed to a relatively high resistance the module capacity and, therefore, the battery capacity was limited to three quarters of the rated capacity. The battery power was much less affected and it was possible to discharge the battery at close to the predicted power levels (Fig. 13.7). Maximum power (half open circuit voltage) was extrapolated, from 200 A pulses lasting 40 s, to be 60 kW.

After bench testing the battery was cooled to room temperature, the failed cells were replaced and the eight modules were packed tightly into an evacuated thermal enclosure which had a 2.5 cm thick double wall containing an advanced glass insulation board to support the thin steel walls under atmospheric pressure (Fig. 13.8). Heat loss through the walls of this evacuated vessel was about 175 W but additional losses through the cover of conventional insulation and the feedthrough increased this to a total of 300 W.

The battery was again heated to its operating temperature preparatory to bench testing but unfortunately an internal short circuit occurred before testing could begin. The battery was again cooled to room temperature and dismantled

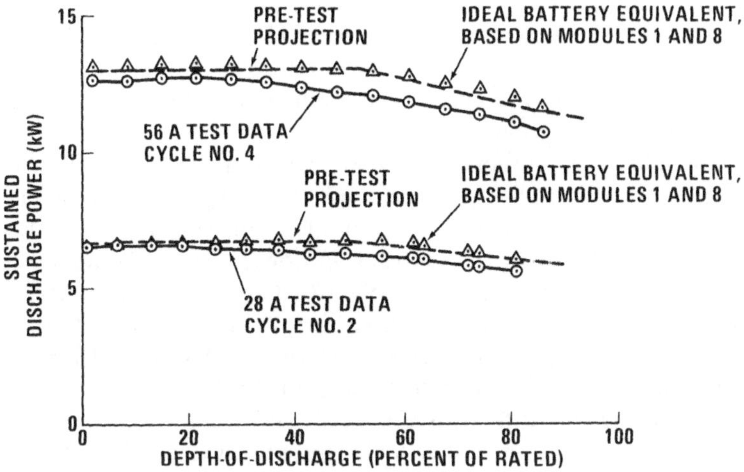

Fig. 13.7 Performance of Ford car battery on bench test.

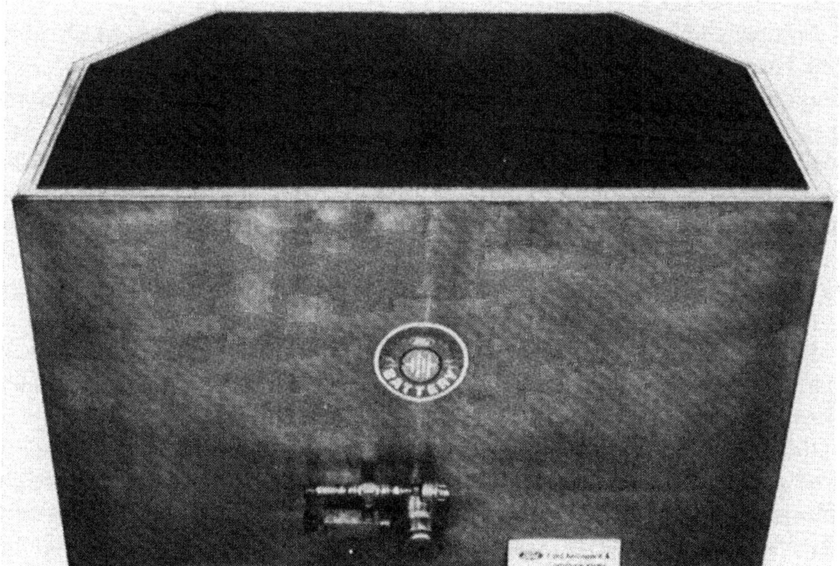

Fig. 13.8 Ford car battery case.

when it was found that a thermocouple wire had caused two modules to short out and overdischarge. The battery had over 400 instrument leads and this incident illustrates the point made in Chapter 11 concerning the advisability of keeping instrumentation of the battery to a minimum for maximum reliability. At the time of writing the battery has been reassembled with the six good modules and testing is about to be resumed.

In late 1983 the authors and their colleagues at Beta Research & Development Ltd. installed a sodium sulfur battery in an electric car. The battery was similar in construction to the British Rail battery described previously; the cell used was a central sulfur design but with annular sodium storage and a capacity of 95 Ah. Each of the four series chains comprised thirty-six cells, giving a nominal battery voltage of 75 V compared with the 79 V of the previous battery. The gravimetric specific energy of the cell (145 Wh kg^{-1} at the five hour rate) was slightly lower than that of the BR30 cell in which the sodium is stored in a reservoir at the end of the cell. The overall specific energy of the battery was 63 Wh kg^{-1} at the five hour rate compared with 71.5 Wh kg^{-1} for the first battery. The volumetric specific energy was higher at 43.6 kWh m^{-3} compared with 35.2 kWh m^{-3} at the same discharge rate.

The electric car in which the battery was installed was a stretched version of the PGE in which the previous battery had been tested (Fig. 13.4). This car had a thyristor chopper controller with regenerative braking and had a maximum speed of 60 km h^{-1}.

Because the battery voltage on load was significantly less than the 72 V for which the controller was designed, the top speed was limited to 45 km h^{-1}. The range of the car was approximately 60 km compared with 40 km when powered by a lead acid battery of slightly larger capacity. The reasons for this were twofold. First, the capacity quoted for both batteries was at the five hour rate, but whereas the sodium sulfur battery still delivered all the rated capacity at the one hour rate, the lead acid battery capacity was considerably reduced. Second, the sodium sulfur battery at 180 kg was significantly lighter than the lead acid battery (360 kg), resulting in a lower energy consumption during acceleration and hill climbing.

To increase the top speed of the car to 60 km h^{-1} the battery voltage was increased by connecting a 12 V lead acid battery in series. This made the vehicle quite traffic compatible.

Although the car was being driven around the Derby area it was felt that any problems arising from the use of the sodium sulfur battery would only manifest themselves if the car was used regularly in a normal driving cycle. Consequently, one of us (J. L. Sudworth) decided to use the car for commuting from his home in Burton-on-Trent to Derby, a distance of 10 miles (16 km) on terrain which although mainly flat included a number of hills along the Trent valley escarpment, the steepest of which had a gradient of 1:9 and a length of 0.3 miles (0.5 km).

The battery was charged during the day at 40 V maximum, the current being allowed to distribute itself among the four chains. End of charge was determined by inspection of the chain currents. On a typical journey from Derby to Burton the battery temperature at the start of the journey was 345 °C and the cooling fans were switched on at that point. The journey time was 24 minutes compared with 20 minutes in the two litre petrol engined car normally used. The temperature at the end of the journey was 375 °C. On arrival the battery was parked on the drive and plugged into a temperature controller which maintained the temperature around 350 °C overnight. The return journey normally took the same time despite the battery being in the single phase region for part of the time.

On one occasion the cooling fans were mistakenly not switched on until near the end of the journey when it was noticed that the temperature had risen to 380 °C. The battery temperature at the end of the journey on that occasion was 383 °C. This temperature rise was not surprising considering the current drawn by the motor. Although the current at cruising speed was normally around 100 A (50 A per cell), this increased to 300 A (150 A per cell) on severe gradients or when starting on a gradient.

These journeys were virtually non-stop and there was no chance for the battery to cool down, so some trips were made during the day in which the car was left parked. Thus it was used to collect people from the railway station, which required it to be parked on the forecourt for 15 to 20 minutes, and it was also used to take visitors out to lunch, during which it was left in the car park for about an hour. On these occasions the battery temperature did not drop noticeably.

Adverse weather conditions were encountered during this period – subzero temperatures some nights and heavy rain some days. The car coped admirably with these. On one occasion when negotiating a flooded section of road at speed, water splashed up to roof height but no adverse effects were observed.

One bonus of using a high temperature battery was that even when left out at night in a severe frost there was no icing or misting of windows and the interior of the car was at a comfortable temperature at the start of the journey. Although the cooling air from the battery was ducted inside the car this was not sufficient to maintain a comfortable temperature on very cold days and a petrol heater was used on those occasions.

After two weeks of commuting a problem was encountered: on a journey from Derby to Burton the controller developed a fault. The field weakening part of the circuit operated intermittently, causing the car to decelerate and then accelerate again spasmodically. It was necessary to return to Derby, and as it proved impossible to repair the controller the car was put off the road. However, during the two weeks the car covered over 200 miles (320 km) without incident. The total mileage up to that point was 230 miles (26 cycles) and no cell failures had occurred. During that period at least, the sodium sulfur battery was more reliable than the chopper controller.

As the car had previously been operated with a lead acid battery of the same nominal capacity over the same terrain, a direct comparison of the two batteries was possible. In terms of driving performance there was little to choose between the two systems except that the lead acid battery car's range was reduced in very cold weather, whereas the sodium sulfur battery car's range was not. Whatever capacity (ampere hours) was put into the sodium sulfur battery on charge could be taken out on discharge. This was not the case with the lead acid battery. A useful feature of the sodium sulfur battery was the change in open circuit voltage with depth of discharge in the single phase region. This allowed an accurate estimate to be made of the residual range of the vehicle when the battery was partly discharged.

One disadvantage of the sodium sulfur battery was the need to plug into the mains if the car was going to be standing for more than a few hours, although the range of the lead acid powered car was so limited that every opportunity was taken to put the battery on charge, so that in practice there was little to choose in terms of convenience. Another advantage of the sodium sulfur battery was the zero maintenance requirement. Inspection of electrolyte levels and topping up with water, although not taking a great deal of time, was a definite inconvenience with the lead acid powered car. Similarly, the precautions needed to avoid the build-up of explosive gases on charge, though simple, were an inconvenience not present with the sodium sulfur powered car.

The installation of a 12 kWh battery in the PGE car demonstrated that a sodium sulfur battery vehicle could be used routinely on public roads with no operational problems. The performance of the car, however, was at the lower limit of acceptability and the range was quite low.

In May 1985 a 25 kWh battery of essentially similar design was installed in a Ford Escort van and the vehicle was tested on a 2.9 mile circuit at the Motor Industry Research Association, Nuneaton, Warwickshire. The fastest lap was done at a speed of 54 mph (87 kmh^{-1}) and the vehicle travelled a distance of 64 miles (103 km) at an average speed of 38 mph (61 kmh^{-1}) on one charge. The specific energy of the battery at the three hour rate was 81 Wh kg^{-1} and 54 kWh m^{-3}.

From the user's point of view, the main improvement required in the sodium sulfur battery is an improved thermal management system which will allow him or her to leave the car unattended for several days without worrying about maintaining the correct temperature. This would also eliminate the cost penalty associated with heating the battery.

13.4 SUMMARY

Although not all the problems of sodium sulfur cells and batteries have been solved, the technology is sufficiently mature to permit the construction and testing of prototype batteries by most of the companies developing this battery. The experience gained from operating these batteries has highlighted the remaining problems at the cell level and identified the major battery problems.

Experience of battery operation has shown that the cell is able to withstand the stresses associated with that environment. It is true that the longest period of continuous operation was achieved with a load levelling battery in which the cell sees an environment very similar to the condition obtaining in single cell tests. Nevertheless, experience with vehicle batteries, in which cells are connected together in long series strings, has not uncovered any serious problems during battery operation. It is when a cell fails that problems occur owing to the variability in the resistance of failed cells. At present this problem can only be overcome by changing the failed cell or shorting it out mechanically. If a cell could be developed which failed to a predictable resistance, the battery could be designed so that cells could be left in place after failure without an untoward effect on battery capacity. As the reliability of cells improves and early failures are eliminated, this problem will diminish considerably and it may be acceptable to maintain the battery by physically bypassing the failed cell. For those applications requiring a high volumetric specific energy the cells will have to be packed very closely together and may not be accessible when the battery is hot. Such batteries would have to be serviced cold and the cell would have to have a high probability of surviving cooling to room temperature.

The major outstanding problems associated with the battery are very application dependent. Thus for load levelling batteries the problems are both technical and economic. The unit cell size must be large in order to achieve a low unit cost. This requires highly effective safety features to be incorporated in the cell if the much greater energy content is to be released in a controlled manner when cell failure occurs. Economic considerations also demand a long life for these batteries – ten years and 2000–3000 cycles – and this obviously requires corrosion of the current collector or cell case to be reduced to minimal levels. Accelerated tests (Sudworth, 1985) indicate that a life of ten years should be feasible and no mechanisms have been identified which would place an upper limit on the life of the sulfur electrode or the solid electrolyte.

The number of sodium sulfur batteries tested to date is very small but the results of those tests have been encouraging enough to justify the next phase of development. This will involve the building and testing of large numbers of batteries, not only in the laboratory but also in the field. This is essential if the problems described are to be solved and if this unique battery system with its highly energetic reactants is to be accepted by the eventual users. Only by building up a history of uneventful and effective operation will the consumer be persuaded to adopt the sodium sulfur battery.

13.5 REFERENCES

Churchman, A. T. (1975) *J. R. Soc. Arts.*

Minck, R. W. (1983) 5th DoE/EPRI Sodium Sulfur Battery Workshop, Washington DC.

Reschke, R. (1983) 5th DoE/EPRI Sodium Sulfur Battery Workshop, Washington DC.

Sudworth, J. L. (1985) 6th DoE/EPRI Sodium Sulfur Battery Workshop, Salt Lake City, Utah.

Appendix A: Useful data

A.1 PROPERTIES OF SULFUR

Melting point:
  Monoclinic 119 °C
  Rhombic 110 °C
(Transition temperature rhombic–monoclinic 95 °C)
Boiling point 444.4 °C
Latent heat of fusion 50.1 kJ kg^{-1}
Latent heat of vaporization at the boiling point 326 kJ kg^{-1}
Volume increase on melting (monoclinic) 8.5%

Density

Solid rhombic sulfur	2.07 g cm^{-3}
Solid monoclinic sulfur	1.96 g cm^{-3}

(Meyer, 1977)
Liquid sulfur

$$\rho(\text{g cm}^{-3}) = a - bT \qquad (T = \text{absolute temperature})$$

values of parameters a and b:

Temp. (°C)	*a*	*b*
121–155	2.1365	8.46×10^{-4}
178–445	2.0445	6.10×10^{-4}

Density of liquid sulfur

Temp. (°C)	*Density ($g\ cm^{-3}$)*	*Temp. (°C)*	*Density ($g\ cm^{-3}$)*
127	1.7982	267	1.7149
147	1.7813	307	1.6905
167	1.771	347	1.6661
187	1.7637	387	1.6416
207	1.7515	427	1.6172
227	1.7393	437	1.6111

(Janz and Rogers, 1983)

Conductance

For $T=217$–$417\,°C$
Conductance κ ($\Omega^{-1}\ cm^{-1}$)$=2.60\times10^{-1}\exp[-22\,265/RT]$
(T=absolute temperature)

T (°C)	$\kappa\times10^{14}$ ($\Omega^{-1}\ cm^{-1}$)	T (°C)	$\kappa\times10^{14}$ ($\Omega^{-1}\ cm^{-1}$)
125	0.9	155	1.8
130	1.2	160	1.5
135	1.4	165	1.8
140	1.7	170	2.3
145	2.0	175	2.7
150	2.4		

(Janz and Rogers, 1983)

Specific heat

T (°C)	C_p ($J\ kg^{-1}\ K^{-1}$)	T (°C)	C_p ($J\ kg^{-1}\ K^{-1}$)
120	966	250	1109
150	993	280	1113
170	1396	300	1118
200	1152	320	1132

(Janz and Rogers, 1983)

Surface tension

γ (dyne cm^{-1})$=102.1-0.105T$ (115–159 °C)
γ (dyne cm^{-1})$=\ 81.2-0.0566T$ (159–434 °C)
(T=absolute temperature)

T (°C)	γ (dyne cm^{-1})	T (°C)	γ (dyne cm^{-1})
117	61.1	227	52.9
127	60.1	277	50.0
147	58.0	327	47.2
177	55.7	377	44.4
207	54.0	427	41.6

(Janz and Rogers, 1983)

Thermal conductivity λ (W m^{-1} K^{-1})

Molten sulfur

T (°C)	λ (Wm^{-1} K^{-1})	T (°C)	λ (Wm^{-1} K^{-1})
115	0.131	165	0.141
130	0.134	190	0.151
150	0.139	210	0.158

(Janz and Rogers, 1983)

Vapour pressure

$\log \rho_S$ (atm) $= A + B/T$
(T = absolute temperature)

Temp. range (°C)	A	B
140–180	9.3614	−4263.7
190–240	8.1246	−3682.3
240–280	7.5549	−3347.6
290–320	7.4553	−3282.5
320–350	7.4254	−3261.7

ρ (atm)	T (°C)	ρ (atm)	T (°C)
2.10×10^{-4}	147	0.128	327
1.74×10^{-3}	187	0.334	377
3.54×10^{-2}	277	0.767	420

(Janz and Rogers, 1983)

Viscosity

η (poise) $= a + bT + CT^2$
(T = absolute temperature)

Temp. range (°C)	$a \times 10^{-3}$	$b \times 10^{-2}$	$c \times 10^{2}$
57–147		$\eta = 3.54642 \times 10^{-5}$	exp(6305.9/RT)
167–187	14.9135	0.3452	—
193–260	31.8865	−1.1359	10.1212
271–316	9.3125	−0.3148	2.6654

Viscosity

T (°C)	η (poise)	T (°C)	η (poise)	T (°C)	η
115.8	0.119	165.8	210	260.2	110
121.3	0.110	171.3	460	271.3	74
126.9	0.100	182.4	890	282.4	52
132.4	0.089	188.0	930	293.5	34
138.0	0.080	193.5	910	304.6	22
143.5	0.074	199.1	820	315.7	18
149.1	0.067	204.6	730	317.0	(14)
154.6	0.065	215.7	570	327.0	(10)
157.4	0.067	226.9	390	337.0	(7)
159.1	0.12	238.0	250	347.0	(5)
163.0	120	249.1	160	357.0	(3.5)

Values in parentheses by extrapolation

(Janz and Rogers, 1983)

A.2 PROPERTIES OF SODIUM

Melting point 97.82 °C
Boiling point 881.4 °C
Latent heat of fusion 113 kJ kg^{-1}
Latent heat of vaporization at the boiling point 3877 kJ kg^{-1}
Volume increase on melting 2.7%

(Foust, 1972)

Density

Solid sodium:
ρ (g cm^{-3}) $= 0.9725 - 20.11 \times 10^{-5} T - 1.5 \times 10^{-7} T^2$
For $T = 0 - 96.6$ °C

T (°C)	ρ (g cm^{-3})
0	0.9725
10	0.9705
20	0.9684
30	0.9663
40	0.9642
50	0.9621
60	0.9599
70	0.9577
80	0.9555
90	0.9532
97.82	0.9514 (extrapolated)

(Thomson and Garelis, 1956)

Liquid sodium:
ρ (g cm^{-3})$=0.952-2.45\times10^{-4}T$
For $T=98$–$440\,°C$

T (°C)	ρ (g cm^{-3})
100	0.928
150	0.915
200	0.903
250	0.891
300	0.879
350	0.866
400	0.854

(Longson and Thorley, 1967)

Resistivity

ρ ($\mu\Omega$ cm)$=a+bT+cT^2+dT^3-eT^4$
For $T=97.8$–$300°C$
$a=6.87$, $b=2.44\times10^{-2}$, $c=2.67\times10^{-5}$, $d=1.07\times10^{-7}$, $e=2.67\times10^{-10}$

T (°C)	r ($\mu\Omega$ cm)
100	9.66
150	11.36
200	13.25
250	15.27
300	17.32

(Addison *et al.*, 1969)

Specific heat

C_p (J kg^{-1} K^{-1})$=1437.08-0.58063T+4.624\times10^{-4}\,T^2$
For $T=$ MP to $900\,°C$

T (°C)	C_p (J kg^{-1} K^{-1})
100	1385
150	1385
200	1341
250	1323
300	1307
350	1293
400	1282

(Foust, 1972)

Surface tension

γ (dyne cm^{-1})$=203-0.08(T$–$98)$ (cold trapped pure sodium)
γ (dyne cm^{-1})$=203-0.11(T$–$98)$ (oxide saturated sodium)
For $T=98$–$303\,°C$

T (°C)	γ (dyne cm^{-1}) (cold trapped pure sodium)
100	202.84
150	198.84
200	194.84
250	190.84
300	186.84

(Longson and Thorley, 1967)

Thermal conductivity

K (Wm^{-1} K^{-1}) $= 918 \pm 13 - 0.49T$
For $T = 100$–$900\,°C$

T (°C)	K (W m^{-1} K^{-1})
100	869
150	844
200	820
250	796
300	771
350	747
400	722

(Evangelisti and Isacchini, 1965)

Vapour pressure

$\log \rho_s$ (atm) $= 6.354 - 5567/T - 0.5 \log T$
For T up to 1150 K

ρ (atm)	T (°C)
1×10^{-6}	232.84
5×10^{-5}	327.66
1×10^{-3}	428.7
1×10^{-2}	533.57
5×10^{-2}	627.98
0.25	748.0
1	881.4

(Foust, 1972)

Viscosity

$$\eta \text{ (cpoise)} = (0.1235 \pm 0.0018)\rho^{\frac{1}{3}} \exp\left(\frac{697 \pm 9}{T}\rho\right)$$

(where ρ is density in g cm^{-3}, T = MP to 500 K)

T (°C)	η (cpoise)
100	0.6802
150	0.5415
200	0.4519
250	0.3900

T (°C)	η (cpoise)
300	0.3450
350	0.3110
400	0.2845

(Foust, 1972)

A.3 PROPERTIES OF SODIUM POLYSULFIDES

Density

		Density ($g\ cm^{-3}$)				
Melt	*wt% S*	*300 °C*	*330 °C*	*350 °C*	*375 °C*	*400 °C*
Na_2S_3	67.7	1.9025	1.8853	1.8740	1.8599	1.8458
$Na_2S_{3.3}$	69.7	1.9225	1.8986	1.8827	1.8628	1.8429
$Na_2S_{3.7}$	72.0	1.9408	1.9244	1.9134	1.8997	1.8861
$Na_2S_{4.4}$	75.4	1.8870	1.8670	1.8537	1.8370	1.8204
$Na_2S_{4.8}$	77.0	1.8953	1.8739	1.8595	1.8416	1.8237

(Cleaver, 1983)

Conductance

	Conductance ($\Omega^{-1}\ cm^{-1}$)					
Melt	*wt% S*	*300 °C*	*340 °C*	*380 °C*	*420 °C*	457 °C
$Na_2S_{2.1}$	60.0	—	—	—	—	1.1040
$Na_2S_{2.9}$	66.1	—	—	0.8331	1.0804	—
$Na_2S_{3.0}$	67.5	0.4261	0.6324	0.8514	1.0738	—
$Na_2S_{3.2}$	70.1	0.3898	0.5868	0.7984	1.0149	
$Na_2S_{3.8}$	72.3	0.3120	0.4812	0.6676	0.8626	—
$Na_2S_{4.2}$	74.8	0.2579	0.4124	0.5848	0.7659	—
$Na_2S_{5.1}$	77.8	0.2100	0.3441	0.4962	0.6579	—

Specific heat

Specific heats of Na_2S_3, Na_2S_4 and Na_2S_5 are temperature invariant (Janz and Rogers, 1983)

		Specific heat		
Composition	*Temperature range (°C)*	*cal K^{-1} mol^{-1}*	*J K^{-1} mol^{-1}*	*J kg^{-1} K^{-1}*
Na_2S_3	247–387	44.8	187.5	1319
Na_2S_4	287–387	55.3	231.4	1328
Na_2S_5	252–387	64.6	270.3	1310

Surface tension

Melt	Surface tension (dyne cm^{-1})				
	300 °C	330 °C	350 °C	375 °C	400 °C
$Na_2S_{3.0}$	175.4	173.6	172.4	170.9	169.37
$Na_2S_{3.3}$	154.4	152.4	151.1	149.5	147.9
$Na_2S_{3.6}$	142.6	141.5	140.9	140.0	139.2
$Na_2S_{3.9}$	133.0	132.0	131.2	130.3	129.5
$Na_2S_{4.3}$	147.7	146.6	145.9	145.0	144.1
$Na_2S_{4.8}$	130.5	127.8	125.9	123.6	121.4
$Na_2S_{5.2}$	116.4	115.2	114.4	113.5	112.5

(Cleaver, 1983)

Latent heat of fusion

Composition	ΔH_f°		
	kcal mol^{-1}	kJ mol^{-1}	kJ kg^{-1}
Na_2S	5.0	20.92	2680
Na_2S_2	1.78	7.45	677
Na_2S_3	3.72	15.56	1094
Na_2S_4	6.87	28.74	1649
Na_2S_5	7.57	31.67	1535

(Janz and Rogers, 1983)

Volume increase on melting

Composition	ΔV (%)
Na_2S_3	3.5
Na_2S_4	7.0
Na_2S_5	5.0

(Janz and Rogers, 1983)

Viscosity (cpoise)

Melt	*wt% sulfur*	*300 °C*	*330 °C*	*350 °C*	*375 °C*	*400 °C*
$Na_2S_{3.1}$	68.4	37.9	23.8	18.4	14.0	11.0
$Na_2S_{3.3}$	70.0	38.4	24.4	19.0	14.3	11.2
$Na_2S_{3.6}$	71.7.	39.1	24.3	18.8	14.2	11.3
$Na_2S_{3.9}$	73.1	37.2	22.8	17.4	13.0	10.1
$Na_2S_{4.3}$	75.1	41.9	24.0	17.9	13.3	10.4
$Na_2S_{4.7}$	76.8	44.0	25.4	18.8	13.8	10.5
$Na_2S_{5.2}$	78.5	55.7	26.1	18.5	13.4	10.5

(Cleaver, 1983)

Coefficient of thermal expansion

All compositions of Na_2S_x have the same thermal expansion coefficient of approximately 35×10^{-6} K^{-1} between ambient and 200 °C (Knoedler and Nicoll, 1984).

EMF/temperature relationships of the sodium sulfur cell (Davies, 1973)

Depth of discharge % wrt Na_2S_3	*Sulfur in melt*		$\frac{\partial E}{\partial T}$	*EMF (V)*			*Thermo-neutral potential*
	mol fraction	*wt fraction*	*(μV K^{-1})*	*330 °C*	*350 °C*	*400 °C*	*(V)*
51.79	0.742	0.802	100	2.077	2.076	—	2.130
59.28	0.7155	0.779	216	2.065	2.061	—	2.195
70.56	0.679	0.748	229	1.994	1.989	1.977	2.132
72.53	0.673	0.743	241	1.988	1.983	—	2.133
77.61	0.658	0.729	231	1.944	1.938	—	2.094
82.21	0.645	0.718	173	1.912	1.908	—	2.016
86.62	0.633	0.707	192	1.891	1.887	1.877	2.007
94.36	0.613	0.689	199	1.836	1.832	—	1.956
95.97	0.609	0.686	199	1.817	1.810	1.794	1.937
99.25	0.601	0.678	388	1.802	1.796	—	2.036
102.2	0.594	0.672	415	1.780	1.773	1.752	2.030

Relationship between EMF and depth of discharge in the single phase region (Davies, 1973):

$$EMF = A(1 - D) + B$$

Temp. (°C)	A	B
300	0.706	1.814
330	0.694	1.795
350	0.696	1.791
400	0.706	1.781

EMF and depth of discharge at point of inflexion on EMF against depth of discharge plot (Davies, 1973):

	Depth of discharge		
Temp. (°C)	*%*	*mol fraction S*	*EMF (V)*
300	58.3	0.720	2.0821
330	56.0	0.728	2.0793
350	55.7	0.729	2.0763
400	55.3	0.731	2.0720

Relationship between EMF and depth of discharge in the single phase region (Gupta and Tischer, 1972):

$$\text{EMF} = A(1 - D) + B$$

Temp. (°C)	A	B
280	0.673	1.805
300	0.688	1.779
330	0.699	1.778
360	0.705	1.766
390	0.731	1.752

EMF and depth of discharge at the point of inflexion in the EMF versus depth of discharge plot (Gupta and Tischer, 1972):

	Depth of discharge		
Temp. (°C)	*%*	*mol fraction S*	*EMF (V)*
280	58.1	0.721	2.079
300	57.8	0.722	2.075
330	57.2	0.724	2.073
360	56.9	0.725	2.071
390	56.4	0.727	2.068

Note: As can be seen from Fig. A.1, the relationship between EMF and depth of discharge departs from linearity, particularly near the composition Na_2S_3. The above relationships are an approximation. Davies's data were numerical, whilst the data of Gupta and Tischer were in graphical form and the numerical values given here were obtained by measurement.

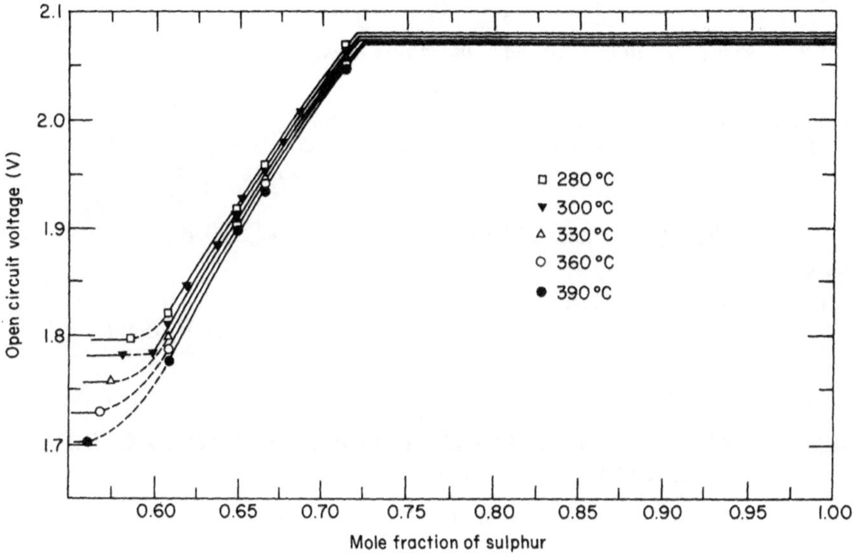

Fig. A.1

A.4 PROPERTIES OF BETA ALUMINA

For more detailed properties see Chapter 2.

Coefficient of linear expansion

Beta″ (0–650 °C) $7.5 \times 10^{-6}\ K^{-1}$ (Barrow, 1980)

Thermal conductivity

Polycrystalline	Temp. (K)	$W\ m^{-1}\ K^{-1}$
	300	3.0
	680	2.9
		(May, 1978)

Specific heat

Polycrystalline	Temp. (K)	$J\ mol^{-1}\ K^{-1}$
	300	1143.0
	680	1323.0
		(May, 1978)

Fracture strength

Beta alumina (fine grain)	230–330 MN m^{-2}	(May, 1978)

Young's modulus

Beta alumina	2.1×10^{11} N m^{-2}	(Tan and May, 1977)

Refractive indices of beta and alpha alumina

	ε	ω
Beta alumina	1,630	1.668
Alpha alumina	1.760	1.768
		(Ridgway *et al.*, 1936)

Dielectric constants of beta and alpha alumina (100 *K*)

Beta alumina	3.22	
Alpha alumina	2.80	
		(Radzilowski *et al.*, 1969)

A.5 PROPERTIES OF TYPICAL ALPHA ALUMINA COLLAR MATERIALS

	Ceramatec A-101	*Smiths FF*
Alumina (%)	99.5	95
Density (g cm^{-3})	3.68–3.9	3.76
Specific heat (J kg^{-1} K^{-1})		
@ 25 °C	850	733
@ 500 °C	1100	—
@ 1000 °C	1250	
Thermal conductivity ($W m^{-1} K^{-1}$)		
20 °C	36	58 (25–40 °C)
500 °C	11	—
1000 °C	6	—
Coefficient of linear expansion		
°C	10^{-6} °C^{-1}	
0–200	—	5.0
0–400	—	6.7
25–500	7.6	—
0–600	—	7.5
25–700	8.1	—
0–800	—	8.2

	Ceramatec A-101	*Smiths FF*
Tensile strength (N mm^{-2})	320	188
Young's modulus (N mm^{-2})	370 000	322 000
Rockwell hardness (R45N)	81	81
Compressive strength (N mm^{-2})	2620	1655

(Data from Ceramatec Product Data Sheet, Ceramatec, Salt Lake City, Utah, USA and Sintox General Information Booklet, Smiths Industries, Rugby, England).

A.6 PROPERTIES OF METAL CELL COMPONENT MATERIALS

A.6.1 NILO K

Composition (%)

Ni	Co	Fe
29.5	17	balance

Density (g cm^{-3}) 8.16
Melting point (°C) 1450

Coefficient of linear expansion

°C	10^{-6} K^{-1}
20–100	6.0
20–200	5.7
20–300	5.3
20–400	5.0
20–500	6.2

Electrical resistivity

°C	$\mu\Omega$ cm
20	43
100	55
200	72
300	88
400	100
500	109

Tensile properties

Temp. (°C)	*Tensile strength* ($N\ mm^{-2}$)	*0.1% proof stress* ($N\ mm^{-2}$)	*Elongation* (%)
20	520	340	42
100	430	260	42
200	400	210	42
300	400	140	45
400	400	110	49

(Data from NILO K data sheet, Wiggin Alloys Ltd, England)

A.6.2 Stainless steel AISI 316

Composition (%)

C (max)	Mn (max)	P (max)	S (max)	Si (max)	Cr	Ni	Mo
0.08	2.00	0.045	0.030	1.00	16.00–18.00	10.00–14.00	2–3.00

Density

°C	$g\ cm^{-3}$
20	7.95
90	7.92
200	7.88
320	7.83
430	7.79

Coefficient of linear expansion (for EN58J equivalent)

°C	$10^{-6}\ K^{-1}$
20–100	15.5
20–200	16.1
20–300	16.8
20–400	17.5
20–500	18.0

Thermal conductivity (for EN58J equivalent)

°C	$W\ m^{-1}\ K^{-1}$
20	16.33
100	16.75
400	20.52

Electrical resistivity

°C	μΩ cm
20	74
90	79
200	87
320	93
430	99
540	104

Specific heat

°C	$J\ kg^{-1}\ K^{-1}$
20	452
90	486
200	528
320	548
430	565
540	573

Tensile properties
(For 0.07% C, 0.26% Si, 0.36% Mn, 8.55% Ni, 18.2% Cr, 2.82% Mo)

Temp. (°C)	*Tensile strength ($N\ mm^{-2}$)*	*0.1% proof stress ($N\ mm^{-2}$)*	*Elongation (%)*
Ambient	678.0	254.8	59
66	603.9	225.5	57.5
100	573.0	194.6	54.25
200	532.8	171.4	45.5
300	531.3	159.1	43
400	522.0	154.4	43
500	503.5	146.7	43

(Data from Woolman and Mottram (1968) and 'Engineering properties at elevated temperatures of the austenitic chromium-nickel stainless steels', an information booklet provided by Inco Europe Ltd).

A.6.3 Mild steel EN 3A

Composition (%)

C	Si	Mn	S (max)	P (max)	Fe
0.15–0.25	0.05–0.35	0.4–0.9	0.06	0.06	balance

Density @ 20 °C 7.86 $g\ cm^{-3}$

Coefficient of linear expansion

°C	10^{-6} K^{-1}
20–100	12.4
20–200	12.8
20–300	13.2
20–400	13.6
20–500	14.0

Thermal conductivity

°C	W m^{-1} K^{-1}
20	51.9
100	51.1
200	48.6
300	44.4
400	42.7

Electrical resistivity

°C	$\mu\Omega$ cm
20	16.9
100	21.9
200	29.2
300	38.1
400	48.7

Specific heat

°C	J kg^{-1} K^{-1}
20–100	485.7
20–200	498.2
20–300	514.9
20–400	535.9

Tensile properties (N870 heat treatment)

(Composition C 0.24%, Si 0.21%, Mn 0.54%, Ni 0.15%)

Temp. (°C)	*Tensile strength* ($N\ mm^{-2}$)	*0.1% proof stress* ($N\ mm^{-2}$)	*Elongation* (%)
RT	526.7	307.3	41
50	518.9	299.6	37
150	674.9	—	16.5
200	678.0	—	23
250	664.1	248.7	30.5
300	633.2	211.6	36.5
350	573.0	193.0	
400	481.9	185.3	36
450	393.8	176.1	30
500	318.2	163.7	33

(All data from Woolman and Mottram, 1968)

A.6.4 Aluminium and aluminium alloys

Pure aluminium

Density @ 20 °C 2.6898 g cm^{-3}
Melting point 660.2 °C
Coefficient of linear expansion (0–100 °C) 23.5×10^{-6} °C^{-1}
Mean specific heat (0–100 °C) 916.9 J kg^{-1} K^{-1}

Thermal conductivity

°C	W m^{-1} K^{-1}
0	236
100	240
300	233
700	92

Electrical resistivity

°C	$\mu\Omega$ cm
0	2.5
100	3.55
300	5.9
700	24.7

Alloys

Material designation and temper	*Melting range (°C)*	*Thermal conductivity 0–100 °C (W m^{-1} K^{-1})*	*Electrical resistivity 20 °C (μΩ cm)*
1080A-0	645	230	2.8
1050A-0	635	230	2.8
6082-TB	570–660	170	4.1

Tensile properties

Alloy 1050A (wire to BS 1475)
Si 0.25%, Fe 0.4%, Cu 0.05%, Mn 0.05%, Mg 0.05%, Zn 0.07%, Ti 0.05% (Aluminium 99.5)

(Condition – as drawn)

Temp. (°C)	*Tensile strength (N mm^{-2})*	*0.1% proof stress (N mm^{-2})*	*Elongation (%)*
20	62.7	28.9	63
100	47.6	24.9	66
150	42.2	24.9	78
200	34.4	19.3	83
250	27.2	16.7	93

Alloy 6082 (bars, tube and sections to BS 1474)
Si 0.7–1.3%, Fe 0.5%, Cu 0.10%, Mn 0.4–1%, Mg 0.6–1.2%, Cr 0.25%, Zn 0.2%, Ti 0.1%, aluminium remainder
(condition solution heat treated and artificially aged)

Temp. (°C)	*Tensile strength (N mm^{-2})*	*0.1% proof stress (N mm^{-2})*	*Elongation (%)*
20	383	339.8	17
100	359.8	322.8	20
150	295.0	256.4	23
200	199.2	173.0	22
250	120.5	97.3	25

All data from 'The properties of aluminium and its alloys' published by the Aluminium Federation, Birmingham, UK.

A.7 PHYSICAL CONSTANTS AND CONVERSION FACTORS

Physical constants

Faraday constant	F 96489 C mol^{-1}
Gas constant	R 8.3166 J K^{-1} mol^{-1}
Boltzmann constant	k 1.3804×10^{-23} J K^{-1}

Conversion factors

1 eV	$= 1.602\,10 \times 10^{-19}$ J
1 calorie	= 4.1868 J
1 kWh	= 3414 BTU
1 hp	= 745.7 W
1 atm	= 14.70 psi
1 atm	= 101 325 Pa
1 Wh l^{-1}	= 0.0948 Wh in^{-3}
1 psi	= 6894.76 N m^{-2}

Equivalents

1 Ah is equivalent to:

0.8578 g Na
0.9905 cc Na @ 350 °C
1.7942 g S (assuming discharge to the composition Na_2S_3).

A.8 REFERENCES

Addison, C. C., Creffield, G. K., Hubberstey, P. and Pulham, R. J. (1969) *J. Chem. Soc. (A)*, 1482–1487.
Barrow, P. (1980) 4th DoE/EPRI Sodium Sulfur Battery Workshop, Derby.
Cleaver, B. (1983) in *The Sulfur Electrode* (ed. R. P. Tischer), Academic Press.
Davies, A. J. (1973) Thesis, University of Southampton.
Evangelisti, R. and Isacchini, F. (1965) *Int. J. Heat Mass Transfer*, **8**, 1303–17.
Foust, O. J. (1972) *Sodium-NaK Engineering Handbook*, vol. 1, Gordon & Breach.
Gupta, N. K. and Tischer, R. P. (1972) *J. Electrochem. Soc.*, **119**, 8, 1033–1037.
Janz, G. J. and Rogers, D. J. (1983) in *The Sulfur Electrode* (ed. R. P. Tischer), Academic Press.
Knoedler, R. and Nicoll, A. R. (1984) *J. Mater. Sci.* (in press).
Longson, B. and Thorley, A. W. (1967) Alkali Metals Conference, The Chemical Society, vol. 22, 155–66.
May, G. J. (1978) *J. Power Sources*, **3**, 1–22.
Meyer, B. (1977) *Sulfur, Energy and Environment*, Elsevier.
Radzilowski, R. H., Yao, Y. F. and Kummer, J. T. (1969) *J. Appl. Phys.*, **40**, 4716–25.
Ridgway, R. R., Klein, A. A. and O'Leary, W. J. (1936) *Trans. Electrochem. Soc.*, **70**, 71–88.
Tan, S. R. and May, G. J. (1977) *J. Mater. Sci.*, **12**, 1058–61.

Thomson, G. W. and Garelis, E. (1956) Chapter 9 in *Sodium: Its Manufacture and Uses* (ed. M. Sittig), Reinhold, New York.

Weast, R. C. (1977) *Handbook of Chemistry and Physics*, 57th edn, CRC Press, Cleveland.

Woolman, J. and Mottram R. A. (1968) *The Mechanical and Physical Properties of the British Standard EN Steels*, BISRA.

Index

GPSR Compliance

The European Union's (EU) General Product Safety Regulation (GPSR) is a set of rules that requires consumer products to be safe and our obligations to ensure this.

If you have any concerns about our products, you can contact us on ProductSafety@springernature.com

In case Publisher is established outside the EU, the EU authorized representative is:

Springer Nature Customer Service Center GmbH
Europaplatz 3
69115 Heidelberg, Germany

Batch number: 10809439

Printed by Printforce, the Netherlands